An Account Of The Danes And Norwegians In England, Scotland And Ireland

Jens Jakob Asmussen Worsaae

Writat

This Edition Published in 2022

ISBN: 9789356396685

TABLE OF CONTENTS

PREFACE

Mr. Worsaae informs us in his Introduction that the following pages were not written solely for the learned. They were designed as a popular contribution to a branch of historical and antiquarian knowledge, which, though highly interesting both to Scandinavians and Englishmen, has been hitherto very imperfectly investigated. The English reader will find in Mr. Worsaae's work not only many facts concerning the early history of this country that are either entirely new to him, or placed at least in a wholly novel light, but he will also meet with many *names* whose form may appear foreign and unfamiliar. It may, therefore, be desirable that on the English reader's introduction to a more intimate acquaintance with that Scandinavian race which has more claims than he had, perhaps, imagined, not only to be regarded as the founders of some of his native customs and institutions, but even to be reckoned among his forefathers, he should be enabled to pronounce their principal names correctly. With this view the following brief remarks are subjoined;—

The double *a* (*aa*), frequently occurring in proper names, must be sounded like the English diphthong *aw*, as in Blaatand, Haarfager.

The *ö*, or *oe*, is pronounced like the French diphthong *eu*.

The *u*, as in German and Italian, is equivalent to *oo* in the English words *cool*, *troop*, &c.; as in Ulf, Huskarl, &c.

C has invariably the sound of *k* (with which, indeed, it is frequently interchanged). The names of Cetel, Oscytel, &c., are to be pronounced Ketel, Oskytel. Where *c* or *k* precedes another consonant, it retains, as in German, its distinct and proper power. In order to represent this power, Latin and English writers have sometimes substituted the syllable *ca* for the initial *c* or *k*; as, for instance, in the name of Canute (*Dan.*, Cnut or Knud). This has led to the very common error of pronouncing the name as if it consisted of two syllables, with an accent upon the first; as Cán-ute, instead of Cănúte.

J has the sound of the English *y*; as in Jarl (*Yarl*, earl), Jorvik (*Yor-vik*, York).

The consonants *th* (the Icelandic Þ[1]) are pronounced like a single *t*. The word *Thing* (assizes, &c.), which the reader will so frequently meet, is sounded like *Ting*. The proper pronunciation is preserved in the word *Hus-ting*, but by altering the spelling. Thus, Thor, Thorkil, &c., must be pronounced *Tor*, *Torkil*.

1. The letter ð has the power of *dh*, or *dth*.

Lastly, the Vikings (*Isl.*, Vikingr, a sea-rover, pirate), who played so great a part during the Danish conquests, were not Ví-kings, but Vik-ings (Veék-ings); so called either from the Icelandic *Vik* (*Dan.*, Vig), a bay of the sea, or from *Vig*, battle, slaughter.

London, Dec. 15th, 1851.

AUTHOR'S PREFACE

In the spring of 1846, his late Majesty Christian VIII. of Denmark determined that an inquiry should be made respecting the monuments and memorials of the Danes and Norwegians which might be still extant in Scotland and the British Islands. His Majesty was the more confirmed in this design as two distinguished British noblemen, his Grace the Duke of Sutherland, and his brother Lord Francis Egerton (now Earl of Ellesmere), had repeatedly stated in their letters to the Royal Society of Northern Antiquaries that, if a Danish archæologist visited Scotland, he should receive all possible assistance, especially in Sutherland, a district so rich in Scandinavian antiquities.

His Majesty did me the honour to intrust this task to me: and the President of the Royal Society of Northern Antiquaries, and of the Royal Committee for the preservation of the national monuments—our present most gracious sovereign Frederick VII.—having, with a lively zeal for the promotion of the inquiry, furnished me with several letters of introduction, I travelled during a twelvemonth (1846-1847) in Scotland, Ireland, and England; where, partly through the personal kindness of the Duke of Sutherland and of the Earl of Ellesmere, and partly by means of their influential names, I invariably met with the best reception and the most valuable assistance in my researches.

The present work contains part of the results of that journey. My aim in it has been to convey a juster and less prejudiced notion than prevails at present respecting the Danish and Norwegian conquests; which, though of such special importance to England, Scotland, and Ireland, have hitherto been constantly viewed in an utterly false and partial light. Whilst writing the work in Denmark, I have but too frequently felt the want of constant access to the well-stored libraries of England; although those literary gentlemen in Great Britain to whom I have written for information, have received my applications with their usual readiness and friendship[2].

2. Amongst the many gentlemen to whom I owe my thanks, I must particularly name: Sir H. Dryden, Bart., of Canons Ashby; C. Roach

Smith, Esq., F.S.A., London; E. Hawkins, Esq., British Museum; J. M. Kemble, Esq.; Professor Cosmo Innes, Edinburgh; Dr. Traill, *ibid.*; C. Neaves, Esq., *ibid.*; R. Chalmers, Esq., of Auldbar Castle; Rev. J. H. Todd, D.D., Trinity College, Dublin; Professor C. Graves; and Dr. G. Petrie, likewise of Dublin.

However, as my work contains the first fully detailed examination of the subject *from the Danish side*, I hope that, notwithstanding all its deficiencies and faults, it may prove of some interest in England, and serve to excite further investigation, which would doubtless throw a clearer light upon a very remote, but not on that account less remarkable, period in the history of England and the North.

J. J. A. WORSAAE.

Copenhagen, April, 1851.

INTRODUCTION
SECTION I

Scandinavia's greatest Memorials.—Those of Denmark and Norway at Sea.—Of Sweden on Land.—The Influence of Climate.

The greatest, and for general history the most important, memorials of the Scandinavian people are connected, as is well known, with the expeditions of the Normans, and with the Thirty Years' War.

In the Norman expeditions the North, mighty in its heathenism, poured forth towards the east, the west, and the south, its numerous warriors and shrewd men, who subverted old kingdoms, and founded new and powerful ones in their place. It was by Danish and Norwegian fleets that Normandy and England were then conquered, and kingdoms won in Scotland, Ireland, and North Holland; whilst Norwegians settled on the Faroe Islands (*Dan.*, Faröerne), and discovered and colonized Iceland. Hence their descendants, having afterwards passed over to Greenland, discovered America, and were in the habit of navigating the Atlantic Ocean centuries before other European nations.

In all these voyages proportionally few Swedes took part. Inscriptions on runic stones in Sweden sometimes speak, indeed, of men who had settled or met their death in the west over in England (Anklant or Inklant). But on the whole the views of the Swedes were at that time, as well as at a later period, mostly directed towards the east. Swedish Vikings, or pirates, harried and established themselves upon the coasts of Finland and of the countries now belonging to Russia; and a tribe of them, the Varæger, even made themselves there the reigning people. Partly in consequence of this, Sweden—and particularly the Island of Gothland, or Gulland—became the centre of the active trade which in ancient times (that is, from the eighth to the twelfth century,) was carried on, through Russia, between Scandinavia and the countries around the Black and Caspian Seas, as well as Arabia.

The Swedes, however, do not appear very prominently either in ancient times or in the early part of the middle ages. They were prevented from playing any considerable part in the distant lands towards the West by the sanguinary intestine disputes which took place between them and the Goths; and it was not till the fifteenth century, and after these disputes were adjusted, that they could appear upon the theatre of the world as a nation. The Swedish Charleses and Gustavuses, by means of the sword, subsequently caused the Swedish name to be feared and honoured; not, however, at sea, but on land, on the plains of Russia, Poland, and Germany. Gustavus Adolphus, in the Thirty Years' War, after the disaster of the Danish-Norwegian king Christian IV., powerfully contributed to uphold Lutheranism, and by that means to establish liberty of conscience for Germany and the rest of Europe.

It was, then, principally at sea that the Danes and Norwegians formerly won a name in the history of the world, whilst the Swedes obtained theirs on land. Indeed, the peculiar nature and situation of the different Scandinavian countries must have necessarily caused the strength and courage which were the common attributes of the Scandinavian race, to be exerted from the first in different directions. Sweden, which towards the west is separated from Denmark only by the Sound and Cattegat, is in like manner towards the east separated from the vast plains of northern Europe by a confined and narrow sea. When, therefore, the thirst of glory and conquest urged the Swedish warriors from their homes, it was only necessary for them to cross over to the opposite shores, or at most to sail along the coasts of the Baltic. In Sweden, forests, valleys, and rivers, are the most prominent natural features, whilst the sea is but a subordinate one. It is scarcely to be expected that such a country should produce good seamen. But in Denmark and Norway the case is altogether different.

Denmark is surrounded on all sides by the sea, which has indented the land with numberless bays and firths, and cut it up into small portions. Nor is it washed only by a confined sea like the Baltic, but also by the more open German Ocean. From the earliest times, therefore, necessity obliged the Dane to put to sea in order to keep up his connections with his friends on the surrounding coasts and islands. Subsequently—when commerce, and more especially when military honour, required it—he was compelled to learn how to navigate the open sea, to struggle with the foaming waves and rapid currents, and to defy the surf—which is still the constant terror of seamen—on the coasts of north and west Jutland.

Thus the Dane early became a bold and daring Viking, and the Norwegian distinguished himself in the same manner. Norway turns her broad and rocky bosom towards the ocean. Her wild and broken coasts, split into deep fiords, or gulfs, bear witness to the never-ceasing and violent attacks of the Atlantic. Towards the east, Norway is separated from Sweden by rocks, forests, and large desert plains. The interior of the country is partly filled with mountains and immense forests, which anciently were still more extensive. The valleys alone, along the banks of rivers, are productive, and capable of cultivation. The greater part of the inhabitants settled therefore originally on the fiords, or in the neighbourhood of the sea, where the pasture land was neither so over grown with wood, nor so sequestered as in the interior, and where also the sea air rendered the climate considerably milder. The weather, however, was variable enough, and the products of the earth being, partly on that account, but scanty, fishing and the chase became important sources of maintenance for the continually-increasing population. The forests supplied them with abundance of timber, the soil was rich in iron; nor were the people wanting in a daring and enterprising spirit. Ships were soon built, capable not only of navigating the fiords, but of venturing beyond their mouths. The first voyages were coasting ones, but subsequently they were extended from the southern part of Norway to the Danish and Swedish shores.

The Norwegian, who had now become skilled in navigating his ship through the mountain waves of the Atlantic and the far more dangerous surfs on the rocks of Norway, no longer dreaded the open sea. When the population had increased to such an extent that the Norwegian rocks could barely afford it a sufficient maintenance; when the reports concerning the rich lands beyond the sea, and their defenceless condition, promised at once renown and booty; and when, lastly, Harald Haarfager's conquests threatened the Norwegians with the loss of their freedom—then thousands of vessels shot out from the fiords of Norway, and steered dauntlessly for the neighbouring western islands. A northern life, and the severe winter's cold, had not only braced the body of the Viking to endure all kinds of hardships, and given him strength to wield the sword with effect; it had also steeled his courage, and taught him fearlessly to face all manner of danger. The clear starry firmament of the North enabled him to observe the course and relative situation of the stars, which were then the only compass by which he steered his ship towards foreign and unknown shores.

Norway must naturally be better calculated to form hardy persevering sailors than Denmark. With the exception of the west coast of Jutland, where there is not a good harbour to be found, and where, consequently, navigation must, in ancient times, have been very limited, Denmark is washed by an enclosed sea with flat coasts. The ocean, on the contrary, washes almost the whole of Norway's rocky shores; where the numerous and deeply-indented fiords resemble so many harbours. There are sufficient indications that anciently the Danes were accustomed to visit only the comparatively neighbouring countries of England, Holland, and France; whilst the Norwegians sailed also towards the north on the wide Atlantic, whose storms and dangers did not prevent them from constantly visiting the Faroe Islands, Iceland, Greenland, and even America. The discovery and first colonization of these countries are, with just reason, the pride of the Norwegians and of their descendants the Icelanders.

A comparison with other European nations will more clearly show how great an influence the climate of the North, and especially the Northern Sea, must have had on the development of navigation among the Danes and Norwegians, and on their whole maritime life. With the exception of England, which, in a still higher degree than Scandinavia, swims in the open sea, and of Holland, which lies as it were half under water, no country in Europe has produced a seafaring people which can be at all compared to the Northmen; and this notwithstanding that Germany, France, and the Spanish Peninsula, have all a very considerable extent of coast. The reason undoubtedly is, that the coasts of those countries are washed by enclosed seas, which naturally cannot be compared with the ocean; whilst the countries themselves, especially Germany and France—and the latter even in spite of its extent of coast towards the Atlantic—have an unmistakeable continental character. It is clear, moreover, that the ocean, as well as the smaller and enclosed seas, have, according to the difference of latitude, an entirely different influence on the people who inhabit their shores. The Mediterranean, surrounded by rich and fruitful, but enervating, countries, has not shown itself capable of producing such seamen as the Baltic, where the climate is more severe, and the gifts of Nature incomparably more sparing. Spain and Portugal, it is true, have a great extent of coast towards the Atlantic, which may almost be compared with the west coast of Norway. But both those countries possess a fruitful soil and a glorious southern climate. Their inhabitants were not, like the Northmen of old, forced to visit foreign shores in order to procure subsistence, and to struggle continually with a raw and severe climate. They preferred to stay at home and enjoy the blessings of their own country; and thus the calm energy and the proud self-reliance which are engendered by a ceaseless struggle with an ungrateful soil and climate, and which are indispensable to a hardy seaman, were not developed in them as in the Norwegians and other inhabitants of the North. This may have been one of the causes why the Spaniards and Portuguese were unable to retain, in later times, their mastery over the new world. They were displaced by the English, a northern seafaring people, who were more at home on the sea.

It was the same quiet energy which, even amid the excitement of passion, so strongly distinguished the northern from the southern races. The inhabitant of the South was more governed, as he now is, by his passions. A torrent of words, an animated play of the features, or even perhaps a violent assault, betrayed the fire that raged within him. The northern man, on the contrary, was of few words. His anger was under the dominion of his cooler reason, and he was capable of concealing the emotions of his soul. But he had a good memory. Years would pass before he revenged himself; and he felt a sort of pleasure in making his preparations, and waiting for the proper opportunity. The revenge of blood, therefore, took place in the cold North, as well as in the fiery South: but in the totally different manner in which it manifested itself we can hardly fail to recognise the influence of Nature.

It must, however, be borne in mind that in every nation, except those situated at the Poles or under the Line, where Nature exerts an almost irresistible and overwhelming force, this influence manifests itself very differently, according to their different degrees of development. In the infancy of a people, and so long as their immediate wants render them entirely dependent on Nature, whose unexplained phenomena appear to them as those of some foreign and unknown power, her influence on their life is naturally strongest. The effect is the same as that which education and the companions with whom he associates produce on an individual. But as nations gradually become more enlightened and refined, they obtain a mastery over Nature, whose influence thus grows weaker and weaker, and at last almost vanishes. It is, indeed, one of the most marked steps in the progress of human development, when man becomes Nature's master, and makes her obedient to his power. Thus when Englishmen, Frenchmen, and others who belong to a people of defined character and perfectly-developed nationality, settle in foreign parts, the influence of Nature, even at the Poles, or under the Line, is scarcely strong enough to produce any great change in their character. And upon the whole, to whatever degree civilization may be carried, most nations will never entirely lose that character which Nature has impressed upon them in the lands which gave them birth.

The influence of Nature upon the Scandinavian people may be traced throughout their history, even down to the present times. In their sanguinary internal wars, the Danes and Norwegians generally gained the victory over the Swedes at sea. Under able leaders they have sometimes been victorious on land also; but here the Swedes have in general been superior. Christian IV. made no progress in the Thirty Years' War. On that occasion he proved himself inferior to Gustavus Adolphus, who, when fighting on land, was in his true element. At sea, on the other hand, Christian IV. signally defeated the Swedish fleet. The chief heroes of the Swedish nation, and those who live most in the memory of the people, are, Gustavus Adolphus, Charles X., and particularly Charles XII.; although that monarch, by his rash wars in Russia, Poland, and Germany, inflicted deep wounds upon Sweden, which took a long time to heal. But the favourite heroes of the Danes and Norwegians are seamen; as Christian IV., Niels Juel, Hvitfeld, and especially Tordenskjold, who, singularly enough, was contemporary with Charles XII. The difference between the people is clearly expressed in the opening lines of two of the most favourite national songs. The Danish—formerly the Norwegian also—runs thus:

"Kong Christian stod ved höien Mast

I Rög og Damp,"

("King Christian stood by the high mast, enveloped in mist and smoke"), where there is an allusion to a fight at sea. But the Swedish lines,

"Kung Karl den unge hjelte

Han stod i rök och dam,"

("King Charles the young hero, stood in smoke and dust"), allude to battle and victory on land. Even to the present day it may with good reason be asserted that the Danes and Norwegians feel more inclination than the Swedes for a seafaring life. But as the battle in Copenhagen Roads (April 2, 1801) maintained the ancient reputation of the Danes at sea, so also recent events have shown, that both the Danes and Norwegians of the present day can fight on land with distinguished bravery.

SECTION II

The Great Memorials of Sweden in their Relation to those of Denmark

and Norway.—Danish-Norwegian Memorials in the British Isles.

Russia, Poland, and particularly Germany, were, as we have seen, the theatre of the greatest victories of Sweden. The glory of Denmark and Norway, on the contrary, was founded in the West, over the sea, in America, Iceland, the British Isles, and France. Denmark's conquests of the southern and eastern coasts of the Baltic in the twelfth and thirteenth centuries, under the Waldemars, terminate, however, the times of the Vikings. The victories of Sweden are of a modern date, and since the last two centuries; but those of Denmark are of the ninth, tenth, eleventh, and twelfth centuries. The remembrance of the Swedish sabre-cut yet remains fresh among the Russians, Poles, and Germans; nay, in some places, the Swedish name is still a terror to the common people.

It is often made a subject of complaint against the great achievements of Denmark and Norway that they are of such remote antiquity; and that, instead of promoting the freedom and spiritual advancement of mankind, like Sweden's struggles in the Thirty Years' War, they rather caused an immense retrograde step in civilization, since the heathen Vikings acted with unbridled ferocity, burnt and destroyed churches and convents, and rudely trampled upon everything that bore the mark of a higher intellectual development. Thus foreigners, and particularly the German historians, usually assert, for instance, that the Danish and Norwegian Vikings brought nothing but misfortune upon the British Isles; whilst, on the contrary, everything great and good in England is mainly attributable to the Saxons, or Germans. This, however, is not to be wondered at, since these critics were obliged to judge of situations for whose right estimation they were entirely without the necessary knowledge, namely, that of the more ancient history of the North.

It would certainly not be gratifying to the national feelings of the Danes and Norwegians if the progress and settlements of the Vikings in foreign lands were marked only by acts of violence, murder, and incendiarism. Nor would it be a whit more pleasing or refreshing if it were necessary to dig up as it were out of the earth the memorials of those deeds, after they had lain for centuries in oblivion, or if we were obliged carefully to revive them and procure their acknowledgment in the countries which were once compelled to bow before the power of the northern warriors.

But what if the Danish name, and the remembrance of the exploits of the Danes and Norwegians, in spite of the many centuries that have passed since they were performed, still live as fresh in the memory of the people of the western lands as the Swedish name in Germany, nay, perhaps even fresher? What if we found that, by means of monuments, the popular character, public institutions, and other traits, a constant powerful and beneficial influence could be traced from the expeditions of the Vikings or Northmen, so that the natives of the lands which they subdued accounted it an honour to descend from the bold natives of the North? Would not the Northman in that case have a double right to be proud of his forefathers? Or would he, upon the whole, any longer have reason to complain?

It is the object of the following pages to convey, partly in the form of travelling impressions, a picture of the memorials of the Danes and Norwegians, as they exist in the monuments and among the people of those countries which in former times most frequently witnessed the victories of the Danes and Normans—namely, the British Islands. It is, however, by no means the exclusive, or even special, design of them, to present to scholars and persons of science detailed and critical observations on every individual ancient monument in those islands, which may be said to be of Danish or Norwegian origin. Their aim rather is to describe the more general, and consequently more appreciable, features of actually existing Scandinavian monuments; in doing which a distinction will, as far as possible, be drawn between the Danish and the Norwegian memorials; and in general between the influence of the Danes in England, and of the Norwegians in Scotland and Ireland.

THE DANES IN ENGLAND SECTION I

Nature of the Country.—Earlier Inhabitants: Britons, Romans, and Anglo-Saxons.

The greater part of England consists of flat and fertile lowland, particularly towards the southern and eastern coasts, where large open plains extend themselves. Smiling landscapes, with well-cultivated fields, beautiful ranges of forest, and small clear lakes everywhere meet the eye. One would often be led to fancy oneself in some Danish province, if the splendid country seats, with their extensive parks, the numerous towns, the smoking factories, and the locomotive engines, with their trains darting continually to and fro, did not remind one of being in that land, which, with regard to riches and commerce, stands first in Europe. The plains are watered by noble and smooth-flowing rivers, which receive in their protecting embraces the thousands of ships which from all quarters seek the coasts of England. The winter is considerably milder than in our northern regions; and the sea air, not permitting the snow to lie for any length of time, renders the climate, on the whole, warmer. In summer the fields are clothed with the most luxuriant verdure. The leafy woods, with their numerous oaks, are filled with singing birds. The charm that is extended over English scenery, united with that freshness of life that stirs itself on all sides, cannot fail to make a deep impression on every foreigner. One feels in its full extent that the nature of the country presents all the requisites for greatness to a powerful and undegenerate people; and one no longer requires an explanation why it was not till after a desperate struggle that the ancient Britons relinquished it, or why, in after times, various nations strove with their utmost efforts for the possession of such a land.

The farther one travels towards the north or west of England, the mountains become higher, the valleys narrower, and the streams more rapid. In the north, however, the mountains rather resemble high hills. They do not tower in broken masses like the granite cliffs of Scandinavia. Their forms are softer and more undulating, and they are, too, clothed with a rich vegetation, and frequently overgrown with wood. In Cumberland and Westmorland are inwreathed those charming lakes whose beauties constantly attract a number of tourists. Even the ridge of the Cheviot Hills is not much more than about two thousand feet above the level of the sea: but stretching from east-north-east to west-south-west, with the river Tweed on one side, and the Solway firth on the other, they form a natural boundary between England and Scotland.

Farthest towards the west rise the mountains of Wales, England's real highland. The valleys here are short and narrow, yet the country has not the wildness of mountain tracts. Although it contains England's highest mountain, Snowdon, whose summit is nearly three thousand five hundred feet above the sea, still it unites the charms of plain and mountain. The whole of Wales may be regarded as a knot of mountains opposed by nature to the enormous waves of the Atlantic Ocean and Irish Sea. The middle is the highest part, whence rivers flow towards the east and west; the latter of which, after a short and foaming course, discharge themselves into the sea. The extent of the country, both in length and breadth, is, on the whole, inconsiderable.

This little mountain tract, which, in comparison with England, is poor as regards fertility, but all the richer in natural beauties, contains the last remains of the former masters of England, the Celtic Britons. By its remote situation, its rocks and narrow mountain passes, the characteristics of its former inhabitants have been preserved to our times. The people speak the ancient Welsh language, a branch of the Celtic stock; and have also inherited no small share of that burning hatred which their forefathers nourished against the English, who gained possession of their original fatherland by force.

Wales was united to England as early as the close of the thirteenth century; yet for ages later the Britons knew how to keep their country almost closed against the intrusion of strangers; whilst the harpers, by their ancient songs, kept alive the remembrance of past exploits and past disasters, and thus, as it were, still more hedged in and protected the language and nationality of the people. It was not till later times, when high roads, and at present railroads, began to open a more frequent intercourse between Wales and England, that the tones of the harp became almost entirely mute. The Welsh language gave way more and more to the English, and the time can hardly be far distant when the Celtic will become entirely extinct in Wales, as it has long been in Cornwall.

The people, whose scanty remnant thus spend the last days of their old age among the Welsh mountains, formerly belonged, both by possessions and kinship, to the most powerful in Europe. Not only were the Scotch and the Irish of the same origin with them, but on the other side of the channel, throughout Gaul, or France, Spain, and the middle and south of Europe, dwelt tribes of the Celtic race. Until about the time of the birth of Christ there was no people north of the Alps, which, with regard to power, agriculture, commerce, skill in the arts, and civilization in general, could equal, much less surpass, the Celts. Yet they were not strong enough to clip the wings of the Roman eagle, when it began to extend them over the Alps. The superior military skill and higher civilization of the Romans, triumphed over the various Celtic tribes, which were torn by internal dissensions, and could not once, even under the danger which menaced them, faithfully unite together. Shortly after the birth of Christ, therefore, the Roman hosts had already gained a footing in Britain, and, notwithstanding the violent and repeated attacks of the natives, soon made themselves masters of the country. They even fought their way to Scotland; where, however, the wild highlands, and their brave inhabitants, the Caledonians, arrested their victorious march. The Romans were now obliged to erect walls, ramparts, and towers, in order to prevent the highland Scots from uniting with the Britons, and to avert the speedy loss of the land which they had already won. Throughout Britain they laid the foundations of a civilization till then unknown there. They promoted agriculture, commerce, and trade; they made roads, and built towns and castles; and, as they had not immigrated in any great multitudes, they left the inhabitants in tolerably quiet possession of the soil of their forefathers.

But the Roman power fell in turn. It was natural that their dominion in so distant and sequestered a land as Britain should decay sooner and more easily than elsewhere, especially as the British chiefs did not fail immediately to revive the old disputes. Their rude neighbours in Scotland, the Picts and Scots, no longer restrained by fear of the Romans, made serious and devastating inroads upon the northern provinces of England, where no slight degree of riches and splendour already prevailed. The Britons, moreover, under the dominion of the Romans, had, like their kinsmen across the channel, already begun to grow cowardly and effeminate. Long oppression had given the power of the Celts a death-blow: and they were consequently unable to withstand the powerful and undegenerate tribes of Germany, which now, in the great tide of emigration from the east and north of Europe, rushed into the old Celtic countries, and made themselves new abodes, either, for the most part, putting the ancient inhabitants to death, or reducing them to a state of thraldom.

In the fifth century Angles, Saxons, and Jutes, from North Germany and the peninsula of Jutland, invaded Britain. The unfortunate Britons, when they would not submit to their conquerors, were persecuted with fire and sword, and were at last driven to the remote mountain districts in the West of England, particularly Cumberland (the land of the Cymbri or Celts), Wales, and Cornwall. After a sanguinary war, which lasted more than a hundred and fifty years, all their fine fruitful plains fell into the hands of their foreign conquerors, who continually brought more and more of their countrymen over, to build up again and inhabit the burnt or destroyed towns and houses, and to cultivate the neglected fields. The Angles settled principally in the north of England, the Saxons in the south and south-west, and mingled amongst both dwelt the Jutes, who do not appear to have been numerous enough to occupy large districts of their own. Under the common name of "Anglo-Saxons," the descendants of these nations continued for several centuries to be the reigning people, although the Britons did not cease to make harassing invasions on the frontiers of their hereditary enemies. For the rest, the Saxons successfully continued what the Romans had begun, with regard to the improvement of the land, and the promotion of civilization among the people. They were, it is true, divided into several tribes and smaller kingdoms, which not unfrequently warred against each other. But Christianity soon began to extend itself, and about the time of its introduction the separate kingdoms were united into one. Churches and convents rose with surprising rapidity throughout the country, and the pursuits of peace, science, and art, throve luxuriantly. Every plant, though foreign, flourished vigorously in the English soil.

In the first ages, however, Christianity produced among the people, as was the case in other countries besides England, a sort of degeneracy and weakness. Instead of the din of battle of the heathens there were now heard songs and prayers, which, joined with the constantly-increasing refinement, made the people dull and effeminate, so that they willingly bent under the yoke of their masters, both spiritual and temporal. In the ninth, tenth, and eleventh centuries the Anglo-Saxons had greatly degenerated from their forefathers. Relatives sold one another into thraldom; lewdness and ungodliness were become habitual; and cowardice had increased to such a degree, that, according to the old chroniclers, one Dane would often put ten Anglo-Saxons to flight. Before such a people could be conducted to true freedom and greatness it was necessary that an entirely new vigour should be infused into the decayed stock.

This vigour was derived from the Scandinavian north, where neither Romans nor any other conquerors had domineered over the people, and where heathenism with all its roughness, and all its love of freedom and bravery, still held absolute sway.

SECTION II

The Danish Expeditions.—The Danish Conquest.

A fate similar to that which the Anglo-Saxons had formerly brought upon the Britons, now partly became the lot of the Anglo-Saxons themselves. The same sea, the North Sea, or, as the old inhabitants of Scandinavia called it, "England's Sea," which in the fifth century had borne the Anglo-Saxons to England, and which had afterwards served to maintain the peaceful connections of trade, and the intercourse between kinsmen in England and in their northern fatherland, now suddenly teemed with the numberless barks of the Vikings, which, from the close of the eighth century, constantly showed themselves in all the harbours and rivers of England. For about three centuries the Danes were the terror of the Anglo-Saxons. They generally anchored their ships at the mouths of rivers, or lay under the islands on the coasts. Thence they would sail up the rivers to the interior of the country, where they frequently mounted on horseback, and conveyed themselves with incredible speed from one place to another. Their frightful sabre-cuts resounded everywhere. Their progress was marked by the burning of churches and convents, castles, and towns; and great multitudes of people were either killed or dragged away into slavery. In a short time they began to take up their abode in the country for the winter, and in the spring they renewed their destructive incursions. The terrified inhabitants imagined they beheld a judgment of God in the devastations of the Vikings, which had been foretold in ancient prophecies.

Not even the remote and poorer districts of Wales were spared. It is true that it was extremely difficult for the Danes to force an entrance on the land side, and, in order to do so by sea, it was necessary to make a troublesome and dangerous voyage round the long-extended peninsula formed by the modern Cornwall and Devonshire. In general its rivers were not large or navigable, and the number of good harbours was but small. Nevertheless, the Northmen seem to have known Wales well, as the old land of the Britons; since it was always called "Bretland," to distinguish it from England. Palnatoke, the celebrated chief of the Jomsvikings, is said to have married there, during one of his warlike expeditions, Olöf, a daughter of the Bretland jarl, Stefner, whose Jarledömme (earldom) Palnatoke afterwards possessed. The Sagas often make mention of *Björn hin Bretske* (Bear the Briton) as being among his men; and it is said that when he assisted at the funeral solemnities which his foster son, King Svend Tveskjæg[3], held in honour of his father, King Harald Blaatand[4], the half of his suite were Britons. Svend himself had ravaged Bretland; and it was there, as is well known, that the Icelander, Thorvald Kodransön, surnamed Vidförle (the far-travelled), delivered him by his noble disinterestedness from a perilous imprisonment.

3. Split-beard.

4. Blue-tooth.

The expeditions of the Danes to Bretland seem, however, to have been confined to the tracts bordering on the north bank of the Severn, and to the Isle of Anglesey; which latter was not unfrequently visited by the Norwegians in their piratical voyages to the Hebrides and Ireland. At least the Sagas mention it as "the southernmost region, of which former Norwegian kings had made themselves masters;" and it was probably here that Palnatoke had his kingdom. The very name of the island recalls a close connection with the inhabitants of the north. Anciently it was called "Maenige;" but the Danes and Norwegians, with regard, clearly, to its situation by the land of the Angles (England), gave it the name of "Öngulsey," or Angelsöen, whence the present form Anglesey may, doubtless, be said to have been derived.

The connections of the Danish Vikings with Bretland were, however, far from being always unfriendly. For as the Britons in Wales and Cornwall constantly nourished a lively hatred against the Anglo-Saxons, on whose lands they continued to make war, the Danes often entered into an alliance with them against their common enemies. The Danish and British armies were either combined, or else the Britons attacked from the west and south, whilst the Danes invaded the eastern coasts. These deep and well-laid plans show that the views of the Danes were no longer confined to robbery and plunder, with a view to gain booty, or to overthrow the churches and convents which threatened their ancient gods with destruction, but that they now seriously thought of conquering for themselves new tracts of country; nay, if possible, of subjugating or expelling the Anglo-Saxons throughout England.

Already in the ninth century the Anglo-Saxons had receded considerably before the Danes, who had obtained possessions on the east coast, where they quickly spread themselves, and where fresh arriving Vikings always found reception and assistance. The Anglo-Saxon king, Alfred the Great, was driven from his throne, and wandered about a long time in the forests, whilst the Danes held the sovereignty in his dominions. He succeeded, indeed, at length in regaining the crown; but in the mean time the possessions of the Danes on the east coast had been extended, and their power continually increased by the arrival of fresh emigrants, who settled in different parts of the country, and married the native women. Alfred, it is true, built fleets for the protection of the coasts; but the militia-men instituted in his time, in order to repel the frequent attacks of the Danes, now went over to them, accounting them their kinsmen. In Northumberland especially, the Danes, and a considerable number of Norwegians, had settled themselves securely under their own chiefs. Here they had sought a refuge against the new order of things which was now about to make itself felt in the mother countries, Denmark and Norway.

Partly as a result of the expeditions of the Vikings, and the frequent contact into which they were thus brought with Christian States, Christianity began, towards A.D. 900, to spread itself in the countries of Scandinavia. About the same time occurred there, as in the rest of Europe, a union of many small kingdoms under a single sovereign: and the Scandinavian tribes were subjected to the kings of Denmark, Norway, and Sweden. Some powerful and malcontent ones had indeed migrated beyond the sea; but, nevertheless, there were materials enough left for dissension in the new kingdoms, before Christianity could be generally introduced, and the power of the kings firmly established. A time arrived when the internal struggles in Denmark and Norway scarcely allowed the inhabitants to send any availing support to their friends in Northumberland, or to the other Danes on the coasts of England. Towards the middle of the tenth century, therefore, the hitherto almost independent Danish provinces in England were compelled to submit to the Anglo-Saxon kings, whose sovereignty, however, was but of short duration; for after the year 980 Danish and Norwegian Vikings again swarmed throughout England. Nor was it now, as formerly, merely the petty kings, who, with a comparatively inferior force, conducted these warlike expeditions. By degrees the Danish and Norwegian kings' sons, and even the kings themselves, endeavoured, with large fleets and well-appointed armies, to wrest the sceptre from the hands of the feeble Anglo-Saxon monarchs. It was in vain that the latter strove against them. They laid a tax on the whole land, called *Danegelt*, in order to defray the great expenses which the defence of the country against the Danes occasioned. But the money thus raised it was often necessary to expend in buying off the Danes, or in supporting their victorious hosts whilst they wintered in the country. The Anglo-Saxon king, Ethelred, after seeing his kingdom harried and fearfully devastated by the Danish king, Svend Tveskjæg, in conjunction with Olaf Trygvesön, the son of the king of Norway, first succeeded in making peace with Olaf in 995, and with Svend in 1002, after paying immense sums as Danegelt, and agreeing to many humiliating conditions.

As a last resource against the daily-increasing number and power of the Danes, Ethelred determined secretly and cruelly to murder these who were settled in England. The massacre took place on St. Bridget's eve, the 13th of November, 1002. Old and young, women and children, were murdered with the most frightful tortures. Not even the churches could protect the Christian Danes against the fury of the Anglo-Saxons. The slaughter was, however, confined almost exclusively to the south of England; since towards the north, and particularly in Northumberland, the population was chiefly of Danish and Norwegian extraction.

No sooner did the news of Ethelred's perfidious and sanguinary act reach Denmark, than a strong fleet was fitted out, and in the following year (1003) the Danish flag waved on the coasts of England. After numerous sanguinary battles, the Anglo-Saxons were compelled to submit to Svend Tveskjæg and Canute. What could not be conquered by force of arms was obtained through prudence and cunning. The Danish conquest of England was completed, and for about one generation Danish kings wore the English crown.

SECTION III

The Thames.—London.

London, and its wealthy neighbourhood, was naturally the main object of the Danish attacks in the south-east part of England. Under the Romans it had already become considerable as a commercial mart; but afterwards, under the Anglo-Saxons, it increased so much in wealth and importance, that it was, if we may use the expression, the heart of England. It was for this reason that the old northern bards used the term "Londons *Drot*" in their songs about the kings of England. From the first London is undoubtedly indebted for its greatness chiefly to its situation on the Thames, which opened an easy communication both with the opposite shores of the Continent and with the interior of England. In our days it is certainly a remarkable sight to observe the numberless ships that assemble there from all parts of the world, and to mark the activity that everywhere prevails on the beautiful shores of the river. But it becomes doubly remarkable when we recollect that this spectacle is neither a new one, nor has arisen under a single people; but that it has been repeated, in a somewhat altered form, for about two thousand years, under the most different circumstances: namely, under the dominion of the Britons, the Romans, the Anglo-Saxons, the Danes, and the Normans. In this respect there is no river whatsoever that can be compared with the Thames. Had it not been one of the most, or indeed quite the most, favourably situated stream in Europe for commerce, the greatest commercial city in the world would hardly have risen on its banks.

But just as the Thames brought, in the olden times, numerous merchant vessels, and, along with them, wealth and prosperity to the south of England, so must it also have frequently drawn down ruin on the surrounding districts, since it attracted thither almost all the Vikings who sought for booty and conquest. Nature herself has cut a deep bay into the eastern coast of England, at the mouth of the Thames, and thus pointed out to the Vikings the way they should pursue. The ships of the Danish Vikings constantly swarmed at the mouth of the Thames. "When they were not strong enough to sail up the river and attack London, or when the winter approached, they anchored under the coast, in places where they could lie in wait for and seize the merchantmen, and whence they could easily reach the open sea, if attacked by too superior a force. Some of their most important stations were under the Isle of Thanet, in Kent, and the Isle of Sheppey, (*Anglo-Saxon*, Sceapige, or the Sheep Island,) which lies at the mouth of the Thames. Thus these islands, whose remote situation rendered them sufficiently dangerous before, suffered doubly from the ravages committed by the Vikings on the coasts. Another place near the Thames, where the northern Vikings and conquerors generally landed when they harried the south of England, and where they often wintered, was the present Sandwich, in Kent. As it was an important landing-place even in the times of the Romans, they had already fortified it. Sandwich (*Ang.-Sax.*, *Wic en Stad*) became in the mouths of the Northmen "Sandvic," or the sandy bay; an appellation which perfectly agrees with the nature of the place. We find the same name for places in Orkney and the Shetland Isles, in Iceland, and Norway. From Sandwich it was but a few miles to Canterbury (in the northern tongue "Kantaraborg"), which, being a rich bishopric, was on that account exposed to remorseless plunder. In the year 1011 especially, the Jarl Thorkel the Tall, visited it with fire and sword. Christchurch, the principal church in England, was burnt down; the monks were put to death, and only one in ten of the citizens spared. Many, and among them Archbishop Elfeg, who was afterwards cruelly murdered, were cast into prison.

To the south of Canterbury, on the channel, lies "Dungeness;" and at the mouth of the Thames, "Foulness," and "Sheerness." The termination *ness*, in these names, seems to be neither Saxon, nor Celtic, but plainly the Danish and Norwegian *Næs* (a promontory, or lofty tongue of land, running out into the sea).

The nearer we approach London by the Thames, the more memorials we find of the Danes. Just before we reach the metropolis, we sail past Greenwich on the left, called by the northmen "Grenvik" (nearer, perhaps, "Granvigen," the pine-bay), whose celebrated hospital contains in our days a little host of England's superannuated seamen, who have fought in defence of her honour, and who, supported by the public, enjoy an old age free from care. In the eleventh century Grenvik was also for a long time the resting-place of a host of naval warriors, who were supported at the public expense; but that was a host of bold Danish Vikings, who, after having fearfully devastated England under their chief, Jarl Thorkel the Tall, had now, in 1011, allowed themselves to be bought off for an immense sum of money, and to settle down peaceably in the service of the English king Ethelred. From this time it became the custom for the English monarchs to have continually a standing army, composed mostly of Danes, "Huskarlene," or "Thingmen," as they were called (Þingmannalið), whose duty it was to keep the country quiet, and to defend it against foreign invasion; whence they sometimes came to fight against their own countrymen. King Athelstan (925-941) had, however, almost a century earlier, made use of Danish warriors to suppress revolt in his kingdom; for which purpose it was ordered that one of these men should be maintained in every house, in order that they might be always ready for the king's service. The Thingmen were to the English kings much what the Varangians were to the Greek emperors in Constantinople. They had certain rights and privileges, and later, in particular, two places were assigned to them for their headquarters—London in the south of England; and in the north, Slesvig (Nottinghamshire). Under King Canute, they played, as is well known, a considerable part.

The name of Canute the Great is connected not only with the town of Brentford (Brandfurda), on the Thames, near the western parts of London, and with Ashingdon (Assatun), in Essex, to the north-east of London, and, as the legend says, to the north of "Daneskoven" (the Danish forest), in which places he fought bloody battles with Edmund Ironsides, before he subdued England; but it is also connected in the closest manner with London itself.

When I sailed up the Thames for the first time, and when at length, above a forest of masts, the gray turrets of the Tower appeared on one side, and London Bridge in the distance, I was involuntarily led to recall the time when King Canute long lay in vain with his ships before the fortress and bridge of the metropolis, whilst a great part of the rest of England submitted to his sway. London Bridge was defended by three castles, one of which stood on the bridge itself. The Danes attempted to dig a canal round the foot of the bridge; and though Canute, who was well supported by Thorkel the Tall, and by Erik Jarl, the Norwegian, is said to have resumed the siege several times, yet it was by negociation alone that he seems to have obtained possession of London.

Even amid the varied impressions created by the metropolis of the world, I could not forget—and what Dane could?—that it was chiefly here that for a long period the Northmen found, as it were, another home, from which they returned to their native land enriched by fresh knowledge, and on the whole with a higher degree of civilization, which they afterwards turned to account in the north; that it was here that not a few of the most zealous promoters and defenders of Christianity in Scandinavia, and amongst them particularly the Norwegian king, Olaf Trygvesön, had dwelt before they began the work of conversion; that it was here, lastly, that several Danish chieftains, and especially Canute the Great, had played the sovereign, and held their court, surrounded by the *Thingmen* and the bards, who in those times usually accompanied the northern kings. On surveying London, its proud river, and beautiful uplands, one cannot help doubly admiring the power of that king, who, at a distance from his native land, was not only able to command all this, as well as the whole of England, but Norway and Denmark in addition. One feels the truth of the words of the Saga about Canute: "Of all kings that have spoken the Danish tongue, he was the mightiest, and the one that reigned over the greatest kingdoms."

Although London was at that time one of the most considerable towns in Europe, it was of course but very small compared with what it is at present. The walls inclosed only that proportionally small part of modern London called the "City," and which forms the centre of its busy commerce. Close by lay a castle (whence the Northmen's name for London, "Lundunaborg"), and undoubtedly on the same spot where, not long after Canute's time, William the Conqueror built the Tower. Somewhat higher up the Thames, on an island which, from the many thorns growing there, obtained the name of Thorney (*Anglo-Saxon*, Thornege), or the Thorn Island, stood another castle, said to have been inhabited at different times by Canute. This island, in whose name we find both the Anglo-Saxon *ege*, and afterwards the northern *ey* (island), and which is therefore sometimes very incorrectly called Thorney *Island*, has now lost both its ancient name and appearance. Under the name of Westminster, it forms at present a continuous part of London.

The Dane who wanders through this immense city, will not only be reminded by such names as "Denmark Court," "Denmark Street," and "Copenhagen Street," and by monuments in St. Paul's and Westminster Abbey, of the sanguinary battles which have taken place in modern times between England and Denmark, as well as of the older ties of friendship, which for a long time found increased support by means of the relationship and reciprocal marriages which occurred in the reigning families of the two countries; but he will also find traces even to this day, of the power and influence which his forefathers, both before and after King Canute's time, possessed in the most important commercial city of wealthy England.

Approaching the city from the west end, through the great street called "the Strand," we see, close outside the old gate of Temple Bar, a church called St. Clement's Danes, from which the surrounding parish derives its name. In the early part of the middle ages this church was called in Latin, "Ecclesia Sancti Clementis Danorum," or, "the Danes' Church of St. Clement." It was here that the Danes in London formerly had their own burial-place; in which reposed the remains of Canute the Great's son and next successor, Harald Harefoot. When, in 1040, Hardicanute ascended the throne after his brother Harald, he caused Harald's corpse to be disinterred from its tomb in Westminster Abbey, and thrown into the Thames; where it was found by a fisherman, and afterwards buried, it is said, "in the Danes' churchyard in London." From the churchyard it was subsequently removed into a round tower, which ornamented the church before it was rebuilt at the close of the seventeenth century.

It has, indeed, been supposed by some that this church was called after the Danes only because so many Danes have been buried in it; but as it is situated close by the Thames, and must have originally lain outside the city walls, in the western suburbs, and consequently outside of London proper, it is certainly put beyond all doubt, that the Danish merchants and mariners who, for the sake of trade, were at that time established in or near London, had here a place of their own, in which they dwelt together as fellow-countrymen. Here it should also be remarked, that this church, like others in commercial towns, as, for instance, at Aarhuus in Jutland, at Trondhjem in Norway, and even in the city of London (in East Cheap), was consecrated to St. Clement, who was especially the seaman's patron saint. The Danes naturally preferred to bury their dead in this church, which was their proper parish church.

The Danes and Norwegians also possessed an important place of trade on the southern shore of the Thames, opposite the city—in Southwark, as it is called, which was first incorporated with London, as part of the city, in the middle ages. The very name of *Southwark*, which is unmistakably of Danish or Norwegian origin, is evidence of this. The Sagas relate that, in the time of King Svend Tveskjæg, the Danes fortified this trading place; which, evidently on account of its situation to the south of the Thames and London, was called "Sydvirke" (Sudrvirki), or the southern fortification. From Sudrvirki, which in Anglo-Saxon was called Suð-geweorc, but which in the middle ages obtained the name of Suthwerk or Suwerk, arose the present form, Southwark, through small and gradual changes in the pronunciation. The Northmen had a church in Sydvirke dedicated to the Norwegian king, Olaf the Saint. Olaf, who fell in the battle of Stiklestad, in 1030, was so celebrated a saint that churches were built in his honour, not only in Norway, where he became the patron saint of the kingdom, and in the rest of Scandinavia, but also in almost every place where the Northmen established themselves; nay, even in distant Constantinople the Varangians had a church called after him. There is still a street in Southwark, close by London Bridge and the Thames, which bears the significant name of Tooley Street, a corruption of St. Olave's Street. On the northern side stands a church, called St. Olave's Church, and which is found mentioned by that name as early as the close of the thirteenth century.

Within the city, in what may be strictly called ancient London, where the Sagas already mention a St. Olaf's Church, there are to be found at this day no fewer than three churches consecrated to St. Olave: namely, in Silver Street; at the north-west corner of Seething Lane, Tower Street; and in the Old Jewry (St. Olave's Upwell). The two last-named stand in the eastern extremities of the city, yet within its ancient boundaries. In the same neighbourhood, near London Bridge, there is also a church dedicated to St. Magnus the Martyr, which likewise undoubtedly owes its origin to the Northmen, either the Norwegians or Danes. St. Magnus was a Norwegian jarl, who was killed in the twelfth century in Orkney, where the cathedral in Kirkwall is also consecrated to him.

That so many churches in London should be named after these Norwegian saints, Olaf and Magnus, who, moreover, were not canonized till after the death of Canute the Great, and the overthrow of the Danish dominion in England, furnishes no mean evidence of the influence of the Northmen in London. It confirms in a remarkable manner the truth of the old statements, that the Danes who dwelt in London could at times even turn the scales at the election of a king; as, for instance, after the death of Canute the Great. An English chronicler, speaking of the power of the Danes at that period, adds, that the citizens of London had, by reason of their frequent intercourse with "*the barbarians*" (the Danes), almost adopted their manners and customs. And it was, indeed, natural that the long voyages of the Northmen, and the important commerce carried on between the countries of Scandinavia and England, should have long secured to the northern merchants an influential position in a city like London, which was in the highest degree a commercial city, and particularly when these merchants had once been established there in great numbers.

But the most striking and remarkable memorial of the early power of the Danes and other Northmen in London is this—that the highest tribunal in the city has retained to our days its pure old northern name "Husting." The word *Thing*, whereby, as is well known, both deliberative and judicial assemblies were designated in the north from the earliest times, does not seem to have been employed by the Anglo-Saxons in that signification, or at all events not before the Danish expeditions and Danish immigrations into England. The Anglo-Saxons used in that sense the term *gemót*, as in "Witena-gemót," which was the name of their parliament. Husthings are also especially mentioned in the Sagas as having been held in the north, particularly by kings, jarls, and other powerful individuals. The Husthing in London was originally established in order to protect and guard the laws and liberties of the city and the customs of the courts of judicature; and the principal magistrates were judges. In the Latin of the middle ages it is said of a person who attended there—"Comparuit in Hustingo." A similar Husting was also formerly found in the Isle of Sheppey, at the mouth of the Thames.

London, beneath whose walls and gates the Danes have fought numerous battles with various success, contains within it memorials both of their greatest power and of the decay of their dominion. On the same side of the Thames as Sydvirke, or Southwark, but somewhat higher up, lies Lambeth (formerly Lambythe, Lambgathre), which is now a part of London, and the residence of the Primate of England, but which in olden times was a village outside the capital. At a country-house there a Danish jarl celebrated his marriage in the year 1042. King Hardicanute, with a number of his followers, was present at the banquet; but just as he was drinking to the bride, he suddenly fell to the ground, in a fit of apoplexy, and shortly afterwards breathed his last at the age of only twenty-six years.

Hardicanute was the last Danish king in England.

SECTION IV

Watlinga-Stræt.—South England.—Legends about the Danes.—The graves of Canute the Great and Hardicanute.

In the heart of the city of London, near St. Paul's Cathedral, is a street called "Watling Street." Anciently it was connected with the great high road of the same name (or more properly Watlinga-Stræt), which had been made by the Britons from the Channel and London through the midst of England to the north-east of Wales, Chester, and the Irish Channel. On account of the importance of this road, as communicating with the interior of England as well as with Ireland, the Romans improved it. But, like most of the high roads of ancient times, it was carried over heights, with the constant view of avoiding streams which would require the erection of bridges. It followed, as nearly as possible, the natural division of the watercourse in England, or the ridge of the land watershed whence rivers take their course in all directions.

About the year 1000 this road not only showed the natural boundary between the northern and southern river-valleys, but likewise indicated in the clearest possible manner a political boundary between the inhabitants of different extraction, and different manners and customs. The districts to the north and east of this road belonged for the most part to the so-called "Dena-lagu," or "Dane-lagh," that is, the Dane's community (from *lag*, whence in the north itself, in Norway, for instance, *Thröndelagen*, and in Sweden, *Roslagen*). For here the Danes, and other conquerors or immigrants of Scandinavian origin, had gradually subdued and expelled the Anglo-Saxons, and here the Danish laws, habits, and customs, chiefly prevailed.

In the districts to the south, on the contrary, the repulsed Anglo-Saxons had concentrated the last remnants of their former power. A great number of wealthy and leading Danes were indeed also settled here, either in the country, or, with a view to commerce, in the principal towns on the coast; as in Winchester, which, like London, long had its "Husting;" Exeter, where a church was in later times dedicated to St. Olave; and Bristol. But, out of London, the Danes scarcely formed at that time any really strong and united power in the south of England. The predominating people was the Anglo-Saxon, and in general the old Saxon characteristics had been preserved.

To the south of Watlinga-Stræt, which had already often been agreed upon between the Danish conquerors and the Anglo-Saxon kings as the boundary between the Danish and Anglo-Saxon kingdoms, Edmund Ironsides received his share of England by agreement with Canute. It was in these districts that the Anglo-Saxon kings had always found their truest and most numerous adherents, and they had therefore generally been the theatre of the more important battles between the Anglo-Saxons and the Danes. Near Wareham, in Dorsetshire, Alfred purchased peace with a host of the latter, who swore on their armlets to observe it; but, though this oath was regarded by the Danes as very sacred, they are said to have broken it immediately. During his exile Alfred concealed himself for a long time at Athelney, in Somersetshire; and near Eddington he again beat the Danes. In the neighbourhood of Athelney, Alfred also induced Gudrun (Gorm), the king of the Danish Vikings, to receive baptism. The oppressed inhabitants were in these parts scarcely ever free from the devastating attacks of the Vikings and conquerors. The Danes frequently established themselves in castles near the coast, as at Exeter, in Devonshire; Dorchester and Wareham, in Dorsetshire; Winchester, in Hampshire; and Chichester, in Sussex. At Southampton, in Hampshire, and under the Isle of Wight, they generally wintered with large fleets. Thence they made incursions into the land of the Anglo-Saxons; and if they could not entirely expel them, and colonize the south of England in their stead, they at least endeavoured to weaken and exhaust it as much as possible.

On the whole, it would not have been very easy for the Danes to settle themselves entirely in any parts of the south, or south-west, of England; not even on the coasts near the harbours, though regularly visited by the ships of the Norwegian Vikings. The inhabitants in these parts were mostly of pure Saxon descent, and consequently already prejudiced against the Danes, on account of the old disputes between the Scandinavian and Saxon races; at all events, they somewhat differed from the Danes in character, manners, and customs. These districts were, besides, too remote from Denmark; and in case of an attack from the Anglo-Saxons, which might naturally be expected to take place, assistance might come too late. The Danes were not so safe there as on the east coast of England, which lay opposite to Jutland, and where, if any danger threatened them, a ship could easily be sent with a message to their friends over the sea, so that, with a tolerably favourable wind, a strong fleet could be speedily brought within sight of the Anglo-Saxons. The Angles, whose descendants inhabited these eastern and northern districts, seem too, with regard to language and national manners, to have borne a greater resemblance to the Danes than the inhabitants of any other part of England, so that it was by no means difficult for the Danes speedily to amalgamate with them. In addition to this, the eastern coasts offered much the same allurements to the Danes as the more southern provinces. They were remarkable for their fertility and for the riches of their inhabitants, acquired as well by agriculture as by trade with Saxony, Belgium, and Gaul. Precisely on the east coast, indeed, were situated at that time some of the largest commercial towns in England.

It is not surprising, therefore, that, with the exception of London and its environs, there are not found in the south of England, as is the case farther north, many names of places of well-defined Danish or Norwegian origin, which have preserved the old northern forms down to the present day, and which thus clearly testify that a genuine Scandinavian population must long have lived there. It is only at the extremities of the coasts that an occasional promontory, or "Næs," and small islands whose names end in *ey* and *holm*, remind one of the Northmen; as Flatholmes (*Dan.*, Fladholmene) and Steepholmes in the Severn, where there are said to be remains of Danish fortifications; Grasholm (*Dan.*, Græsholm), to the west of Pembrokeshire; Bardsey, west of Caernarvonshire; Priestholm (*Dan.*, Præsteholm), near the northern inlet of the Menai Straits; and several others.

In the south of England one cannot discover any striking resemblance to the Danes either in the language, features, or frame of body of the people. What they have chiefly left behind them here is a name, which will certainly never be entirely eradicated from the people's memory. Centuries after the Danish dominion was overthrown in England, the dread of the Danes was handed down from one generation to another, and even to this day they occupy a considerable share in the remembrance of the English nation. Throughout England the common people—nay, even a great number of the more educated classes—know of no other inhabitants of the north of Europe than "the Danes;" and as they include under this name both Swedes and Norwegians, the idea of the unity of Scandinavia has unconsciously taken root amongst them. That they have so implicitly awarded the first place in Scandinavia to the Danes, has not originated solely from the fact that, anciently, the Danes were really regarded as the leading people in the north—whence also the old Norwegian language was often called "dönsk tunga" (Danish tongue); nor because the Danes at that time undoubtedly exercised a more important influence on the British Isles than the other inhabitants of the north; it may, likewise, have arisen from the circumstance that, partly in consequence of its situation, Denmark has continued to stand, even down to our time, in much closer relations both of peace and war with England, than Sweden has; and that the separation of Norway from Denmark is still too recent an event to have completely penetrated to the knowledge of the less informed part of the English people. Even had the remembrance of the Danes in England lain slumbering there, such events as the battle in Copenhagen roads in 1801, and the seizure of the Danish fleet in 1807, must at once have brought all the old tales respecting the doings of the Danes in England to the lips of the English people.

Legends about "the Danes" are very much disseminated among the people, even in the south of England. There is scarce a parish that has not in some way or another preserved the remembrance of them. Sometimes they are recorded to have burnt churches and castles, and to have destroyed towns, whose inhabitants were put to the sword; sometimes they are said to have burnt or cut down forests; here are shown the remains of large earthen mounds and fortifications which they erected; there, again, places are pointed out where bloody battles were fought with them. To this must be added the names of places; as, *the Danes-walls*, *the Danish forts*, *the Dane-field*, *the Dane-forest*, *the Danes-banks*, and many others of the like kind. Traces of Danish castles and ramparts are not only found in the southern and south-eastern parts of England, but also quite in the south-west, in Devonshire and Cornwall, where, under the name of *Castelton Danis*, they are particularly found on the sea coast. In the chalk cliffs, near Uffington, in Berkshire, is carved an enormous figure of a horse, more than 300 feet in length; which, the common people say, was executed in commemoration of a victory that King Alfred gained over the Danes in that neighbourhood. On the heights, near Eddington, were shown not long since the entrenchments, which, it was asserted, the Danes had thrown up in the battle with Alfred. On the plain near Ashdon, in Essex, where it was formerly thought that the battle of Ashingdon had taken place, are to be seen some large Danish barrows, which were long, but erroneously, said to contain the bones of the Danes who had fallen in it. The so-called dwarf-alder (*Sambucus ebulus*), which has red buds, and bears red berries, is said in England to have germinated from the blood of the fallen Danes. It is therefore also called *Daneblood* and *Danewort*, and flourishes principally in the neighbourhood of Warwick; where it is said to have sprung from, and been dyed by, the blood shed there, when Canute the Great took and destroyed the town.

Monuments, the origin of which is in reality unknown, are, in the popular traditions, almost constantly attributed to the Danes. If the spade or the plough brings ancient arms and pieces of armour to light, it is rare that the labourer does not suppose them to have belonged to that people. But particularly if bones or joints of unusual size are found, they are at once concluded to be the remains of the gigantic Danes, whose immense bodily strength and never-failing courage had so often inspired their forefathers with terror. For though the Englishman has stories about the cruelties of the ancient Danes, their barbarousness, their love of drinking, and other vices, he has still preserved no slight degree of respect for Danish bravery and Danish achievements. "As brave as a Dane" is said to have been an old phrase in England; just as "to strike like a Dane" was, not long since, a proverb at Rome. Even in our days Englishmen readily acknowledge that the Danes are "the best sailors on the Continent;" nay even that, themselves of course excepted, they are "the best and bravest sailors in all the world." It is, therefore, doubly natural that English legends should dwell with singular partiality on the memorials of the Danes' overthrow. Even the popular ballads revived and glorified the victories of the English. Down to the very latest times was heard in Holmesdale, in Surrey, on the borders of Kent, a song about a battle which the Danes had lost there in the tenth century.

Amidst the many memorials of "the bloody Danes," the name of Canute the Great lives in glorious remembrance amongst the English people. It is significant that later times have ascribed to Canute the honour of important public undertakings for the common benefit, which, however, at most, he can only have continued and forwarded. In the once marshy districts towards the middle of the east coast of England, there is a ditch several miles long, called the Devil's dyke (in Cambridgeshire), the formation of which is by some attributed to Canute, although it existed in the time of Edward the Elder. Canute's name is also given to a very long road over the morasses near Peterborough (Kinges or Cnutsdelfe), although it was made before his reign. Canute's name is also preserved in Canewdon (Canuti domus), near London, and close by the battle-field of Ashingdon, in Essex, where he is said to have frequently resided. In like manner a bird, said to have been brought into England from Denmark, has been called after him *Knot* (*Lat.*, Tringa Canutus seu Islandica).

It may be asserted, with truth, that not many English kings have left a better name behind them than Canute. He does not owe this only to the favour he showed the clergy, the authors of most of the chronicles of ancient times. He acquired it by his numerous and excellent laws, by the power he exerted in restoring order and tranquillity in the kingdom, by his wisdom in suppressing the ancient animosities between the Danes and Anglo-Saxons, as well as by the care he took to promote the knowledge and piety of his people. He issued severe laws against heathenism, and endeavoured to wipe out the traces of his forefathers' devastations by re-building convents and churches. He even caused the corpse of Archbishop Elfeg, so cruelly murdered by the followers of Thorkel the Tall, to be conveyed with great solemnity from London to Canterbury, and deposited in the cathedral. To these traits may be added his many excellent personal qualities, his sincere repentance for the acts of violence which he committed in the heat of passion, and his profound humility before

God. The story of his shaming some of his courtiers, who flattered him when walking on the seashore whilst the tide was flowing, is, if possible, still better known in England than in Denmark. It would be difficult to find any one who is not acquainted with all the particulars of it, and who has not heard it stated that Canute, from that very day, placed his golden crown on the altar of Winchester cathedral, and never wore it more. This is one of those traits of true nobility and greatness of soul that are imperishable in all times and ages.

Canute was first buried in the old convent of St. Peter's at Winchester; but his body was afterwards removed into the grand choir of the cathedral, where both his and his son Hardicanute's tombs are still to be seen. Over Hardicanute's, in the wall that surrounds the middle of the choir, was placed (1661) a stone, on which a ship is carved, and the following inscription:—

Qui jacet hic regni sceptrum tulit Hardicanutus;

Emmæ Cnutonis gnatus et ipse fuit.

In hac cista Lo. 1661. Obiit A.D. 1042.

Or, "Hardicanute, who lies here, and who was a son of Emma and Canute, bore the kingdom's sceptre. He died in the year of our Lord 1042, and was placed in this coffin in 1661."

[+ +] Hardicanute's Tombstone, ship

The form of the ship on the tombstone shows it to be of no older date than the seventeenth century; but it was possibly carved there because a ship of war had previously adorned the tomb of Hardicanute. At all events, it indicates his relationship with the powerful Scandinavian sea-kings, and his descent from those Northmen who for centuries were absolute on the ocean.

Above the before-mentioned wall, in the grand choir, there stands to the left of the entrance a rather plain wooden coffin, decorated with a gilt crown, half fallen off, with the inscription:—

"In this and another coffin, directly opposite, repose the remains of Kings Canute and Rufus, of Queen Emma, and of the Archbishops Winde and Alfvin."

[+ +] Canute's Tomb

In Cromwell's time, the coffins of the kings in the grand choir of Winchester cathedral were broken open, and the bones dispersed; but they were afterwards collected together, as far as this could be done, and again placed in the grand choir in coffins like the one just mentioned. Thus Canute the Great, whose ambition could not be bounded even by three kingdoms, has not retained so much as a grave for himself and his beloved Emma. The presentiment of the perishableness of all earthly power that seized him when he deposited his golden crown in the same place has, in truth, been fulfilled!

The other royal coffins that surround the grand choir in Winchester contain the bones of several old Saxon kings. That the Danish kings Canute and Hardicanute should be entombed among them, in the midst of Anglo-Saxon south England, is a sufficient proof of the immense change that had taken place with regard to the Danes in England since their first appearance there as barbarous heathen Vikings. Instead of their kings seeking renown by the destruction of churches and convents, and by murdering or maltreating the clergy; instead of their despising any other kind of burial than that in the open fields, on hills under large cairns, or monumental stones, their successors were now regarded as the benefactors and protectors of the Church, and as such worthy to repose in the most important ecclesiastical edifices, even in the principal district of their former mortal enemies. Nay, the clergy there were indefatigable in handing down their glory to the latest ages; and thus a statue of Canute the Great was long to be seen in the cathedral of Winchester.

But this also affords a striking proof that the Danes and Anglo-Saxons no longer regarded each other so much in the light of strangers, or with such mutual feelings of enmity as before; and that Canute had thus happily broken through the strong barrier which had hitherto separated Saxon south England from Danish north England.

SECTION V

The Wash.—The Five Burghs.—The Humber.—York.—Northumberland.—Stamford

Bridge.

The Thames certainly brought many Danes in ancient times to the country south of Watlinga Stræt; but the large bay on the eastern coast of England, called the "Wash," and the rivers Humber, Tees, and Tyne, attracted still more of them to the eastern and northern districts. The Wash especially seems to have been one of the landing places most in favour with them. Whether it were its situation, directly opposite to Jutland on the one side, and on the other, on a line with the fruitful midland districts of England; or whether it were rather the rapid current which sets in there that attracted the ships of the Vikings, is a point that we must leave undecided. This much, however, is certain, that the first and richest settlements of the Danes were around this bay; and from it afterwards extended itself quite up to the frontiers of Scotland, the so-called "Danelagh;" which was a district so considerable as to comprise fifteen of the thirty-two counties, or shires, then existing in England, and amongst them the extensive county of Northumberland.

South of the Wash, and extending towards the Thames, lay East Anglia (Norfolk and Suffolk); which, a century after the commencement of the Vikings' expeditions, was already in the hands of the Danes. Alfred the Great was compelled to cede it, together with several adjacent tracts of country, by formal treaty, to the Danish King Gudrun, or Gorm. It is certain that it had at that time, like Kent, received many Danish settlers, particularly from the neighbouring Jutland, and their number continually increased. Yet in East Anglia they seem to have been scarcely more in a condition to compete with the Anglo-Saxons, in regard to population and power, than in Kent. It was only on the coast, and indeed only on that of Norfolk, that they had any settlements, as the Scandinavian names of places still preserved there show. These districts lay too near to the main strength of the Anglo-Saxons. The Saxon inhabitants did not easily suffer themselves to be expelled, and the Danish dominion there could not, consequently, become of permanent importance.

But to the north and west of the Wash the Danes obtained a very different footing. In the province called Mercia (or the Marches), which formed the centre of England, and in that of Lindisse (or, in old Norsk, Lindisey), which extended from the Wash to the Humber, they were not only in possession of a great number of villages and landed estates, which they had selected to settle on, but had likewise made themselves masters of several towns, and particularly the five strong fortresses of Stamford, Leicester, Derby, Nottingham, and Lincoln. These places, which as early as Alfred's reign belonged to the Danes, and which were distinguished by their size, their commerce, and their wealth, obtained the name of "The Five Burghs" (Femborgene). They formed, as it were, a little separate state, and possessed in common their own courts of judicature, and other peculiar municipal institutions. The hostile and dangerous neighbourhood of the Saxons naturally compelled them to coalesce together as much as possible; and for a very long period they formed the chief support of the Danish power in England. Protected by them from all attacks from the south, the Scandinavian settlers were enabled securely to continue establishing themselves in the more northern districts. To arrest the sudden attacks of the Britons in the west, the Danes also had, on the north-eastern frontier of Wales, the city of Chester, whose name (*Anglo-Saxon*, Lægeceaster, from the Latin castra, a camp) shows that it had been a fortified place still earlier, under the Romans.

Chester formed one of the principal entrances from Wales into the midland parts of England, as well as into what was then called Northumberland: under which name was comprised, at least by the Danes and Norwegians, all the country to the north of the rivers Mersey and Humber, from sea to sea, and up to the Scottish frontier. Covered by the "Five Burghs," it was here that the greater part of Danish England lay. It was a country filled, particularly in the north-west, with mountains, and intersected by numerous rivers. Near these, valleys opened themselves in every direction, of which the largest and most considerable lay around the tributary streams of the Humber, in what is now Yorkshire. A separate kingdom had existed here from the oldest times; and here the Danes, like the Britons, the Romans, and the Anglo-Saxons before them, possessed the most important city in the north of England. Built on the river Ouse, which falls into the Humber, it carried on an extensive trade; and, as the principal seat of the Northumbrian kings and chiefs, was doubly important. The Britons called it "Caer Eabhroig," or "Eabhruc," the Romans "Eboracum," the Anglo-Saxons "Eoforwic," and the Danes "Jorvik;" whence it is plain that the form "York," now in use, is derived.

The Humber and York were for the north of England much what the Thames and London were for the south. It is not therefore surprising that York came to possess within its walls the largest and most splendid cathedral in England, which still towers aloft, a proud and awe-inspiring monument of the power and religious enthusiasm of the middle ages; nor that the history of York comprises, so to speak, the whole of that of Northumberland.

The soil of south England received the dust of the Christian Danish kings, and of Canute the Great, the hero of Christendom. But the north of England held the bones of many a mighty Danish chieftain, who had never renounced his belief in the ancient gods; and, in the neighbourhood of York, one of the most renowned of heathen heroes, King Regner Lodbrog, met his death. The names of Regner and his sons were reverenced and feared in England from their earlier Viking expeditions. When about to invade England, he suffered shipwreck, and together with only a few of his men saved himself on the coast of Northumberland. The Saxon king, Ella, advanced against him from York; a battle ensued, and, after the bravest resistance, Regner was overcome and made a prisoner. With true northern pride he would not make himself known to Ella, who caused him to be thrown into a pen filled with snakes; and it was not till the dying Regner had sung his swan's-song, "Grynte vilde Grisene, kjendte de Galtens Skjebne" (How the young pigs would grunt if they knew the old boar's fate), that Ella too late observed to his terror that he had exposed himself to the fearful vengeance of the king's sons; who, guided by the shrewd Ivar Beenlöse, had long been silently preparing for the conquest of

Ella's kingdom. Ella was vanquished and made prisoner; and, according to the Norwegian legend, Regner's sons, to avenge their father's miserable death, caused a blood-eagle to be carved on Ella's back. The place of Ella's death is said by some to have been near the town of "Ellescroft," or Ella's Grave. The English accounts make Regner's sons, Ingvar and Ubbe, revenge their father's death in the year 870, by murdering in a most horrible manner King Edmund (who was afterwards canonized) at the castle of Æglesdon, in East Anglia. They shot at him as at a mark, then cut off his head, and lastly laid the body among thorns, in the same forest where their father had been put to death.

Ivar Beenlöse (the Boneless) succeeded to the kingdom of Northumberland after Ella; where also such names of subsequent kings as Sigtryg, Regnald, Godfred, Anlaf (Olaf), and Heric (Erik), unmistakably show their Scandinavian origin. In Olaf's time, at the beginning of the tenth century, the Anglo-Saxon king Athelstane (Adelsteen) succeeded in subjecting Northumberland, whilst Denmark and Norway, as before mentioned, were prevented by internal distractions from sending any effectual assistance to the Danes in England. Olaf fled to Ireland, and Godfred to Scotland, to assemble the Scandinavian warriors in those parts, and Athelstane in the mean time destroyed the Danish castle in York. It is related that Olaf returned with more than six hundred ships, and again took possession of York. He had with him a great number of Northmen and Danes from Ireland and Scotland, together with a great many Celtic Cymri and Britons, and the Scottish King Constantine was also in his army. Athelstane and this brother Edmund arrayed a mighty force against them at Brunanborg (Bromford?), where, in the year 937, a battle was fought; which, though unfavourable to the Danes, afforded the old northern bards matter for enthusiastic song, of which the Sagas have still preserved some remains. Subsequently a treaty with King Edmund, in 941, gave Olaf the dominion over the country east and north of Watlinga-Stræt; but the dispute soon broke out afresh. After the death of the Northumbrian King Erik in 951, Northumberland ceased to be a kingdom. From this time it became an earldom (Jarledömme), which was, however, for the most part, almost entirely independent of the Anglo-Saxon kings, and governed by Norwegian chieftains. For a long time it constantly received fresh inhabitants from the mother countries, Denmark and Norway. Many Norwegians came over; nay, even the King Erik just mentioned may possibly have been the renowned Norwegian King Erik Blodöxe, a son of Harald Haarfager, the first absolute sovereign of Norway. After the death of Harald, Erik became chief sovereign in Norway; but he and his queen, the notorious Gunhilde, ruled here with so much cruelty, that the Norwegians gave Erik the surname of Blodöxe (Blood-axe). Driven from his kingdom, he at length repaired to Northumberland, where King Athelstane is said to have made him a tributary king, and where, after many vicissitudes of fortune, he met his death.

Between the Northumbrian Jarledömme—whence the dignity of the Northern "Jarls" began to extend itself to the rest of England, which has still preserved it in the title of "Earl"—as well as between the Danish part of England and the proper kingdom of the Anglo-Saxons in general, disputes must naturally have prevailed of a more or less sanguinary kind. As a necessary consequence of this, the Danish kings, in their later expeditions against the Anglo-Saxons for the purpose of conquest, resorted to, and sought support in, the Danish part of the north of England, in the districts near the Humber. In the year 1013, King Svend Tveskjæg anchored in this river with a powerful fleet, when he came over to conquer England. In conjunction with his son Canute, who afterwards completed the conquest, he had previously lain at anchor at Sandvik (Sandwich), in Kent. From the Humber he anchored in the river Trent, at Gegnesburgh (or Gainsborough), in Lincolnshire; whence he harried the whole of eastern, and part of southern England. The Old Danish land to the north of Watlinga-Stræt was the first to pay him homage; the rest of England soon yielded to him, and King Ethelred was obliged to fly to Normandy. But just as Svend, in the midst of his victorious career, had returned to Gainsborough—just as he was fleecing and levying contributions both on laity and clergy—he suddenly fell from his horse at an assize, or *Thing*, in a fit of illness, and died the following night, the 3rd of February, 1014. Monkish chronicles relate that it was St. Edmund who killed him. Ethelred, who now returned to England, in vain ordered a strict search to be made for the body of Svend, with the view of wreaking a cowardly vengeance on the impotent corpse of the man who, when alive, had been so terrible an antagonist to him. But the body had been secretly conveyed to York, where it was kept concealed during the winter (but scarcely in the cathedral, although that church had been founded long before, and was, perhaps, even considerably enlarged by the Norwegian princes who resided at York). Towards the spring it was brought over to Denmark by some Englishwomen, who were probably of Scandinavian extraction, and placed in the cathedral of Roeskilde, in one of the pillars in the grand choir.

Under the Danish rule, the Danish-Norwegian population in the north of England increased considerably, both in strength and numbers; although Christianity, by the wise arrangements of Canute, and particularly by his severe laws against heathenism, was almost completely disseminated there. Even after the Danish dominion had come to an end by the death of Hardicanute in 1042, and the Anglo-Saxon kings had again taken the helm, the old warlike spirit of the north continued, in spite of Christianity, to stir in the Northumbrian people. The successors of the Vikings still preferred, to a natural death, a glorious one on the field of battle; but Christian tenets no longer permitted them to be marked, when on the bed of sickness, with the point of a spear, in order to consecrate themselves to Odin, according to the heathen custom. The mighty Danish jarl Sivard (Sigeward or Siwerd) reigned over them at that time, who had fought in many battles both in England and Scotland, whereby his name became immortalized in Shakspeare's "Macbeth." When the news was brought to him that his son had fallen in battle, he inquired whether he had received his death wound in front or behind. Being answered, "Before;"—"In that case," he exclaimed, "I have reason to rejoice, for no other death was befitting my son, or me." When Siward himself afterwards lay on his death-bed, and felt the approach of dissolution, an old chronicler (Henry of Huntingdon) represents him as breaking out into sorrowful complaints, and exclaiming, "How shameful it is for me, that I have never been able to meet death in my numerous battles, but have been reserved to die with disgrace like an old cow. Clothe me at least in my impenetrable armour, gird me with my sword, cover my head with my helmet, place my shield in my left, and my gilded axe in my right hand, that I, the bold warrior, may also die like one." Attired in full armour, he passed gladly to his fathers in the year 1055, and doubtless with the secret hope of enjoying in Valhalla a continuation of that proud martial life for which there would soon have been no longer room either in Northumberland or in the parent lands of Scandinavia.

Shortly after the death of Siward, the country near York also became the theatre where one of the last celebrated Vikings of the north fell. Harald Haardraade was indeed a Christian, and a king in Norway; but with him, as with many of his cotemporaries, Christianity dwelt only on his lips. In his heart he was still the bold Viking, who valued Hildur's bloody game more than holy psalms, and who preferred conquest on foreign shores to the peaceful government of an hereditary kingdom. Whilst still young he had distinguished himself in expeditions in the East, and in the Greek Empire. It seemed to him disgraceful that those lands, particularly in the north of England, which had once belonged to his forefathers, should for ever be wrested from Norway. He therefore agreed to assist Toste Godvinsön against his brother, the English King Harald Godvinsön; but on the condition that he himself, if he succeeded in conquering Harald, should have the dominion of England, whilst Toste was to have the half of it as jarl, or earl. They landed in the Humber; but in the battle which shortly afterwards took place (in 1066) at Stamford Bridge, a little to the east of York, both Toste and Harald fell. Thus the latter gained no more of England's soil than the English King Harald had offered him before the battle, namely, "seven feet of earth, or as much as he was taller than other men."

This was one of the last serious attempts on the part of Denmark or Norway to reconquer England; and in the same year the Normans, after the battle of Hastings, in which King Harald fell, seized the kingdom which their Danish kinsmen had formerly possessed. William the Conqueror went in person against the Northumbrians; but before he disembarked he is said to have broken up the tumulus on the coast (by the Humber?) in which, according to the legend, Regner Lodbrog's son, Ivar Beenlöse, had ordered himself to be buried, in order to avert the attacks of foreigners. William had to combat long before he could reduce Northumberland; but, as we shall afterwards see, he never succeeded in subduing that spirit of freedom and independence which the Danes and Norwegians had planted there.

SECTION VI

Danish-Norwegian Memorials in the North of England.—Coins.—The

Raven.—The Danish Flag.

If even the old Saxon south England is distinguished by its richness in legends and still-existing memorials of the Danes, it is natural that they should be met with in still greater numbers in the old Danish districts to the north and east of Watlinga-Stræt.

Here also the Norwegian saint, "St. Olave," has been zealously worshipped, both in the country and in the towns. In Norfolk (East Anglia) there is a bridge called "St. Olave's Bridge." In itself it is a remarkable monument of a time when bridges over rivers were regarded as such considerable and important structures that, like churches, they were named after, or dedicated to saints; in ancient Scandinavia they even built bridges, as several runic stones testify, "for their souls' salvation." In the city of Chester, on the northern frontier of Wales, there is to be found in the southern outskirts, opposite the old castle and close to the river Dee, a church and parish which still bear the name of St. Olave. By the church runs a street called "St. Olave's Lane." In the north-west part of York there is likewise a St. Olave's church, said to be the remains of a monastery founded by the powerful Danish Jarl Siward, who was himself buried there in the year 1055. There can be no doubt that similar churches dedicated to St. Olave were scattered about in other towns of north England, where further researches might possibly yet discover at least some of them.

These traces of the importance formerly conferred on St. Olave in the towns of north England lead one to conjecture that, even after the Danish ascendancy in England was annihilated, a great number of Northmen must have continued to reside there, as was the case in London. This is so much the more natural, as, long before the Norman Conquest, the Northmen preponderated in many, perhaps in most, mercantile towns of the north of England, and particularly in the fortified towns occupied by the Danes. At the time of the

Conquest, the population in some of the largest and most important cities towards the east coast, such as Lincoln and York, is said to have been almost exclusively of Scandinavian extraction; hence it was that Lincoln and York, at least, preserved their original Scandinavian "husting" throughout the middle ages, and even later.

In and about the last-named city, which was the chief place in Danish north England, are numerous Scandinavian memorials. The names of several streets in York end in *gate*. In London, where the same termination of the names of streets frequently occurs, some have, indeed, endeavoured to derive this *gate* from the gates which these streets adjoined; and, as far as regards London, this explanation may probably in most cases be correct. But in York, where formerly there were at least a score of such streets, it is certainly by no means a probable conjecture that twenty gates existed from which their names were derived; and it therefore becomes a question whether these *gates* should not be derived from the old Scandinavian "*gata*" (a street), particularly when they appear in compound names, such as Petersgate (Petersgade), Marygate (Mariegade), Fishergate (Fiskergade), Stonegate (Steengade), Micklegate (from the old Scandinavian "mykill," signifying great); which have a striking resemblance with Scandinavian names of streets; nay, there is even a legend respecting Godram, or Guthramgate, that it was named after a Danish chieftain, Guthrum or Gorm, who is said to have dwelt there. The historical accounts of the number and influence of the Northmen in York cannot but strengthen these suppositions in a high degree.

North-east of York, on the coast towards the German ocean, is a promontory called "Flamborough-head." It is separated from the main land by an immense rampart said to have been raised by the Danes, and called on that account "the Danes' Dyke," behind which they intrenched themselves on landing. At no great distance, near Great Driffield, is "the Danes' Dale," and "the Danes' Graves," where remains of the Danes who fell in a battle are said to have been dug up. South of York, on the Humber, between Richal and Skipwith, human bones and pieces of iron have likewise been found in several barrows, or tumuli, ascribed to the Danes. It is supposed that the Danes and

Norwegians landed in this neighbourhood at different times, when proceeding up the Humber on their warlike expeditions.

The popular legend of the bloody battle by Stamford Bridge, or, as it was afterwards called, "Battle Bridge," is not yet obsolete. A piece of ground near the bridge over the river Derwent is called "Battle-flats," and in the surrounding fields, where, for about a century after the battle, large heaps of human bones were to be seen, joint-bones, together with iron swords and other weapons, have been ploughed up, as well as horse-shoes that would be suitable for the small Norwegian horses. The English chronicles which describe this battle are lavish in their praises of a Norwegian, who, in the midst of the fight, stood quite alone on the bridge over the Derwent, and for several hours kept Harald Godvinsön's whole army at bay, until at length a man glided under the bridge and ran him through from below with a spear. The inhabitants of the village of Stamford Bridge have to the present day kept up the custom of celebrating this deed at an annual festival, by making puddings in the form of a vessel or trough; for, as the legend states, it was in a trough that the slayer of the Norwegian passed under the bridge. It is certain, however, that the river Derwent hereabouts has only lately been made navigable.

It would lead us too far to relate, even in an abbreviated form, all the legends, or to reckon up all the numerous memorials, which, to the north of Watlinga-Stræt, are connected with the Danes. It is not only the common people in England who in general ascribe every ancient monument of any importance to the Danes; there was a time, and no very distant one, when many learned men were but too much inclined to do the same. In proof of this it suffices to remark that the celebrated circle of stones at Stonehenge, on Salisbury Plain, in Wiltshire—the most superb monument of its kind in the British Islands, or even in the whole of northern Europe—was also at one time described by the learned as a Danish place of sacrifice, although it is clearly distinguished, both by its structure and whole appearance, from the ancient monuments of Scandinavia; and although, on the contrary, the highest degree of probability proclaims its having originated from the older inhabitants of England, the ancient Britons. It is undoubtedly true, that want of adequate experience and knowledge was generally the real cause why the learned were never able to distinguish, with certainty, between what ancient monuments were really Danish and what were not. Nevertheless they would assuredly never have given the Danes credit for so many monuments, at the expense of their own countrymen and ancestors, had they not acknowledged that the immigration and settlement of the Danes in England was of the most widely-extended importance.

Even in our days English antiquarians are not disinclined to ascribe British, Roman, or Anglo-Saxon antiquities to the Danes; as well as to suppose, on the whole, that there are more monuments of the Danes extant in England, than, strictly speaking, that people can validly claim.

At first sight it might indeed appear that the Danes, who so early, and for so long a period, had extensive possessions in the north of England, must have left there a great number of tumuli, stone circles, and cairns; as well as, in consequence of their numberless fights and battles, a considerable quantity of entrenchments. It is sufficiently known how careful the old Northmen were to hand down to posterity the memory of a hero, and of his deeds. The doctrines of Odin even commanded it, as a sacred duty, to erect bauta stones in memory of the brave; which is one of the principal reasons why Scandinavia is distinguished, even down to modern times, by such a striking abundance of ancient monuments.

But with regard to England, we must not forget that the inhabitants of the central and northern parts had for centuries been Christians when the heathen Danes began to make conquests there. Among the Danes, as among the Northmen in general, the belief in their ancient gods had been weakened, and faith in their own power and strength had frequently usurped its place. Living among Christians in a foreign land, and doubtless, also, often marrying native females, they easily adopted, at least in form, the novel doctrines of Christianity, and with them the customs which they brought in their train. They soon renounced the usage of placing the dead in mounds, after the heathen manner, and of providing them with the weapons and ornaments which were dearest to them when alive. The bodies were buried in churchyards, or in the churches themselves; and the precious things which were formerly thought to secure for the hero an honourable seat in Valhalla, now for the most part remained above ground, where they generally found their way into the pocket of the monk, in order that he might deliver the deceased from purgatory by masses for his soul, and procure him an easy entrance into the kingdom of heaven. By degrees, as the Danes abandoned themselves to the influence of the higher civilization of England, they must also have adopted the most essential parts of the English dress, or at all events English ornaments; and consequently, even if only some few of these were deposited in the barrows, it became almost impossible to decide, when these graves were opened after a long lapse of time, whether it were Danes or Anglo-Saxons who had been originally interred in them.

Thus it is easily explained why but, proportionally, very few really Danish or Scandinavian barrows and monumental stones are to be found in England. We must not ascribe it to the progress of agriculture alone that, even in the north of England, we may search the fields in vain for stones, which, by runic inscriptions in the ancient language of Scandinavia, have preserved the remembrance of some distinguished warrior from the eastern lands beyond the sea. It is but rarely that one can even fancy that he has met with a Scandinavian runic stone; but a closer inspection will soon show that both the runes, and particularly the language in which the inscriptions are couched, betray a foreign, and especially an Anglo-Saxon, origin. The most important runic stone in these northern districts is found near the English border, in the Scotch town of Ruthwell, on the other side of Solway Firth. It is of considerable height, and is ornamented with a number of carvings of biblical scenes, mingled with figures of leaves, birds, and animals. Besides Latin inscriptions indicating and explaining these Christian carvings, there is a runic inscription on the stone which was long considered, both by British and Scandinavian archæologists, to be Danish, or at least to contain remnants of the old Scandinavian language. But it is now shown to be derived neither from the Danes nor Norwegians, but from the Anglo-Saxons, as the supposed Scandinavian inscription includes some verses of an old devotional Anglo-Saxon poem. The whole appearance of the stone, also, is rather Saxon than Danish. The runic characters are, in part at least, different from those of Scandinavia, and the words are not, as in them, separated by points. Ornaments with similar so-called Anglo-Saxon runic inscriptions are not altogether uncommon in England, particularly in the north. But as not a few ornaments, as well as runic stones with inscriptions in the selfsame character, are also found in the countries of Scandinavia, both in Denmark and Norway, and particularly the latter, and the west and south of Sweden (and there mostly in Bleking), it may be a question whether this runic writing was not originally brought over to England by Scandinavian emigrants. It would otherwise be inexplicable that they should have used entirely foreign runic characters in Scandinavia, whilst they possessed a peculiar and genuine Scandinavian runic writing of their own. The true state of the matter will not, however, be brought to light till antiquarians succeed in explaining, in a satisfactory manner, the inscriptions with Anglo-Saxon runes that are found in England as well as in Scandinavia, and which, for the most part, have not hitherto been deciphered.

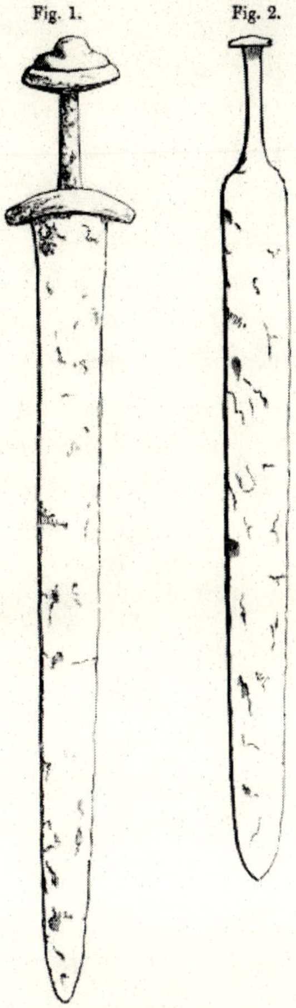

[+ +]Swords -
Fig. 1. Scandinavian
and Fig. 2. Saxon

It is a matter of course that arms and ornaments should be at times dug up in England that belonged to Scandinavian Vikings, who found either death or a new habitation on the English shore. In the rivers on the eastern coast, where the Vikings' ships showed themselves so regularly, and where remains of these ships are supposed to be now and then discovered, iron swords have been found, as for instance in the Thames, of undoubted Scandinavian origin. (Fig. 1.) They are in general longer and heavier than the Saxon sword (Fig. 2.), and are superior to them from having a guard, and a large, and commonly triangular, knob at the hilt. On the other hand, they are exactly of the same kind as our Scandinavian swords of what is called "the iron age;" that is, they belong to the latest period of heathenism. The Vikings, who often had to combat from their ships, and who, being few in number, were so much the more obliged to depend on their arms and the strength of their weapons, were necessarily compelled to have them both long and good. "Danish battle-axes" are usually mentioned in the old English and Frankish chronicles as excellent and dangerous weapons of attack. Nay, even from the distant Myklegaard, or Constantinople, where the Northmen, under the name of Varangians, served for a long series of years as the Greek Emperor's bodyguard, stories have reached us of the particular kind of battle-axes which they wielded with such strength. These axes, like the swords, were frequently inlaid with silver or gold, and were of excellent workmanship. It is also related by Giraldus Cambrensis that the Irish procured their battle-axes from the Northmen. The Danes in England, at least towards the latter part of their sway, are likewise said to have used shirts of mail, or chain armour, in which, however, the rings were not interlaced, but sewed on by the side of each other; helmets, with iron bands that covered the nose; and lastly, large pointed triangular shields. Some are even of opinion that these coats of mail were commonly black, and that this gave rise to the Danes being sometimes called "the black Danes." Others derive this surname from the colour of their hair and skin, which must at that time have been in general considered darker than the Norwegian complexion; whilst others, again, infer that the Danes generally used black sails for their ships, and the Norwegians white. The Scotch and Irish distinguish clearly between "Dubgall" or the black stranger (whence the present name Dugal), and "Finngall," or the fair stranger. Old Irish authors also call the inhabitants of Denmark "Dublochlannoch" (dark Lochlans), and the inhabitants of Norway "Finn-Lochlannoch" (fair Lochlans). Lochlan is with them the usual appellation of Scandinavia.

Besides their arms, the ornaments and decorations of the Danes and Norwegians were also of a peculiar kind; at least they are in general clearly different from the Anglo-Saxon ornaments now discovered in graves in England. As the Danish and British antiquities of the earlier, or what is called the bronze period, betray a considerable and well-defined difference, so also a comparison between the corresponding antiquities of the iron period will clearly show, that even if Roman taste formed the basis of art both among the Anglo-Saxons and the Danes and Norwegians during the last-named period, yet that each people followed its own independent course. That the Northmen, consequently, were not exclusively indebted to England for all that fresh development of taste which predominated at the close of heathenism and commencement of Christianity, but that they had themselves, before the Conquest of England, already made a great step in advance, was however no more than what one might expect from a people capable of building ships that crossed the Atlantic, and who were acquainted with, and frequently used, a peculiar sort of writing, the Northern runes.

But though, at present at least, it is scarcely possible to point out in England proper a single runic memorial of undoubted Danish or Norwegian origin, still there are found at times, particularly in north England, certain antiquities, with inscriptions that perfectly supply the want of those illustrations which the runic stones would otherwise afford, respecting the influence and settlements of the Northmen in England. These are small silver coins struck by Danish-Norwegian kings and jarls during their dominion there. I do not allude, of course, to coins of such kings as Canute the Great, Harald Harefoot, and Hardicanute; for as these princes held a confirmed dominion in England—and that at a time when coining was general in Europe, and when on the whole the light of history begins to shine clearer—there would be nothing strange, nor particularly instructive in an historical point of view, that they also had coined money. I refer to coins of Danish-Norwegian chiefs, whose deeds in England the chronicles have related either sparingly or not at all, and who lived more than a century before the Conquest by Canute the Great.

A short stay would easily have sufficed to erect a runic or bauta-stone; and great and imminent indeed must have been the danger which threatened the Northman of the olden time if he omitted, even on a foreign soil, to perform the last honours for a fallen friend or relative. But a coin was not so quickly minted. The countries of

Scandinavia had not a mintage of their own before the year 1000, or thereabout; when the Danish king, Svend Tveskjæg, having brought home with him from his expedition into England, a quantity of Anglo-Saxon coins, began to have them imitated. The Scandinavian Viking, to whom coining was a strange and unknown art, had enough to do, during a short and dangerous expedition for conquest, to procure a footing and support for his army; and if he failed in conquering a kingdom, he was glad to bring home as booty some pounds of foreign money. It was only when he had made himself king or jarl over a considerable district, and when he had begun to exchange his wild warrior's life for the milder occupations of peace, that he could have leisure to reflect that he also, like other princes in England, should promote his people's welfare and his own advantage by ordering those coins to be minted which are so important for trade and commerce. The older the dates of such Danish-Norwegian coins struck in England—the rarer the minting of coins in general, even in the more enlightened countries—so much the more clearly is the existence proved of well-established Scandinavian kingdoms, where works of peace were already capable of thriving.

Some few years ago (1840), a highly remarkable and very ancient treasure of silver was discovered near Cuerdale in Lancashire, within the boundaries of the ancient Northumberland. It consisted of bars, armlets, a great number of pieces of broken rings and other ornaments, as well as about seven thousand coins, all of which were inclosed in a leaden chest. To judge from the coins, which, with a few exceptions, were minted between the years 815 and 930, the treasure must have been buried in the first half of the tenth century, or almost a hundred years before the time of Canute the Great. Amongst the coins, besides a single Byzantine piece, were found several Arabic or Kufic, some of north Italy, about a thousand French, and two thousand eight hundred Anglo-Saxon pieces, of which only eight hundred were of Alfred the Great. But the chief mass, namely, three thousand pieces, consisted of peculiar coins, with the inscriptions, "Siefredus Rex," "Sievert Rex," "Cnut Rex," "Alfden Rex," and "Sitric Comes" (jarl); and which, therefore, merely from their preponderating number, may be supposed to have been the most common coins at that time, and in that part of north England where the treasure had been concealed. Cnut's coins were the most numerous, as they amounted to about two thousand pieces of different dies; which proves a considerable and long-continued coining.

Not only are the names Sitric (Sigtryg), Alfden (Halvdan), Cnut (Knud), Sievert (Sivard), and Siefred (Sigfred), visibly of Scandinavian origin, but they also appear in ancient chronicles as the names of mighty Scandinavian chiefs, who in the ninth and tenth centuries ravaged the western lands.

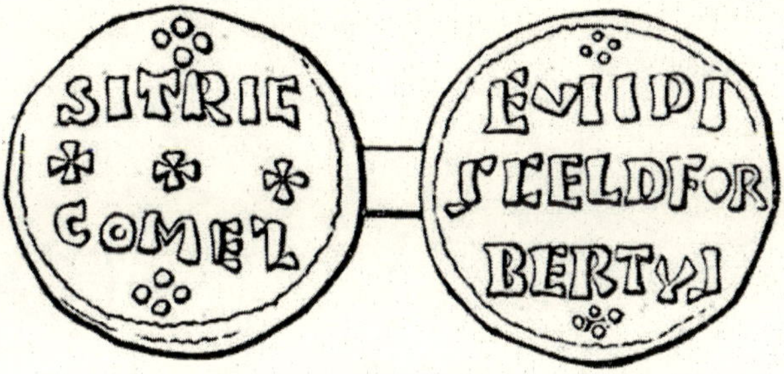

[+ +] Coin: Sitric Comes

Sitric Comes is certainly that Sitric Jarl who fell in a battle in England about the year 900. Alfden is undoubtedly the same king "Halfden," who at the close of the ninth century so often harried south England,—where he even besieged London—till he fell in the battle at Wednesfield in 910. Cnut, whose name is found inscribed on the coins in such a manner that one letter stands on each of the four arms of a cross, whilst the inscription R, E, X. (Rex) is inclosed between them, is probably he whom the Danes called "Knud Daneast" (or the Danes' Joy), a son of the first Danish monarch Gorm the Old; as it is truly related of him that he perished in Vesterviking (or the western lands). Sigfred must either have been the celebrated Viking king for whose adventurous expedition France, and its capital Paris in particular, had to pay dearly; or that Sigefert, or Sigfred, who in the year 897 ravaged the English coasts with an army of Danes from Northumberland.

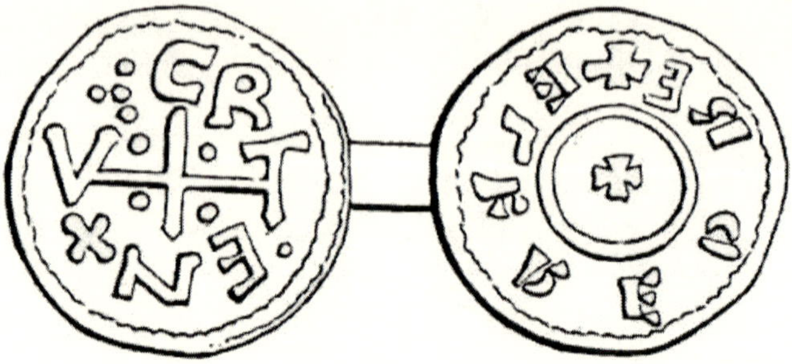

[+ +] Coin: Cnut

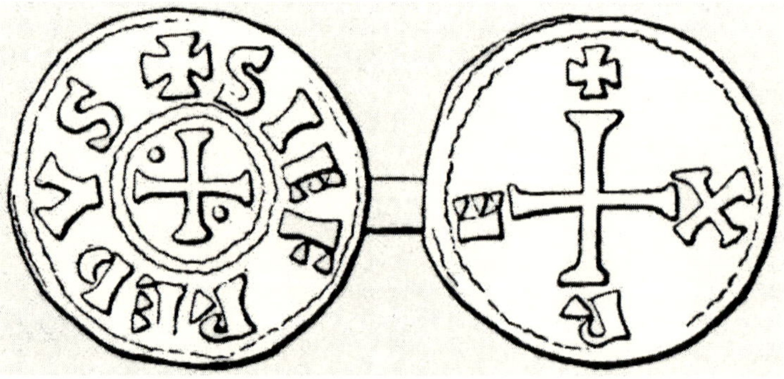

[+ +] Coin: Cnut reverse

The steady connection which the Vikings in England maintained with France affords a natural explanation why their coins were imitations both of contemporary English, or Anglo-Saxon, and of French coins. Thus on the reverse of Cnut's coins just mentioned, we sometimes find, as on that engraved above, the inscription "Elfred Rex," which is purely Anglo-Saxon; and sometimes the particular mark for Carolus, or Charles (Karl), which otherwise is only found on the French Carlovingian coins.

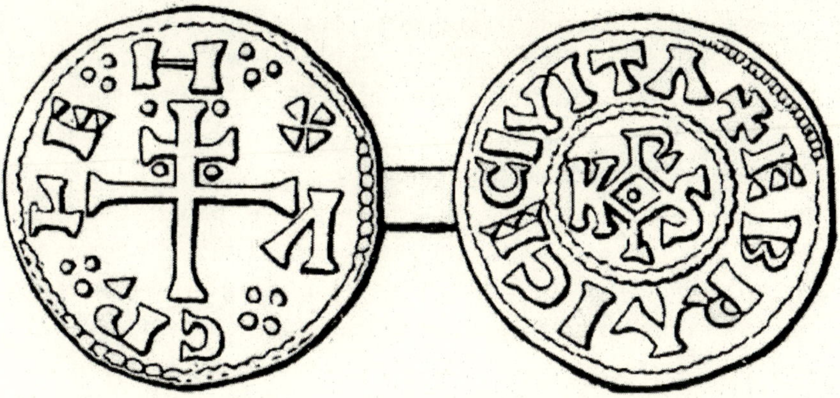

[+ +] Coin: Ebraice

A very frequent inscription on the Scandinavian coins here alluded to, as for instance in the last engraving, is "Ebraice Civita," or "The city of York;" whose ancient name "Eabhroig," and in the barbarous Latin of the time "Eboracum," was converted into "Ebraice." On other contemporary coins struck at York, namely on some of what is called St. Peter's money, York is also called "Ebracec" and "Ebraicit." For the Cuerdale coins, in order to express the name "Ebraice," coins of French kings of the city of "Ebroicas," or Evreux, in Normandy, seem to have been particularly chosen as patterns; for by a slight change of a few letters this Ebroicas could be converted into Ebraice; which was the easier process at a time when the art of stamping coins was not much practised. An additional proof that these coins were really minted by Scandinavian kings in Northumberland, and in the city of York, is, that none such have been found in any other part of England; whilst, on the contrary, one of Canute's coins, which have been so frequently mentioned, was dug up, together with English and French coins of the same kind as those found at Cuerdale, at Harkirke near Crosby, also in Lancashire; and consequently at places whose names ending in *kirke* (church) and *by* (town), bear witness no less than that of Cuerdale (from *dal*, a valley) to the dominion of the Northmen in those parts.

Should any doubt still exist that, so early as the ninth century, Danish-Norwegian kings and jarls minted a considerable number of coins in York, in imitation of contemporary Anglo-Saxon and French coins, it is at all events certain that the Northumbrian kings Regnald, Anlaf or Olaf, and Erik, who resided in York during the first half of the tenth century, caused coins of their own to be minted there, and which agree exactly with the historical accounts. Regnald, who reigned from about 912 to 944, was a son of King Sigtryg, and brother to the Olaf before mentioned, who fought the battle of Brunanborg; Erik (+ 951) is either King Erik Blodöxe of Norway, or a son of King Harald Blaatand of Denmark, who is said to have ruled in Northumberland about the same time.

In the main points these coins are also imitations of the Anglo-Saxon, but are distinguished from them by various and very striking peculiarities, which show them to have been coined both by Danes, or Norwegians, and by conquerors. Erik designates himself on them by the Latin title "Rex," as was usual at that time even among the Anglo-Saxons; but Regnald and Anlaf use the pure Northern title "Cununc;" or, in the Icelandic mode of writing, *Konungr*, the ancient Scandinavian word for *King*. Some of these coins have martial emblems which do not appear on the Anglo-Saxon coins of the same period, and which, therefore, were clearly intended to be in honour of the warlike qualities and victories of the Northmen. Erik's coins have a sword of the peculiar Scandinavian form, with a triangular pummel at the end of the hilt.

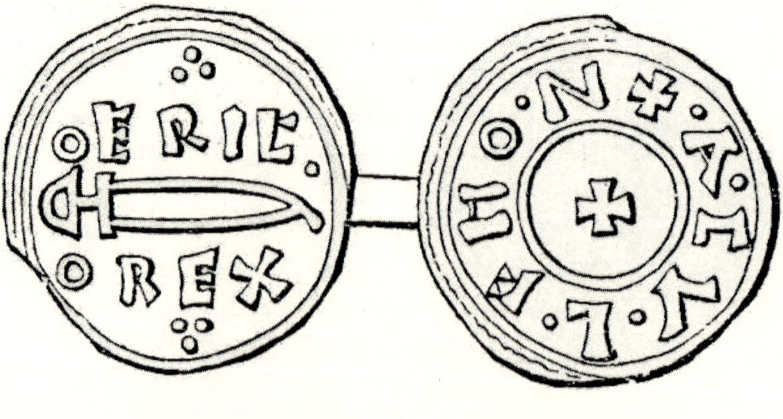

[+ +] Coin: Erik Rex

Similar swords are also seen on the St. Peter's money before mentioned, coined at York during the rule of the Scandinavian kings. One of these coins represents a bent bow with the arrow on it, and on the reverse a sledgehammer, or battle-axe.

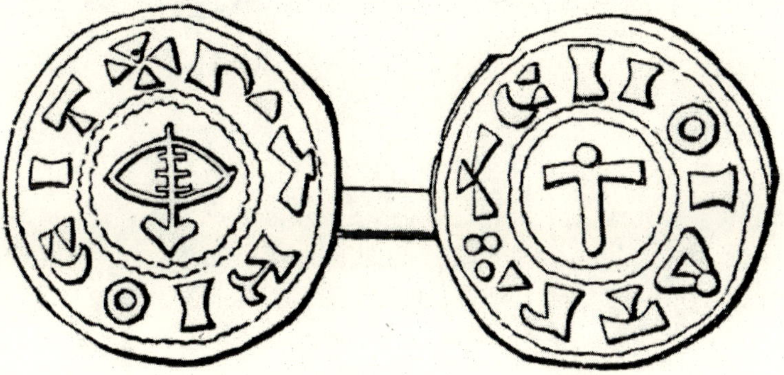

[+ +] Coin: Olaf

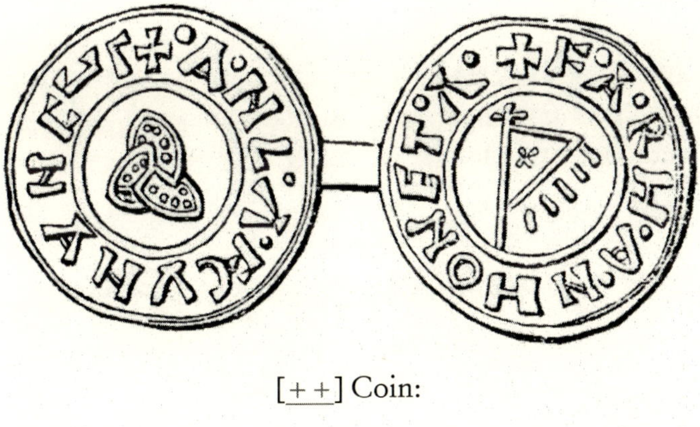

[+ +] Coin:

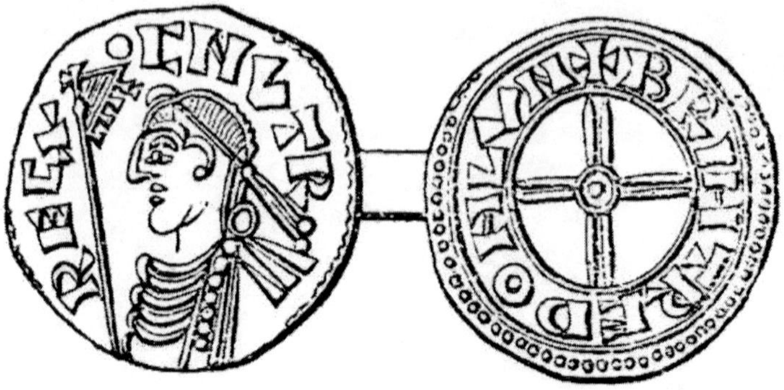

[+ +] Coin: Cnutr. Recx

Regnald's and Anlaf's (or Olaf's) coins, with the Scandinavian legend "Cununc" instead of "Rex," are ornamented with shields placed together (an emblem which may have been transferred from them to the later coins of Harald Haardraade and other Norwegian kings); as well as with flags of a triangular form, with hanging fringes. It is remarkable enough, that though such flags are not to be found on contemporary English coins, a piece of the Danish-English king's, Canute the Great, has lately been found on which the king's bust is represented, and before it a striped triangular flag with hanging fringes, of the same form as the flags on the coins of the Danish-Norwegian kings in north England. The legend on one side is, "Cnutr. Recx;" and on the other, "Brihtred on Lun;" which shows that the coin was minted in London.

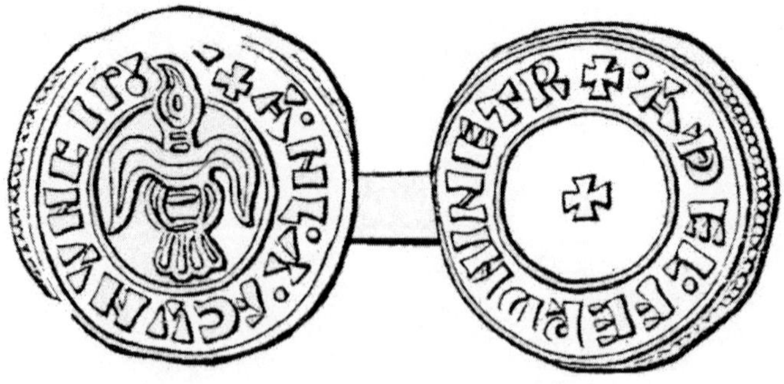

Coin: [+ +] Anlaf Cunune

Thus the coins, in conjunction with the chronicles, contribute to prove that flags were important emblems with the northern conquerors, which was indeed quite natural with a people like the ancient Scandinavians. The old Sagas in particular contain frequent accounts of the great value that the Northmen set on these flags, or, as they were then called, "mærker" (marks). Thus the Norwegian chief Harald Haardraade, before he became king of Norway, and after his return from his many expeditions into the Greek Empire, sitting and conversing one evening (according to the nineteenth chapter of his Saga) with King Svend Estridsen of Denmark at the drinking table, Svend asked him what precious things he had that he set most value on? He answered, his banner, called Landöde (or, the land-ravager). Svend then asked what qualities this banner had, since he esteemed it so precious a thing? Harald replied, "They say that he before whom this banner is borne always gains the victory; and such has constantly been the case since I possessed it."

The class of coins before alluded to as minted by Danish-Norwegian sovereigns in England not only presents a remarkable view of the importance, as well as appearance, of the old Scandinavian flags, or marks, but also serves in a high degree to confirm the repeated accounts of the English chroniclers, that "the Danes," during their conquests in the western lands, often bore a common standard, or national flag; a point about which the Danish chronicles or Sagas are silent. A coin of Anlaf, or Olaf, king of Northumberland, is particularly illustrative of this. It has the legend, "Anlaf Cununc," and represents a bird with extended wings, in which English antiquarians have very justly recognised the raven, the chief ensign, or emblem, of the ancient Danes.

From the most ancient times, and almost since the period that war was first waged, certain ensigns were undoubtedly known and used, around which the warriors rallied in battle. This had its origin, indeed, in necessity, in order that, in the tumult of battle, the combatants might always be able to discern where their fellow-warriors were; and such a rallying point was particularly of the greatest importance when an army was thrown into disorder, or began to fly. To this it may be added, that the commander, or the principal leaders, were generally near the ensign; which thus became a signal where the battle was usually hottest, and a point to rally round in order to protect the chief when in danger.

But these ensigns, which doubtless were originally boughs of trees or other simple things easy to be recognised at a distance, obtained by degrees a religious importance, and must thus have still more excited the courage of the combatants. For ensigns those figurative images were principally chosen under which men were accustomed to represent to themselves their principal gods, or to which a peculiar religious faith was attached. In the course of time these ensigns were adopted by whole tribes as national ones. The eagle, Jupiter's sacred bird, served the Romans for a warlike ensign, and animated the legions on their distant and universally-celebrated expeditions. With them, however, it did not flutter in a banner, but was cast in metal and fixed on the end of a staff. The national ensign used by at least a great part of the Gallic tribes in the south of France about the time of the birth of Christ, was of a similar kind. According to a few still-existing representations of it on monuments, it presented the image of a hog, fastened, like the Roman eagle, at the end of a staff. Among the Gauls the hog was a sacred animal, whence it is afterwards found frequently represented on the old Gallic coins.

Among the German and Scandinavian races, on the contrary, we cannot point with certainty to any such early national ensigns. These people, as it is well known, formed, for several centuries after the birth of Christ, a number of petty and independent kingdoms, which were, besides, often divided amongst several powerful chiefs. It was customary for every chief to have a peculiar sign, often an animal, delineated on his shield; and which was likewise represented on the banner that he carried with him into battle. This banner, or mark, was generally borne before him in the combat by his "marksman;" and at sea it waved on the prow of his ship. It was not, like that of the Romans and Gauls, of cast metal, but of variegated cloth.

It was not till the time that the Danes and Norwegians began to invade the countries of the west, and to make great conquests there, and consequently not till the ninth century, that we find the oldest traces of the Danes, or rather perhaps the Danish-Norwegian Vikings, having fought under one flag; which was not, like the earlier ones, that of a single chief, but rather an established national ensign. We must remember that they were heathens, making war upon a Christian land, and fighting for Odin and Thor against White[5] Christ. Regardless of their former contests in the north itself, the Vikings were now united on these foreign shores by the ties of mutual interest and a common religion; and nothing, therefore, was more natural than that the ensign which conducted them in battle should be consecrated to Odin, or, as he was called, the father of victory, in whose presence they expected at some time to assemble and enjoy the delights of Valhalla. The eagle had been consecrated to Jupiter by the Romans; among the Northmen the raven was Odin's (or, the Father-of-all's) sacred bird. One of Odin's names was therefore "Ravne-gud" (raven-god). The ravens Hugin and Munin sat on his shoulders, and only flew away to bring him intelligence of what happened in the world. The ancient Northmen had consequently an especial confidence in the omens of Odin's bird. When the Viking Floke Vilgerdesön set out from Norway to discover Iceland, he consecrated at a sacrifice three ravens, which he wished to take with him, to show him the way. He was therefore called Ravnefloke. The Northmen, also, made prognostications from the scream and from the flight of the raven; and the warriors, in particular, regarded it as a good omen if a raven followed them as they marched to battle.

5. An epithet applied by the Northmen to our Saviour.

As Jupiter's eagle had been the war sign of the Romans so was Odin's raven the chief mark of the Danes in the heathen ages. An old chronicler (Emma's Encomiast) relates, that in the time of peace no image whatever was seen in the flag, or mark, of the Danes; but in time of war there waved a raven in it, from whose movements the Danes took auguries of victory or defeat. If it fluttered its wings, Odin gave them a sign of conquest; but if the wings hung slackly down, victory would surely desert them. From the few historical accounts that remain to us of this raven's mark we are not, however, justified in believing that it was so long or so generally adopted among the Danes as the eagle was among the Romans. We find it expressly mentioned only during the Danish conquests in the British Islands; yet, remarkably enough, at such different times and under such peculiar circumstances, that we may with good reason assert that the raven's mark was really a common flag of battle and conquest for the Danes and Norwegians.

It is mentioned for the first time in the year 898, consequently nearly a thousand years ago; that is to say, about the time of the banner-coins before described, and especially of that coin of Anlaf, or Olaf, on which is seen the bird with extended wings. At that time, it is said, the Danish chiefs suffered a great defeat in South England, in which they lost their war-ensign, or banner (*Anglo-Saxon*, guð-fana), which they called "the raven" (*Anglo-Saxon*, ræfen v. hrefn. v. hræfen). Another account adds, that these chiefs were sons of Regner Lodbrog, and that the flag, or mark, was cunningly woven by Regner's daughters. The raven borne upon it was thought to forbode either victory or defeat.

This ensign is again spoken of a century later, in the time of Canute the Great. It is mentioned in the great battle of Clontarf, in Ireland (1014), when Sigurd, the Norwegian Jarl of Orkney, bore a raven-standard against the Irish. Two years afterwards, in the sanguinary battle at Ashingdon in Essex (1016), which partly decided Canute's conquest of England, the Danish army had begun to give way; when the jarl, Thorkel the Tall, shouted to the warriors, as he pointed to the flag, that the raven fluttered its wings, and predicted a glorious victory. The Danes took fresh courage, and victory crowned their efforts. The mighty Danish jarl Sivard, or Sigurd, surnamed "Digre" (the stout) (+ 1055), who ruled the earldom of Northumberland somewhat after Canute's time, and after the Danish dominion in England had ceased, also bore a raven ensign, which was called "Ravenlandeye," or the raven that desolates the land. ("*Corvus terræ terror.*") There seems to have been many legends among the people, both as to the manner in which Sigurd procured this ensign, and as to its supernatural power.

After the time of Canute the Great and Sigurd Digre, there is scarcely any coin to be found bearing the image of the raven; but fortunately there is a representation of another kind, belonging to the eleventh century, which in no slight degree proves that raven-ensigns were actually borne by the successors of the Danes and Norwegians in the west of Europe until about the year 1100.

It is known that Scandinavian Vikings, and particularly Normans and Danes, conquered the French province afterwards called from the Northmen (Normænd) Normandy; and that the successors of Rollo, or Rolf (Ralph), continued to govern that land as dukes. From Normandy, Duke William, surnamed the Conqueror, passed over in 1066 into England, which he conquered by the battle of Hastings. The whole expedition, together with this battle, is represented in the old and extremely remarkable piece of tapestry, preserved in the cathedral of Bayeux, in Normandy, and said to have been worked by William the Conqueror's own consort, Matilda; at all events it was made shortly after the conquest of England. There can, therefore, be no question about the fidelity of the figures represented, at all events, as far as regards the Normans. It is here seen that the Norman chiefs, after the old Scandinavian fashion, had each his ensign or banner of party-coloured cloth cut out into tongues or points, and fastened to the pole of a lance. But where William is represented on the Bayeux tapestry advancing to the battle of Hastings, the chief banner is borne by a mounted knight clad in chain armour, who rides before another knight, likewise clothed in armour, and having on his lance an ensign or flag with five tongues or points, and with a cross in it.

[++] Bayeux tapestry: Two Knights

On the chief banner, the only one of that form among the many flags in the tapestry, but which in its whole shape and pendant fringes bears a striking likeness to the old Danish flags before mentioned, there is seen in the middle the figure of a little bird, which may, with the greatest probability, be taken for Odin's raven. For it is very natural that the Scandinavian Vikings, or Normans, who had achieved so many and such famous conquests under Odin's raven, should continue to preserve this sign, even after they had adopted Christianity; and that thus the Normannic dukes in Normandy should also long bear their forefathers' venerable ensign with them as a Palladium in the combat.

After the conquest of England by the Normans, however, the Norman kings abandoned the old Scandinavian raven-mark, and adapted themselves more to the English customs. Probably each king had his own mark or flag, after the custom of that time, until the national banner afterwards received a settled form. But the remembrance of the Danish raven by no means became obsolete among the English nation. Whilst the raven-flag has almost been erased from the memory of the Danish people, the remembrance of it still exists freshly in the British islands; and both poets and artists who represent, however simply, the ancient combats of the Danes with the Anglo-Saxons, the Scotch, and the Irish, seldom neglect to make "the enchanted raven" wave in the Danish ranks.

On the often-mentioned Bayeux tapestry is also represented the fall of the English king, Harald Godvinsön, at the battle of Hastings. The king's flag-bearer, or marksman, who, as well as the king, is on foot, bears a flag-staff, on which is fixed a figure, probably of cloth, cut in the resemblance of a dragon, which was the royal mark of the Anglo-Saxon king. Close before him lies a fallen knight, by whose side is seen a lance with the point downwards, and on which hangs a similar dragon.

[+ +] Bayeux tapestry: Harald Godvinsön

This fallen knight is without doubt the king. From the form of his flag, or mark, we may conclude that the Danes' raven-mark probably consisted at times of the figure of a raven fixed to a shaft, and cut out or sewed in a similar manner.

What colours were used for the raven-mark can now hardly be decided. The bird, or raven, on William the Conqueror's war-flag appears to have been of a blue-black on a pale yellow, or light, ground. This colour in the tapestry may, perhaps, have been accidental; and the account of an English chronicler would lead us to suppose that the ground of the Danish flags, or marks, was, at least in time of peace, white. But the colours were certainly different at different times. There can be no doubt that the ground was often red; for, from the most ancient times, red was a very favourite colour in the north, especially in time of war. The old inhabitants of the north, when they came as friends, used to show a white shield, but when they appeared as enemies it was red; then "they raised the war-shield." In Norway red seems to have been the national colour from an early period; and it was even ordered in Gulething's laws, that every man who possessed six silver marks[6] should have a red shield. Something similar was probably the case in Denmark. An old legend preserved by the Scotch historians relates that, in a battle in Scotland about eight hundred years ago, the Danes wore red and white tunics. That red and white appear so prominently on the Danish national colours ever since the thirteenth century is certainly owing to an ancient predilection among the people for these colours. It is perhaps, therefore, most probable that the banners, or marks, of the ancient Danes were, in time of peace, of a light colour, but in war time of a blood colour, with a black raven on the red ground.

In the ninth, tenth, and eleventh centuries the raven, the Danebrog of heathenism, waved victoriously in the western lands. It was with Canute the Great at Ashingdon, with the Norman William at Hastings, and was thus present at two conquests of England. But the battle of Hastings was the last important battle that the raven won. Heathen Scandinavia had exhausted its strength by numerous and far-extended conquests. Christianity, and with it a new and a higher civilization, advanced with a power not to be checked even among the ancient followers of Odin. The raven, Odin's mark, to which the heathen Danes had attached themselves with all the strength of religious faith, no longer inspired them as before when the warriors had lost the hope of the joys of Valhalla. If they now fought, it was mostly against heathens who would not bow before that cross on which Christ bled and suffered for the sins of mankind. In order to inspire the combatants, it was necessary that the banner which they followed should be an expression of the spirit which stirred among the people, of that living hope which animated them respecting the manner of their existence in another world. The raven, the symbol of heathenism, paled by degrees, as antiquated and meaningless, and at last quite gave place to the symbol of Christianity, the holy cross.

6. A mark was half a pound of silver.

The same representations on ancient coins and tapestry, which exhibit the raven, and the old flags, also show the sign of the cross. The flag on Olaf's and Regnald's coins (p. 53) has a figure in the middle resembling the cross. This is still more distinct on the Bayeux tapestry, where William's chief banner is borne (p. 59), for immediately after the raven follows a flag with the cross. This last, moreover, certainly represents the identical consecrated banner with the figure of a cross, which the Pope sent to William on the occasion of his expedition against England.

The sign of the cross must by degrees have naturally superseded the raven, not only among the descendants of the Danes and Norwegians in England, but also, though perhaps somewhat later, in the north itself. If we may not assume that the present "Danebrog," with its white cross on a red ground, became the Danish national flag immediately after the introduction of Christianity, it is at least certain that the Danish kings, in the first two centuries after that event, bore flags with crosses as their personal banners, or marks; and particularly in the twelfth century, when the crusades against the heathen Wends began. An old Saga, or legend, relates, that during one of the crusades of King Waldemar the Victorious in Livonia, in 1219, the "Danebrog" fell from heaven among the Danish army. This much, however, is certain—that it is not till after these crusades that the "Danebrog" appears as the established national flag of the Danes; and ever since that time, for more than six centuries, it has continued to wave unchanged in the Danish fleets and armies. It is remarkable that, as the flag of the fleet, and of all fortified places, and as the royal flag, it is split; and it can scarcely be doubted that this form must have originated from the fringes and tongues, or points, with which the old Danish and Scandinavian flags were ornamented in the tenth and eleventh centuries. The Scandinavian people is the only one which from remote antiquity has uninterruptedly borne this split flag; and it is possible that Sweden, as well as Norway, obtained theirs, which is of comparatively late origin, by imitating the old Danebrog.

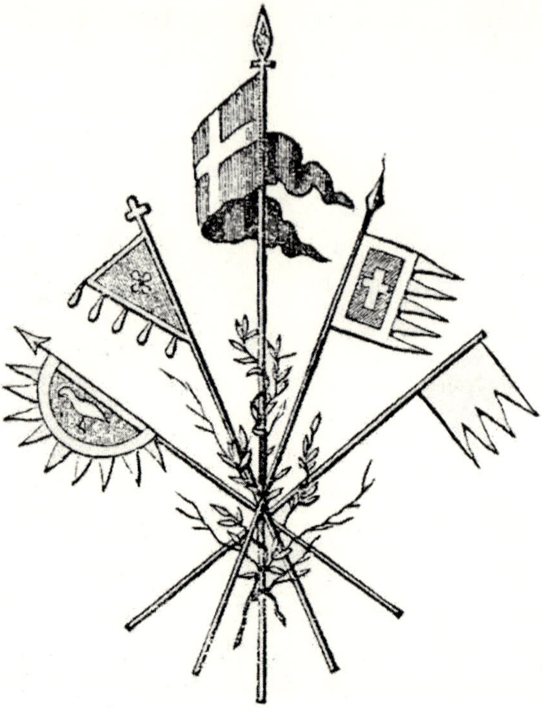

[+ +] Flags and Ensigns

Other European countries also derived from the crusades flags with crosses as their national banners; as, for instance, England the St. George's banner, which was white with a red cross; and Scotland a blue flag divided by a white St. Andrew's cross. About the same time the different kingdoms began to adopt a fixed national coat of arms. Thus Denmark assumed that still in use,—three blue leopards, or lions, on a golden shield, strewed with red hearts; which was originally the family arms of the royal house. It has, however, undergone a few slight changes. With regard to this subject, it is remarkable that three leopards were also borne by the Norman dukes, who were of Norwegian descent, and who, after the conquest, introduced the leopards, or lions, into the arms of England. Generally the lion was not, nor is indeed at present, found on coats of arms in England and France, whereas it appears very frequently in those of the north. Sweden has, besides others, the Gothic lion; the Norwegian national coat of arms is a lion with a halberd; and Denmark has, besides the proper national arms, the Cymbric lion, and the two Sleswick lions. But the lion is so peculiarly Scandinavian that it does not even cross the Eider; Holstein, which is German, has an entirely different coat of arms—a nettle-leaf. There is also this similarity between the Danish and English lions, that they are represented standing, whilst those on the other national arms are depicted springing. Would it, therefore, be quite groundless to trace, even in the armorial bearings of England, one of the many proofs of the influence which the Northmen, and the Scandinavian elements, still continued to exert there at the time when the national arms were adopted, and when the foundations of an entirely new and superior social system had already been laid?

SECTION VII

Danish-Norwegian Names of Places.

On the extremity of the tongue of land which borders on the north the entrance of the Humber, there formerly stood a castle called Ravnsöre (raven's point—in old Scandinavian, Hrafnseyri), and afterwards Ravnsere. *Öre* is, as is well known, the old Scandinavian name for the sandy point of a promontory. Ravn (or Raven) may possibly have been either the name of the man who first conquered the surrounding district and built the castle; or, what is certainly far more probable, the Northmen, on erecting this important castle on one of their first landing places on the greatest river in north England, named it after the bird sacred to Odin, which fluttered in their banner, and prognosticated to them victory in the fight. In that case it was a singular coincidence that Harald Haardraade's son Olaf should, after the battle of Stamford Bridge, have embarked at Ravnsöre for the Orkneys and Norway with the feeble remnant of the Norwegian army. The very place which had before so often seen multitudes of Northmen, intoxicated with victory, land with Odin's raven-flag, now beheld the flight-like departure of their successors, after they had combated in vain under that celebrated banner "Landöde" (the land-ravager), which had accompanied Harald Haardraade in his expeditions to the East, against the Saracens and other enemies of Christianity. It was one of the many proofs that "White Christ" was not yet for the Northmen, at least in battle, what Odin had been previously.

It is, however, at least certain that the name "Ravenspurn" (Ravnsöre) is derived from the Scandinavian conquerors. An Icelandic Saga, written a hundred and fifty years after the conquest of England by the Normans, or after the battle of Hastings (1066), says that "Northumberland was mostly colonized by Northmen; for after Lodbrog's sons, who conquered the country, had again lost it, the Danes and Norwegians often harried it; and there are still many places to be found in the district that have names taken from the

Scandinavian tongue, such as Grimsby, Hauksfliot, and numerous others."

Old English chroniclers also state that many towns in England had new names given to them by the Northmen; for instance Streaneshalch came to be called Whitby, and Northweorthig was named in the Danish language "Deoraby."

A surer and more decisive proof than all written historical accounts of the Danish-Norwegian settlements and diffusion in the midland and northern districts of England is, that the above-named places, namely, Grimsby ("the town of Grim"), Whitby (Hvidby, "the White town"), and Deoraby Dyreby ("town of deer"), contracted to Derby, are to be found to this day in Lincolnshire, Yorkshire, and Derbyshire; and also that in these old Danish districts there is, moreover, a very considerable number of towns with names of just as undoubted Danish origin. A close inspection of even a common map of England will soon show that there are not a few names of places in the north of England, whose terminations and entire form are of quite a different kind from those of places in the south.

The greater number of names of places in the south of England end in ——ton, ——ham, ——bury, or ——borough, ——forth or ——ford, ——worth, &c. These, which are of Anglo-Saxon origin, and which also serve still further to prove the preponderating influence of the Anglo-Saxons in that part, are, it is true, also spread over the whole of the north of England. But, even in the districts about the Thames (in Kent, Essex, Suffolk, and Norfolk) they already begin to be mixed with previously unknown names ending in ——by (*Old Northern*, býr, first a single farm, afterwards a town in general), ——thorpe (old Northern Þorp, a collection of houses separated from some principal estate, a village), ——thwaite, in the old Scandinavian language Þveit, tved, an isolated piece of land, ——næs, a promontory, and ——ey, or öe, an isle; as in Kirby, or Kirkby, Risby, Upthorpe and others. As we approach from the south the districts west of the Wash, such as Northamptonshire and Warwickshire, the number of such names constantly increases, and we find, among others, Ashby, Rugby, and Naseby. As we proceed farther north, we find still more numerous names of towns and villages having in like manner new terminations;

such as, ——with (*i.e.* forest), ——toft, ——beck, ——tarn (*Scandinavian*, tjörn, or tjarn, a small lake, water), ——dale, ——fell (rocky mountain), ——force (waterfall), ——haugh, or, how (*Scand.*, haugr, a hill), ——garth (*Scand.*, garðr, a large farm); together with many others. The inhabitants of the north will at once acknowledge these endings to be pure Norwegian or Danish; which is, moreover, placed beyond all doubt by the compound words in which they appear.

It is not of course very easy to point out the meaning of every name of a place that has a Danish or Norwegian termination; the original form having been partly corrupted by later differences of pronunciation, and partly changed, by the ancient Scandinavians having often merely added a Scandinavian ending to the older names, or at most re-modelled them into forms that had a home-like sound to their ears. Still there are names enough of places whose signification is quite clear. To instance some derived from the situation or nature of the place: Eastby (*Dan.*, Ostby; *Eng.*, the eastern village), Westerby (*Eng.*, the western village), Mickleby (*Dan.*, Magleby; *Eng.*, the large village), Somerby, Markby (*Eng.*, the field village), Newby (*Dan.*, Nyby; *Eng.*, the new village), Upperby (*Dan.*, Overby; *Eng.*, the upper village), Netherby (the lower village), Langtoft (the long field), Kirkland (church-land), Stainsby (the stone village), Haidenby (*Dan.*, Hedeby; *Eng.*, the heath village), Raithby (*Dan.*, Rödby, from *rydde*, to clear away), Dalby (village in the dale), Scawby and Scausby (village in the wood), Scow, Askwith (*Dan.*, Askved, or Askeskov, *i.e.* Ashwood), Storwith (*Dan.*, Storved, or Storskov; *Eng.*, the large wood), Lund (Danish for *grove*), Risby (the beech village), Thornby (the thorn village), Birkby (*Dan.*, Birk; *Eng.*, the birch village), Ings (*Dan.*, Enge; *Eng.* meadow), Brackenthwaite (Bregentved, from Brackens), Northorpe (*Dan.*, Nörup; *Eng.*, north village), Millthrop (*Dan.*, Möldrup; *Eng.*, mill-village), Staindrop (*Dan.*, Stenderup; *Eng.*, stone village), Linthorpe (*Dan.*, Lindrup; *Eng.*, lime-tree village), Stonegarth (*Dan.*, Steengaard; *Eng.*, stone farm), Dalegarth (*Dan.*, Dalsgaard; *Eng.*, valley farm), Fieldgarth (*Dan.*, Fjeldgaard; *Eng.*, rocky farm), with others. A village on the river Eden in Cumberland is called Longwathby (from a long ford, or wading place; *Danish*, at vade); and north and south of the Humber, at a spot where there is a ferry over the river (*Dan.*, Færge), lie north and south Ferriby! Almost all these names, to which a great number of similar ones might be added, answer to names of places still in use in Denmark, only with this difference, that *thwaite* has there passed into *tvede*, or *tved*, and *thorpe* into *trup*, *drup*, or *rup*.

The following examples may be cited of Danish-Norwegian names of places in England, called after animals: Codale (Cowdale), Swinedale, Swinethorpe, Hestholm (*Eng.*, Horse-holm), Calthorpe, and Hareby.

Names of places containing personal names are, however, beyond comparison far more numerous, and were probably taken from the first Scandinavian conquerors; as, for instance, Rollesby (Rolfsby), Ormsby (Gormsby), Ormskirk, Grimsdale, Grimsthorpe, Haconby, Gunnerby, Aslackby, Swainby, Swainsthorpe, Ingersby, Thirkelsby, Asserby, Johnby, Brandsby, Ingoldasthorpe, Osgodby, Thoresby, and several others.

Among this species of names of places are found such as Tursdale, Baldersby, Fraisthorpe, and Ullersthorpe. Now it is certainly probable that these were only derived from men named Thor, Balder, Freyer, and Uller, or Oller; yet we cannot avoid thinking of the old gods who bore these names, particularly as it was a common custom among the ancient Scandinavians to name towns and estates after them. In England also are found Asgardby, Aysgarth (or Asgaard, in Yorkshire), as well Wydale and Wigthorpe, or Wythorpe; which two names have undoubtedly the same origin as the old sacrificial and assize town Viborg, in Jutland (from Vébjörg, or the holy mountains); namely, from *vé*, a sacred place. Even the name of one of the most important sacrificial places in the Scandinavian north, is to be found in Yorkshire, in Upsal (from Upsalir, the high halls). The names of places in England which have preserved traces of the Danes after they had become Christians, may all the more assure us that we are not mistaken in regarding the names just mentioned as remarkable remains of the short period of their domination when heathens. The names of Bishopsthorpe (Bispetorp), Nunthorpe (Nonnetorp), Kirkby, Crosby, and Crossthwaite, sufficiently prove that Christian had succeeded to sacrificial priests, and that church and cross were now erected where heathen altars and temples had formerly stood.

The name of the village of Thingwall[7] in Cheshire affords a remarkable memorial of the assizes, or *Thing*, which the Northmen generally held in conjunction with their sacrifices to the gods; it lies, surrounded with several other villages with Scandinavian names, on the small tongue of land that projects between the mouths of the rivers Dee and Mersey. At that time they generally chose for the holding of the *thing*, or assizes, a place in some degree safe from surprise. The chief ancient *thing* place for Iceland was called like this Thingwall, namely Thingvalla (originally "Þingvöllr," "Þingvellir," or the *thing-fields*).

7. Wall, *Dan.*, Vold, a bank or rampart.

The before-mentioned names Bishopsthorpe and Nunthorpe apply to estates that belonged to the church; the following ones, viz., Coningsby, Coneysthorpe, Coneysby, Kingthorpe, and Kingsby, denote property belonging to the kings, or destined for their maintenance. Some towns are named after the trade or business of the original inhabitants as Smisby (Smithby) Weaverthorpe, and Copmanthorpe (Kjöbmandsthorpe, *i.e.*, merchants-thorpe); others point to the descent of the inhabitants, such as Romanby, Saxby, Flemingsby, Frankby, Frisby and Fristhorpe (but this possibly came from "Freyr"), Scotby, Scotsthorpe, Ireby, Normanby, Danby or Denby, and Danesdale.

It also deserves to be mentioned that many of these names of places have by degrees become family ones, which are constantly heard in England; for instance, Thoresby, Ashby, Crosby (whence again Ashby and Crosby Streets in London), Thorpe, Sibthorpe, Willoughby, Scoresby, Derby, Selby, Wilberforce, &c.

In order, lastly, to convey an idea of the abundance of Scandinavian, or Danish-Norwegian, names of places, which occur in the midland and northern districts of England, a tabular view of those most frequently met with is here subjoined from the English maps. This list, which is principally drawn up for the use of those readers who have not a comprehensive map of England at hand, will, with all its deficiencies, clearly and incontestably prove the correctness of the historical accounts, which state that the new population of Danes and Norwegians that immigrated into England during the Danish expeditions, settled almost exclusively in the districts to the north and east of Watlinga-Stræt, and there chiefly to the west and north of the Wash. Norfolk, Northamptonshire, and Lancashire, have each only about fifty names of places of Scandinavian origin; Leicestershire has about ninety; Lincolnshire alone, nearly three hundred; Yorkshire above four hundred; Westmoreland and Cumberland each about one hundred and fifty. The colonization has clearly been greatest near the coasts, and along the rivers; it had its central point in Lincolnshire (the Northmen's "Lindisey"), and in the ancient Northumberland, or land north of the river Humber. Yet it was not much extended in Durham and the present Northumberland, each of which contains only a little more than a score of Scandinavian names.

A TABULAR VIEW OF SOME OF THE MOST IMPORTANT DANISH-NORWEGIAN

NAMES OF PLACES IN ENGLAND.

(*Extracted and collected from "Walker's Maps," London, 1842.*)

Part A

Names ending in	by	thorpe	thwaite	with	toft	beck	næs
In Kent, north-east of Watling Street	1	.	.	.	.	.	4
In Essex	2	3	.	.	.	.	3
-Bedfordshire	.	3	.	.	1	.	.
-Buckinghamshire	1	2	.	.	.	.	.
-Suffolk	3	5	1	.	.	.	1
-Norfolk	17	24	2	.	.	1	.
-Huntingdonshire	1	.	.	.	.	.	.
-Northamptonshire	26	23	.	.	3	.	.
-Warwickshire	2	1	.	.	.	.	.
-Leicestershire	66	19	.	.	1	.	.
-Rutland	.	7	.	.	.	.	.
-Lincolnshire	212	63	.	1	4	8	1
-Nottinghamshire	15	20	.	.	.	1	.

-Derbyshire	6	4	.	.	1	.	.
-Cheshire	6	.	.	.	.	.	.
-Yorkshire:							
--East Riding	35	48	1	6	3	1	1
--West Riding	32	29	6	8	2	4	.
--North Riding	100	18	2	6	1	7	.
-Lancashire	9	.	14	2	.	.	2
-Westmorland	20	6	14	1	.	17	1
-Cumberland	43	1	43	.	.	12	2
-Durham	7	7	.	.	.	.	.
-Northumberland	.	1	.	.	.	1	.
In all	604	284	83	24	16	52	15

Part B

Names ending in	ey.	dale	force	fell	tarn	haugh	Total
In Kent, north-east of Watling Street	1	.	.	.	.	.	6
In Essex	3	.	.	.	.	.	11
-Bedfordshire	.	.	.	.	.	.	4
-Buckinghamshire	.	.	.	.	.	.	3
-Suffolk	.	.	.	.	.	.	10
-Norfolk	.	.	.	.	.	.	44
-Huntingdonshire	.	.	.	.	.	.	1
-Northamptonshire	.	.	.	.	.	.	52
-Warwickshire	.	.	.	.	.	.	3
-Leicestershire	.	1	.	.	.	.	87
-Rutland	.	1	.	.	.	.	8
-Lincolnshire	.	3	.	.	.	.	292
-Nottinghamshire	.	.	.	.	.	.	36
-Derbyshire	.	.	.	.	.	.	11

-Cheshire	.	.	.	.	.		6
-Yorkshire:							
--East Riding	.	12	2	.	.	.	109
--West Riding	.	12	.	15	2	.	110
--North Riding	.	40	4	7	1	.	186
Lancashire	2	13	.	7	.	.	49
Westmorland	.	36	6	42	15	.	158
Cumberland	.	16	1	15	9	.	142
Durham	.	5	2	2	.	.	23
Northumberland	.	3	.	7	.	10	22
In all	6	142	15	95	27	10	1373

Besides many other names ending in -holm, -garth, -land, -end,

-vig, -ho (how), -rigg, &c., c.

The same table still further shows that the names ending in by, thorpe, toft, beck, næs, and ey, appear chiefly in the flat midland counties of England; whereas, farther towards the north, in the more mountainous districts, these terminations mostly give place to those in thwaite, and more particularly to those in dale, force, tarn, fell, and haugh. This difference, however, is scarcely founded on the natural character of the country alone; it may also have arisen from the different descent of the inhabitants. For although in ancient times Danish and Norwegian were one language, with unimportant variations, so that it would scarcely be possible to decide with certainty in every single case whether the name of a place be derived from the Danes or from the Norwegians; yet it may reasonably be supposed that part at least of the last-mentioned names are Norwegian; namely, those ending in ——dale (as Kirk-dale, Lang-dale, Wast-dale, Bishops-dale); in ——force (as Aysgarth-force in Yorkshire, High-force, and Low-force, in the river Tees, and in the stream called "Seamer Water"); in ——fell (old Norwegian, fjall; Mickle-fell, Cam-fell, Kirk-fell, Middle-fell, Cross-fell); in ——tarn (*Old Nor.*, tjörn, or tjarn, a small lake); and in ——haugh (as in Northumberland, Red-haugh, Kirk-haugh, Green-haugh, Windy-haugh). Exactly similar names are met with to this day in the mountains of Norway; whilst they are less common, or altogether wanting, in the flat country of Denmark. That Norwegians also immigrated into England, even in considerable numbers, both history and the frequently occurring name of Normanby in the north of England, clearly show; but they appear to have betaken themselves chiefly to the most northern and mountainous districts, which not only lay nearest to them, but which in character most resembled their own country. In this respect it deserves to be noticed, that places whose names end in *tarn*, and are consequently pure Norwegian, are found only in the most northern counties; and that those in *haugh*—although there are names of places in Denmark ending in *höi* (hill)—must also, from the form, be Norwegian. They are found exclusively in the present Northumberland, and within the Scotch border.

We may, however, venture to set down the greater part of Scandinavian names of places in England as Danish. The terminations in *thwaite* and *thorpe*, indeed, are to be met with in Sweden, Norway, and Iceland, as well as in the Saxon and Frisian districts of North Germany; yet as the corresponding English names are for the most part composed of pure Scandinavian or Danish words, and as they seldom appear either in the tracts conquered by the Norwegians in Scotland and Ireland, or in the southern and south-western, originally Anglo-Saxon, districts of England, but keep strictly within the same boundaries as the rest of the Danish names of places, and particularly of those in *by* (Danish for town or village), these are valid reasons for regarding them in general as Danish.

The names of places in England ending in *by* are only to be found in the districts selected by the Danes for conquest or colonization. With the exception of a single Kirby, or Kirkby, in Kent, not far from London, they are nowhere to be found to the south of Watlinga-stræt (for Tenby, formerly Tenbigh, in Pembrokeshire, is from a different derivation); whilst towards the north, they cease in the most north-eastern county of England, the present Northumberland; in the south-westernmost part of Scotland (Locherby in Dumfries, Sorby in Wigtonshire); and in the Isle of Man (Sulby, Jurby, Dalby). If we except Duncansby in Caithness, and Oreby in the Isle of Lewis, as well as some few villages in Orkney and the Shetland Isles, they do not appear among the many pure Norwegian names of places in the north and west of Scotland, and in Ireland; which, as will be explained in its proper place, have generally quite a different character from the Scandinavian (chiefly Danish) names of places in England. It can hardly be said that this was solely owing to the natural character of the country in England being more favourable for the building of villages than in those districts in Scotland and Ireland which were occupied by the Northmen: first, because the Norwegians seem to have dwelt closely together in many places there, doubtless in order to resist the attacks of the natives; secondly, because the land there, though often separated by nature into many districts, as for instance in Caithness and the Orkneys, by no means prevented them from assembling together in villages; and lastly, because *by* originally denoted only a single estate or farm. In Norway, the Faroe Isles, and Iceland, many names of places are to be found, which indicate the existence both of single farm-houses and collections of them, or villages; but they have this peculiarity, that they generally end in *bœr* or *bö*, far more rarely in *býr* or *by*; whilst, on the contrary, this last form is essentially Danish. Names of places ending in *by* are spread over the peninsula of Jutland quite down to Danevirke and the Eyder; are found in great numbers in the southern boundary of South Jutland, or Sleswick; as well as in the islands and old Danish countries of Skaane, or Scania, Halland, and Bleking; whence they extend themselves over a great part of Sweden, and far into Finland. From the most ancient times down to the present, this difference between the Norwegian form *bœr*, and the Danish *býr* or *by*, seems on the whole to have clearly prevailed; and thus that, as early as the eleventh century, the English towns and villages are written in William the Conqueror's "Domesday-book," with the Danish ending *by* or *bi*, and not with the Norwegian form *bœr* or *bö*, is certainly no slight corroboration of their assumed Danish origin. Besides, as *by* is not found in the names of places south of the Eyder, in Holstein or North Germany, and as it is wholly unknown in the Saxon or German languages, there is consequently so much the greater probability that in England it was derived from the Danes.

For the same reasons, towns whose names end in *by* are most numerous in the counties situated on the coast opposite Jutland; viz., in Leicestershire, 66; Lincolnshire, 212; and the North Riding of Yorkshire, 100. In the two other Ridings, there are altogether about 70 names of places ending in *by*; in Cumberland, 43; and in Westmoreland, 20. For the rest, this termination occurs so frequently throughout the old Danish part of England, that, of 1370 Scandinavian names of places, above 600 (as the tabular view given at page 71 shows) end in *by*, whilst no other names exceed 280; and even this number is reached only by the ending *thorpe*, which also is certainly pure Danish; whilst the most numerous after thorpe fall down to 140. This remarkable preponderance of Danish endings in *by*, will of itself sufficiently prove the important and wide-extended influence of the Danes in the midland and northern counties of England.

The not inconsiderable number (1370) of Scandinavian names of places collected together in the preceding tabular view, could be much increased if we were to include all the Scandinavian appellations used by the common people in many parts of the north of England. A hill, or small mountain, is there called *hoe* or *how* (Höi in Jutland: Höw or Hyv); a mountain ridge, *rigg*; a ford, *wath*; a spring, *kell*; a holm or small island, *holm*; a farm (*Dan.*, Gaard), *garth*, &c., &c. We might thus, on a very low calculation, compute in round numbers the clearly recognisable Scandinavian names of places in England at one thousand five hundred.

That they should have been preserved in such numbers for more than eight centuries after the fall of the Danish dominion in England, and that they should have retained, as it has been shown, the original Scandinavian forms, and that often in a highly-striking degree, completely disproves the opinion that the old Danish-Norwegian inhabitants of the country north of Watlinga-Stræt were supplanted or expelled after the cessation of the Danish dominion (1042), first by the Anglo-Saxons, and afterwards by the Normans from Normandy; for if such had been the case, the names of places would naturally have become altogether changed and impossible to recognise. As the matter stands it is sufficiently proved that Danes as well as Norwegians must have continued to reside in great numbers in the districts previously conquered by them, and particularly in the north; and consequently that a very considerable part of the present population in the midland and northern counties of England may with certainty trace their origin to the Northmen, and especially to the Danes.

SECTION VIII

Resemblance of the People to the Danes and Norwegians.—Proper Names.—Popular Language.—Songs and Legends.

The present English people is certainly composed, as we have seen, of the most heterogeneous elements. The Englishman reckons among his ancestors Britons, Romans, Anglo-Saxons, and Northmen, especially Danes and Normans. All these people, who successively reigned over England for centuries, must naturally have left numerous descendants behind them. But as in ancient times it was a combat of life and death for dominion, the conquered and their posterity could not immediately amalgamate with the conquerors. Long after the Norman conquest (1066) the Britons, Saxons, Danes, and Normans, were still hostilely opposed to each other. These disputes were brought to a close during the middle ages; prejudices vanished; mixed marriages became more frequent; the different races acquired common interests; and at last, with the exception of those Britons who kept themselves aloof in Wales, passed into one great nation. From this time it was no longer usual in marriages to regard family descent; it was only some of the richer sort, and higher lineage, who considered it an honour to preserve the original blood as pure as possible. There are families still to be found in England who pretend that they descend in a direct line from Saxon or Norman ancestors, and who assert that Saxon or Norman features have been transmitted to them. But even these families have in the course of time been considerably mixed with races of an entirely different extraction; nay, even the Britons in Wales have not been able to prevent some of the hated English blood from gradually supplying and deteriorating that which runs in their own veins. Moreover, if we consider what an immense number of Irishmen, Frenchmen, Germans, Jews, and others, have, particularly during later centuries, immigrated into England, where they have settled, and by degrees married natives; and, lastly, if we remember that most foreigners have settled on the east coast, or in the midland and north-eastern districts; we might almost deem it impossible to point out from the features and bodily frame of the inhabitants of these districts, any preponderating degree of descent from Saxons, Danes, or any one race of people that colonized England in times so long past. In this respect we can of course scarcely think of comparing districts of small extent, such as two neighbouring parishes, or two adjoining counties on the east coast of England. Nevertheless, if by taking a survey of such extensive districts as north and south England, we were able to discover a tolerably decided difference in the general appearance of the inhabitants, this would be a weighty corroboration of the assertions of history, and of the proof derived from names, that these districts were originally peopled by inhabitants of entirely different descent.

The Englishman of London, and the rest of southern England, does not in general betray in his exterior any perceptible resemblance to the Danes and Norwegians. On the contrary, he decidedly differs from them. The black hair, the dark eye, the fine hooked nose, and the long oval countenance, remind one either of relationship with the Romans, whose chief seat in England was in the south, or rather, perhaps, of a strong compound between the ancient Britons and the Anglo-Saxon and Norman races, which afterwards immigrated into England. Many of the Britons seem to have been dark-haired; for among their descendants in Wales, as well as among their near kinsmen, the Highland Scots and the Irish, there are still frequently found—and particularly in remote districts, as, for instance, in the Hebrides—dark-haired and generally small people, having on the whole dark complexions. It was, too, in the south and south-west of England that the greatest mixture took place between the original British tribes and those that afterwards came over.

But as we proceed from the southern towards the middle and northern parts of England, we find that by degrees an entirely different physiognomy, which before we only got a glimpse of now and then, and which could scarcely be remarked in the confusion of people in London, becomes more and more the prevailing one. The farther one proceeds towards Northumberland, the more distinct does it become. The form of the face is broader, the cheek bones project a little, the nose is somewhat flatter, and at times turned a little upwards, the eyes and hair are of a lighter colour, and even deep red hair is far from being uncommon. The people are not very tall in stature, but usually more compact and strongly built than their countrymen towards the south. The Englishman himself seems to acknowledge that a difference is to be found in the appearance of the inhabitants of the northern and southern counties; at least one constantly hears in England, when red-haired compact-built men with broad faces are spoken of; "They must certainly be from Yorkshire:" a sort of admission that light hair, and the broad peculiar form of the face, belong mostly to the north-of-England people. On the other hand, little importance must be attached to the circumstance that Englishmen generally attribute the red hair to the immigration of the Danes; for though it is true that many Danes, and particularly many Norwegians, were red haired, yet some tribes of the original Celtic inhabitants of the British Isles also had red hair; and the same feature may likewise be partly ascribed to the Saxons.

In the midland, and especially in the northern part of England, I saw every moment, and particularly in the rural districts, faces exactly resembling those at home. Had I met the same persons in Denmark or Norway, it would never have entered my mind that they were foreigners. Now and then I also met with some whose taller growth and sharper features reminded me of the inhabitants of South Jutland, or Sleswick, and particularly of Angeln; districts of Denmark which first sent colonists to England. It is not easy to describe peculiarities which can be appreciated in all their details only by the eye; nor dare I implicitly conclude that in the above-named cases I have really met with persons descended in a direct line from the old Northmen. I adduce it only as a striking fact, which will not escape the attention of at least any observant Scandinavian traveller, that the inhabitants of the north of England bear, on the whole, more than those of any other part of that country, an unmistakeable personal resemblance to the Danes and Norwegians.

Old Scandinavian national names, such as Thorkil, Erik, Haldan, Harald, Else, and several others, were formerly, at least, not unfrequently used in these districts. Surnames, such as Adamson, Jackson, Johnson, Nelson (Nielson), Thomson, Stevenson, Swainson, and others, all of which have endings in *son* or *sen*, which never appear in Saxon names, still frequently occur. The ending *sön* or *sen* (a son) is quite peculiar to the countries of Scandinavia, whence it was brought over to England by the Scandinavian colonists. It is not, however, confined to the north of England, but is spread over all the British Islands where the Northmen settled; for instance, in Scotland we find Anderson, Matheson, &c. It is very remarkable that the name of Johnson, which, as is well known, is one of the commonest in England, is also, perhaps, in the selfsame form, that which most frequently occurs in Iceland.

The still-existing popular dialect affords an excellent proof that the resemblance of the inhabitants of the northern counties of England to the Danes and Norwegians is not confined to a, perhaps accidental, personal likeness. The pure English language itself includes, both with regard to its vocabulary and inflexions, many Scandinavian elements, the result of the Danish immigration. But, in the north of England, many words and phrases are preserved in the popular language, which are neither found nor understood in other parts, although they sound quite familiar to every Northman. These original Scandinavian terms are not only applied, as I have before said, to waterfalls, mountains, rivulets, fords, and islands, but are also in common use in daily life; as, for instance, *late* (*Dan.*, lede; *Eng.*, to seek), *lite* (*Dan.*, lide; *Eng.*, to rely), *helle* (*Dan.*, helde; *Eng.*, to pour out), *hit* (*Dan.*, hitte; *Eng.*, to find), *clip* (*Dan.*, klippe; *Eng.*, to cut), *forelders* (*Dan.*, Forældre, or Forfædre; *Eng.*, ancestors, forefathers), *updaals* (*Dan.*, opdals; *Eng.*, up the valley), *kirk-folk* (*Dan.*, Kirkefolk; *Eng.*, people going to church), *kirk-garth* (*Dan.*, Kirke-gaard; *Eng.*, churchyard), with many others.

These originally Scandinavian words are now chiefly found in the north-west of England, among the remote mountains of Yorkshire, Westmoreland, Cumberland, and Lancashire, where they have withstood the changes of time. On entering a house there one will find the housewife sitting with her *rock* (*Dan.*, Rok; *Eng.*, a distaff) and *spoele* (*Dan.*, Spole; *Eng.*, spool, a small wheel on the spindle); or else she has set both her *rock* and her *garnwindle* (*Dan.*, Garnvinde; *Eng.*, reel or yarn-winder) aside, whilst standing by her *back-bword* (*Dan.*, Bagebord; *Eng.*, baking-board) she is about to knead dough (*Dan.*, Deig), in order to make the oaten bread commonly used in these parts, at times, also, barley-bread; for *clap-bread* (*Dan.* Klappebröd, or thin cakes beaten out with the hand) she lays the dough on the *clap-board* (*Dan.*, Klappebord). One will also find the *bord-claith* spread (*Dan.*, Bordklæde; *Eng.*, table-cloth); the people of the house then sit on the *bank* or *bink* (*Dan.*, Bænk; *Eng.*, bench), and eat *Aandorn* (*Eng.*, afternoon's repast), or, as it is called in Jutland and Fünen, *Onden* (dinner). The chimney, *lovver*, stands in the room; which name may perhaps be connected with the Scandinavian *lyre* (Icelandic, ljóri); *viz.*, the smoke-hole in the roof or thatch (*thack*), out of which in olden times, before houses had regular chimneys and "*lofts*" (*Dan.*, Loft; *Eng.*, roof, an upper room), the smoke (*reek* or *reik*, *Dan.*, Rög) left the dark (*mirk* or *murk*, *Dan.*, mörk) room. Within is the *bower* or *boor* (*Eng.*, bed-chamber), in Danish, *Buur*; as, for instance, in the old Danish word Jomfrubuur (the maiden's chamber), and in the modern word Fadebuur (the pantry).

Outside, in the *garth*, or yard (*Dan.*, Gaard), stands the roomy *lathe*, or barn (*Dan.*, Lade), which directly shows how fruitful the soil is that belongs to the *garth* (*Dan.*, Gaard; *Eng.*, a manor, farm). The shepherd or herdsman, whose *nowth* (*Dan.*, Nöd; *Eng.*, neat cattle) are restless in the *boose* (*Dan.*, Baas; *Eng.*, stall) and *crib* (*Dan.*, Krybbe; *Eng.*, manger), is about to cleanse the stable, and with a *greype*, or gripe (*Dan.*, Möggreve; *Eng.*, dung-fork), bears out the *muck* (*Dan.*, Mög; *Eng.*, dung) to the midding (*Dan.*, Mödding; *Eng.*, dunghill). If we accompany him to the fields he tells us in a lively tone about the many *threaves* of corn (*Dan.*, Traver, bundles of twenty or thirty sheaves), particularly of *big* (*Dan.*, Byg; *Eng.* barley) that have been got from the poor *ling* (*Dan.*, Lyng; *Eng.*, fern) which covers the sides of the *haughs* or *haws* (*Dan.*, Höie; *Eng.*, hills); of all the *slaa-torns* (*Dan.*, Slaatjörn; *Eng.*, sloes), *lins* (*Dan.*, Lindetræer; *Eng.*, linden trees), *roan trees* (*Dan.*, Rönnetrær; *Eng.*, Scotch rowan trees), and *allars* (*Dan.*, Elletræer; *Eng.*, alders), that grow in yonder little *shaw* (*Dan.*, Skov; *Eng.*, wood), or in that *lawnd* (*Dan.*, Lund; *Eng.*, grove), which is likewise full of *hindberries* (*Dan.*, Hindbær; *Eng.*, raspberries), and which is resorted to by many *gowks* (*Dan.*, Gjöge; *Eng.*, cuckoos). A field farther on, which in its time was acquired by *mackshift* (*Dan.*, Mageskifte; *Eng.*, deed of exchange), has been allowed to *ley-breck* (*Dan.*, ligge-brak; *Eng.*, to lie fallow). Through this field winds a *beck* (*Dan.*, Bæk; *Eng.*, brook), or rivulet well stocked with fish, in which with a *liester* (*Dan.*, Lyster; *Icelandic*, Ljöstr, grains, or a sort of barbed iron fork on a long pole) one may be able to make a good capture.

In the river are the *trows*, or troughs (*Jutland*, trow; *Old Scan.*, Þró), made use of to cross over to the opposite shore. These *trows*, or troughs, are two small boats, originally trunks of trees hollowed out, and held together by a cross-pole. He who wishes to pass over places a foot in each trough or boat, and rows himself forward with the help of an oar. It is said that Edmund Ironsides and Canute the Great rowed over to the Isle of Olney (in the river Severn) in such boats at the time when they concluded an agreement to divide England between them. The original inhabitants of Europe undoubtedly passed the great rivers in the same simple manner.

Amongst the words in the popular language that still remind one of ancient Scandinavian customs, those of *yuletide*, *yuling* (Christmas), *yule-candles* (*Dan.*, Julelys), and *yule-cakes* (*Dan.*, Julekager), deserve particular notice. Christmas was certainly kept as a solemn feast among the Anglo-Saxons, but it does not appear to have had that importance with them which it had with the Scandinavians; of which this is a proof, that the old name of Christmas (*Yule*) is preserved only in those districts in the north that were more especially colonized by the Northmen. Yule, or the mid-winter feast, was, in the olden times, as it still partly is, the greatest festival in the countries of Scandinavia. Yule bonfires were kindled round about as festival-fires to scare witches and wizards; offerings were made to the gods; the boar dedicated to Freÿr (*Dan.*, Sonegalte) was placed on the table, and over it the warriors vowed to perform great deeds. Pork, mead, and ale abounded, and yuletide passed merrily away with games, gymnastics, and mirth of all kinds. It is singular enough that even to the present day it is not only the custom in several parts of England to bring a garnished boar's-head to table at Christmas, but that the descendants of the Northmen, in Yorkshire and the ancient Northumberland, do not even now neglect to place a large piece of wood on the fire on Christmas Eve, which is by some called the *yule-block*, by others *yule-clog*, or *yule-log* (perhaps from the old Scandinavian *lág*, *log*, a felled tree; Norwegian, *laag*). Superstitious persons do not, however, allow the whole log to be consumed, but take it out of the fire again in order to preserve it until the following year. Exactly similar observances of Christmas customs still exist in the Scandinavian North. At Smaaland, in Sweden, a boar's-head, called *julhös* (from *hös*, the skull), is set on the table at Christmas; and in East Gothland a large loaf, called *juhlegalt*, is seen on table throughout the festival, of which, however, nothing is eaten. *Juhlhös* and *juhlegalt*, as well as the boar's-head in the north of England before alluded to, owe their origin unmistakeably to the expiatory barrow-pig, or "Galt," offered up by the old Northmen to Freÿr. The remembrance of the games of the Northmen is also preserved in England in the Scandinavian word *lake* (to play), which is heard only in the ancient Danish districts.

To enumerate all the Scandinavian words in the English popular tongue would, from their quantity, be both a tedious and a superfluous labour. The following selection of a hundred of the most common of them will surely be regarded as sufficient clearly to prove in what a highly remarkable manner "the Danish tongue" has imprinted itself on the north of England, in comparison with other countries occupied by the Normans, as, for example, Normandy; where the Scandinavian language, notwithstanding the very considerable immigrations from Scandinavia, has disappeared to such a degree that but very few traces of it now remain.

A HUNDRED DANISH WORDS, SELECTED FROM THE VULGAR TONGUE,

OR COMMON LANGUAGE, NORTH OF WATLINGA STRÆT.

Provincial English[8].	English.	Danish.
arr	scar	Ar
attercop	spider	Edderkop
awns	beads of corn	Avner
bank	to beat	banke
bairn, bearn	child	Barn
bede	to pray	bede

bid	to invite	byde, indbyde
bide	to stay	bie
big, biggin	to build, building	bygge, Bygning
blend	to mix	blande
boll, or bole	trunk of a tree	Bul (Træ)
brosten	burst	brusten
clammer	to quarrel, grasp	klamres, fast-klamre
claver	to climb	klavre
cluve	hoof	Klov, Hov
dyke, dike	ditch	Dige
elt	to knead	ælte
festing-penny	earnest-money	Fæstepenge
fra	from	fra
frem folks	strangers	Fremmede Folk
full	drunk	fuld, drukken
gainest way	nearest way	Gjenvei
gammon	merriment	Gammen
gants, ganty	to be merry	gantes
gar	to make	gjöre
gar	to hedge	gjerde
glowing (glouring)	staring	gloende
greit, greets	to weep, tears	grœde, Graad
grepen	clasped	greben
grise	young pig	Griis

groats	husked corn	grudtet Korn
hack	to stammer	hakke, stamme
halikeld	holy-well	Helligkilde
hand clout	towel	Haandklæde
handsel	earnest	Handsel
harns, harns-pan	brain, brain pan	Hjerne, Hjerne-skal
heck	hay-rack	Hække (til Hö)
hesp	latch	Haspe (Dör)
hose	stocking	hose
kaam, kem	comb, to comb	Kam, kæmme
kail, kale	cabbage	Kaal
kern-milk	churn-milk	Kjernemelk
kern	to churn	kjerne
kilt	to tuck up	kilte (op)
kitling	young cat	Killing
laid	just froze	logt (Iis)
mauf, meaugh	brother-in-law	Maag, Svoger
mind	to remember	mindes
nab	to catch	nappe
neaf (or neif) neaf-full	fist, handful	Næve, Nævefuld
neb	bill, beak	Næb
nipping	to sip	nippe
pot-scar	pot-sherd	Potteskaar
quern	hand-mill	Qværn

querken'd	suffocated	qværket
raise	a heap of stones, cairn	Rös, Steendysse
read (or rede)	to guess, know fully	raade, udtyde
read	to comb	rede (Haar)
reasty	toasted	ristet
rid	to remove	rydde
rig, riggin	back, ridge of a house	Ryg, Rygning
rip up	to revive (injuries)	rippe op
rise	underwood	Riis (Underskov)
rive	to split, divide	rive (splitte)
sackless	without suit	sageslös
sark	shirt	Særk
scarn	dung	Skarn (Smuds)
schrike (or skrike)	to cry, shriek	skrige
scoll	toast (health)	Skaal (Drikkelag)
sele	to bind, fasten	bind i Sele
skift	to change (clothes)	skifte (Klæder)
slade	sledge	Slæde
sleck	to put out (quench)	slukke
smiddy	blacksmith's shop	Smedie
smooth-hole	hiding-place	Smuthul
smouch	kiss	Smadsk (Kys)
snirp	to pine	snirpe
speer (or spar)	to ask	spörge

spire	young tree	Spire
stee (or stey)	ladder	Stige
steert	point	Stjert
stew	dust	Stöv
stive	to raise dust	stöve
stumpy	short, thick	stumpet
stot	young horse, or bullock	Stod (Hest)
swale	shade	Svale (Skygge)
sype (or sipe)	to drop gently (ooze)	sive
tang	sea-weed	Tang
theaker	thatcher	Tækker
toom (or tuam)	empty	tom
twine	to murmur, weep	tvine
unrid	disorderly, filthy	uredt, urede
uphold	to maintain	holde oppe
wadmal, woadmel	coarse woollen cloth	Vadmel
wan	rod	Vaand
wark	ache, pain	Værk (Smerte)
way zalt	to weigh salt, a game	veie Salt (Leeg)
wong	a field	Vænge

8. Many of these words are Scotch.

These numerous and striking Danish terms, still existing in the north of England almost a thousand years after the destruction of the Danish power there, and after an almost equally protracted struggle with the constant progress of the English language, show that the

Scandinavian tongue must possess no mean degree of durability. These Scandinavian words, moreover, taken in conjunction with the unusually numerous Scandinavian names of places in England, put it beyond all doubt that a Scandinavian population must have been far more diffused, and have taken much deeper root there, than in any other foreign land.

The popular language of the north of England is particularly remarkable for its agreement with the dialects found in the peninsula of Jutland. Several words which are common to the north of England and Jutland, are not to be found elsewhere. For instance, in the north of England, the shafts of the carts used there are called *limmers*, a word clearly of the same origin as the Jutlandish *liem*, a broom; both being derived from the old Scandinavian *limi*, which signifies *boughs*, *branches*. But it is the broad pronunciation in particular that makes the resemblance so surprising. Thus, for instance, we have in the north of England, *sty'an* (*Dan.*, Steen; *Eng.*, a stone), *yen* (*Dan.*, een; *Eng.*, one), welt (*Dan.*, vælte; *Eng.*, to upset), *swelt* (*Dan.*, vansmægte; *Eng.*, overcome with heat and exercise), *maw* (*Dan.*, Mave; *Eng.*, stomach), *lowe* (*Dan.*, Lue; *Eng.*, flame), *donse* (*Dan.*, dandse; *Eng.*, dance), *fey* (*Dan.*, feie; *Eng.*, to sweep), *ouse* (*Dan.*, Oxe; *Eng.*, ox), *roun* (*Dan.*, Rogn; *Eng.*, spawn or roe of fishes), *war and war* (*Dan.*, værre og værre; *Eng.*, worse and worse); with many others of the same kind, which are pure Jutlandish.

On the whole, of all the Danish dialects the Jutland approaches nearest to the English. The West Jutlander uses the article *æ* before words like the English "the," although the Danish language in other provinces does not recognise such an article; and the broad open *w*, which the natives of Funen and Zealand can, after the greatest difficulty, only pronounce with tolerable correctness, is as easy for the Jutlander as for the Englishman. Many Danish words pronounced in Jutlandish become purely English; as, for instance, *foul* (*Eng.*, fowl; *Dan.*, Fugl), *kow* (*Eng.*, cow; *Dan.*, Ko), *fued* (*Eng.*, food; *Dan.*, Fod), *stued* (*Eng.*, stood; *Dan.*, stod), *drown* (*Eng.*, drown; *Dan.*, drukne); besides many others. Many words are even quite common to Jutland and England; such as the Jutlandish *forenoun* and *atternoun* (*Eng.*, forenoon and afternoon; *Dan.*, Formiddag and Eftermiddag), *stalker* (*Eng.*, stalker;

Dan., en Stork), *kok* (*Eng.*, cock; *Dan.*, en Hane), *want* (*Eng.*, to want; *Dan.*, mangle, behöve).

This affords a very important proof of the close connection which must have anciently subsisted between Jutland and England. Although it may be doubtful to what extent the Jutes had tracts specially assigned to them for their settlements in the south of England (as in Kent and the Isle of Wight, at the time of the Anglo-Saxon conquest in the fifth century), it is, at all events, quite certain that, both at that time and at a later period, a number of Jutes settled on the east coast of England, and particularly in the more northern districts. Jutland lies nearer to England than any other part of Scandinavia. The Limfjord, which in remote ages was a roadstead for the Vikings' ships, and afterwards the rendezvous of Saint Canute's fleet when he intended to reconquer England, certainly dispatched numerous Vikings' barks to the British coasts. In legends still existing in Jutland, the old connections with England, and the wars there, are not forgotten; nay, in some places the people tell of battles fought with the English in Jutland itself: of which ancient names of places likewise bear witness, as in the neighbourhood of Holstebro, "Angelandsmoor" (Angelandsmosen), with the adjacent "Prince Angel's barrow" (Prinds Angels Höi), which is surrounded with a number of tumuli. The remembrance of the same old connections with England still resounds in the Jutlandish and other ancient Scandinavian ballads, or heroic songs, in which the scene is frequently laid on the "engelandish strand."

The near relationship of the north Englishmen with the Danes and their Scandinavian brothers is reflected both in popular songs and in the folk-lore. It is well known that the old Northmen were in a high degree lovers of minstrelsy. The Scandinavian kings were generally accompanied on their Viking expeditions by bards, who encouraged and cheered the champions with songs respecting the exploits of former times, and about every glorious deed that had been performed during the expeditions. These historical epics passed from mouth to mouth, and from generation to generation. Nor did the Scandinavian conqueror in foreign lands disdain to be celebrated by the bards of his native country. Canute the Great, who was himself a poet, placed the

Scandinavian bard high in his hall; and numerous lays, which are still partly preserved in the Sagas, sounded his fame over the north. After the warlike life of heathenism had ceased, the poetical and historical talent of the people expressed itself in ballads and heroic songs, which, during the middle ages, succeeded the lays of the ancient bards. The old ballad, in its characteristic form, belongs peculiarly to the countries of Scandinavia; and it is very remarkable that the corresponding English ballads, which often, both in their prevailing tone and in their form—as, for instance, with regard to the burthen—betray a surprising similarity with the Scandinavian, are in England found exclusively in the north. They are, however, heard still more frequently in the Scotch Lowlands, whither great immigrations of Northmen also took place. In the north of England a very peculiar kind of song for two voices was also formerly heard, and which the English themselves ascribed to the Danes.

It is more difficult to adduce pure Scandinavian remains of popular superstitions, as in this respect the Teutonic races have so very much in common; and consequently one is afraid to draw too strong conclusions from the striking agreement usually shown in the phantoms of the imagination among north Englishmen and their Scandinavian kinsmen. Yet it deserves to be mentioned that the Scandinavian name *Nök* (a river-sprite), is not yet forgotten in Yorkshire; although some by "Nick" or "Oud-Nick" erroneously imagine the devil to be meant, instead of the water-sprite. Many little tricks performed by the *nix* (*Dan.*, nisse, a brownîe) are knòwn there, as well as in Scandinavia. Once, in England, the conversation happening to turn on these little beings, I related our Scandinavian legend about a peasant who was plagued and teazed in all possible ways by a *nisse* or brownîe, till at last he could bear it no longer, and determined to *flit* (move house) to another place. When he had conveyed almost all his goods to the new house, and was just driving thither with the last load, he accidently turned round, and whom did he see? Why, the brownîe with his red cap, who sat quietly on the top of the load, and nodded familiarly to him, with the words, "Now we flit." One of the persons present immediately expressed a lively surprise on hearing a legend related as Danish, and that, too, almost word for word, which he had often heard in Lancashire in his youth. The word *flit* was, and still is, used there by the common people.

A natural result of the long-continued and extensive dominion of the Danes in the north of England is, that they also are classed with the invisible mystical beings, which, in the imagination of the people, haunt that district. In certain places among the remote mountains of the north-west, people still fancy that they hear on the evening breeze tones as of strings played upon, and melancholy lays in a foreign tongue. Often, too, even when nobody hears anything unusual, the animals prick up their ears as if in astonishment. It is "the Danish boy," who sadly sings the old bardic lays over the barrows of his once mighty forefathers.

SECTION IX

The Outrages of the Danes.—The Danes and Normans.—Influence of the Danes in England.

It is thus shown, by numerous and incontestable proofs, that the Danes held dominion in England for a short period, and that they also exercised, in conjunction with the Normans, so important and lasting an influence for centuries before and after the time of Canute the Great, at all events in that portion of England lying to the north of Watlinga Stræt, that even a great part of the population there may be safely assumed to be of Danish extraction. Nevertheless, the generally received opinion in England on this subject is expressed in the following passage in a brief History of Denmark lately published in London ("Edda, or the Tales of a Grandmother"), which states that after the suppression of the Danish power in England, "*Both nations [the Danes and English] separated soon after, and in a few years the Danish supremacy had vanished like a vision of the night; so little did it leave any traces in England, or produce any important political benefits to Denmark.*"

It would, however, have been extremely astonishing, nay, utterly inexplicable, if great effects had not manifested themselves in Denmark from the expeditions towards the west, and from the complete conquest of a country like England, which, in regard both to religious and political development, stood so far above Scandinavia. History, also, sufficiently shows of what great importance the conquest of England was, not only for Denmark, but for the whole Scandinavian North. The Christianity of Scandinavia arose, indeed, out of the smoking ruins of the English churches and convents. Scandinavian kings and warriors were frequently baptized during their Viking expeditions; and it was English priests who proclaimed the doctrines of Christianity on the plains of Denmark and in the rocky valleys of Sweden and Norway. Many of the first bishops in the North were of English extraction, and even the style of the ecclesiastical edifices attested the powerful influence of wealthy England. The more advanced cultivation of science and art in general which prevailed there, communicated itself in many directions to the countries of Scandinavia; where it certainly contributed, just as much as the great emigrations, to weaken heathenism, and thus, both in a religious and political point of view, to found a new and better order of things.

But for whatever benefits Denmark and the North received in this manner from England, they did not fail to yield a full equivalent. It cannot reasonably be reproached to the Danes exclusively that, in order to obtain settlements in England, they made their way with fire and sword, for this was no more than all other conquerors, and particularly the Romans and Anglo-Saxons, had done before them. With regard to bloodshed, and acts of violence and destruction, the Anglo-Saxon conquest of England exceeded rather than fell short of the Danish. It annihilated the civilization which had been so widely disseminated there by the Romans, and subjugated or expelled the older inhabitants in the most frightful manner. It is the circumstance of the Danish expeditions having taken place at a far later time, when the monks wrote chronicles, and when on the whole history was more circumstantial, that has alone contributed to place the Danish expeditions in so prominent and so hateful a light.

But even the present age, with its severe views, is scarcely justified in condemning unconditionally the Scandinavian sea-king, who was not instigated solely, or even chiefly, by a savage desire of plunder or murder, but who valued deeds of arms, a glorious name, and the joys of Valhalla, more than his life, and who therefore "went to death with a laugh." Even with him religion was a spur to his achievements in Christian lands. He was combating for his own gods, in whom in general he certainly believed as firmly as most of the Christians of that time did in Christ. The ideas, too, which then prevailed respecting conquest, slaughter, and rapine, were altogether different from ours. If the heathen Viking regarded it as an honour to acquire lands and booty by his sword, the same thought was also cherished not only by the early Christians, but throughout the middle ages; when Christian citizens, noblemen, and princes contended in mortal combat, with fire and sword, for the possession of estates and lands. The Christian Anglo-Saxons of those times felt no hesitation in secretly massacring the Danes who had settled in England; and as many of these had been converted, one Christian thus murdered another! To dismember general history into a number of unconnected events, and then to pass judgment upon these separately according to our moral feelings, would be an infamous act, and more difficult to defend before the tribunal of morality than perhaps all the expeditions of the heathen Danish Vikings put together. Such a method of proceeding would lead to the most confined views of history that can possibly be imagined. No correct conception can be formed of any part of the history of the world if it be not examined in its due connection, whereby both causes and effects become perceptible. Many events, which the moralist would otherwise condemn, find in this manner both excuse and defence in the superior historical necessity that produced them. Viewed in this light, violent devastations, which have for a time, perhaps, arrested the progressive development of a people, will appear to have ultimately founded and educed purer and more wholesome manners and customs. Severe shocks are now and then as useful for the general welfare of a nation as a violent fit of sickness for the health of an individual, or storms for the purification of an oppressive atmosphere.

The germ of a higher civilization was first implanted in the rude and warlike tribes, which then predominated throughout Europe, by the Greeks and Romans. The bold expeditions of the latter, in particular, introduced the arts and sciences into the countries north of the Alps; and it was from the south that even the Christian religion began its progress. But before Christianity could take firm root among the European tribes, before a really Christian state could be founded, it was necessary that an immense revolution should take place. Heathenism and barbarism then collected all their strength in order to destroy Roman power and Roman civilization. The Roman Empire, and with it almost all the older states, was overthrown by the vast national migrations; and a new and different population, with which a fresh civilization was to begin, spread itself over Europe. It was these migrations that brought the Anglo-Saxons into England, after they had abandoned their ancient habitations on the south and south-west shores of the Baltic; whence they were expelled by the advancing Slavonic tribes of the Wends, or Vandals.

Contemporaneously with the diffusion of Christianity in the south and west of Europe, larger Christian states gradually arose. Charlemagne had already, about the year 800, founded an immense kingdom; and, in order to strengthen it both against inward disturbances and outward attacks, had established apparently durable institutions. But as it was too often necessary, in those early times, to force Christianity on the people by dint of arms, without seeking any real support for it in their convictions and belief—a circumstance that rendered prevalent a very great moral relaxation, and even wickedness—they were thus induced to regard the political institutions which sprang from it as something foreign, which neither proceeded from themselves, nor possessed any intrinsic strength. Both Church and State tottered. The whole structure of Christian communities was in its weak and early childhood; and it was not till the people had been convinced of its necessity, by their calamities and sufferings, that Christianity was able to gain a really firm footing.

The Christian States were now attacked at once and on all sides by the enemies of Christianity, the Mahometans and heathens. The Saracens, towards the south; the Magyars, or Madjarers, the forefathers of the Hungarians, towards the east; and the Northmen towards the north and west, all invaded the Christian States. Europe long groaned under this terrible scourge. Meanwhile, however, separate States grew stronger in this combat with their exterior enemies; whilst great tribes of the latter settled in the conquered districts, adopted Christianity, and mingled with the natives. The destructive expeditions which for a time indeed retarded, in certain directions, the commencements of civilization, ended by exhausting all the strength of heathenism, in preparing a complete victory for Christianity, and in producing in Church and State a vigour hitherto unknown in those lands which had long embraced the Christian faith. It was now that a period was put to the throes which had given birth to a new and Christian Europe. The descendants of the lawless Vikings became the most zealous champions of Christianity. The Normans, who by degrees had raised themselves to be the ruling people in several of the western and southern States of Europe, and had thus brought a new and wholesome power to the helm, broke many a doughty lance with the Mahometans and heathens. In these crusades the knight was now accompanied by the troubadour, as the Viking formerly had been by the bard or scald. It was among the Normans in particular that the knightly and feudal system developed itself, which was of such decided importance throughout the middle ages, and the forerunner of the freer and more advanced state of society of modern times.

Under the name of "Normans" are included all those swarms of Swedes, Norwegians, and Danes, which, from the close of the eighth until far into the eleventh century, either laid waste or settled on the eastern and southern coasts of the Baltic, as well as the coasts of the west and south of Europe. "Norman" signifies neither more nor less than a man from the north. The Danish conquest of England was therefore just as fully Normanic as the conquest, by the Norwegians and Danes, of a part of France, called, after them, Normandy. Hence there was a natural reason why the Danish conquerors, and Svend Tveskjæg in particular, concluded an alliance with the dukes of Normandy, in order that they might find a reception among these kinsmen in case they should not be able to make themselves masters of England; and hence, in like manner, Canute the Great obtained the more readily the hand of Emma, the daughter of a Norman (and consequently nearly related) duke. But between the above-mentioned conquests there was this difference, that the Danish conquest of England, together with the Norwegian conquests in Scotland and Ireland, was of far greater extent, and of quite a different and more extensive importance for the British Isles, than the Norwegian-Danish conquest of so small a district as Normandy was for France. Whilst the Northmen principally brought thither only a number of powerful chiefs, who, at the expense of the natives, constituted themselves into an imperious feudal nobility, and who afterwards for the most part went over with William the Conqueror into England, in search of still greater feudal possessions, the Danish expeditions to and conquest of England were, on the contrary, the means of bringing an entirely new population into a very considerable portion, perhaps even the half, of that kingdom.

All accounts attest what proud and energetic men the Norwegian-Danish Normans were who settled in Normandy, and who afterwards became the progenitors and founders of the English nobility. The chronicles of that time cannot sufficiently praise their bravery and contempt of death, whilst at the same time they highly extol their chivalric spirit. In but a short time after their settlement in France they had readily acquired its politer manners; and not only these, but that higher mental cultivation which then raised the southern countries above those of the far north. It was a distinguishing trait of the Normans that they very quickly accommodated themselves to the manners and customs of the countries where they settled; nay, even sometimes quite forgot their Scandinavian mother tongue, without, however, losing their original and characteristic Scandinavian stamp. But what the Normans in particular, with all their French refinement, did not lose, was the ancient Scandinavian feeling of freedom and independence. The descendants of those powerful chiefs who had quitted the hearths of their forefathers because they would not suffer themselves to be enslaved by kings—and who on their arrival in Normandy, when the question was put to them, "What title does your chief bear?" are said to have answered, "None, we are all equal"—continued steadily to maintain their freedom against the Norman dukes, and not least so against the despotic William the Conqueror, even after he had distributed among them the rich estates of conquered England. The later English nobility, whose power and influence William's conquest had thus founded, did not in any way degenerate from their Norman forefathers. From the earliest period of the middle ages the English barons were the stoutest protectors and defenders of freedom against ambitious kings; and it is also their respect for the proper liberties of the people that has alone insured to them the quiet possession of the power which they still continue to retain. The English nobility have in several other ways preserved to the present time traces of their ancient origin. Thus among the English aristocracy we not only find the old Scandinavian title of Jarl, or Earl, which in the North itself has given way to the German one of Graf, or Greve, but a Northman will easily discover many characteristic traits that remind him of his own ancestors. It is truly remarkable that the love of bodily exercises, games, hunting, and horse-racing, not to mention the predilection for daring sea voyages so strongly prevalent amongst them, was likewise manifested, according to the Sagas or legends, by the rich and powerful in Iceland, and the rest of the Scandinavian fatherlands.

Under these circumstances it would, indeed, have been in the highest degree surprising if the Danish-Norwegian Normans, who conquered England at the same period that their near kinsmen, the Norwegian-Danish Normans, conquered Normandy, who had migrated from the north for the selfsame reasons as these kinsmen, and who were subject to the same virtues and vices—if these Normans in England alone, I say, should have been barbarous "robbers and plunderers," trampling on and destroying all that was "great and good," whilst their brothers in Normandy distinguished themselves by an early civilization, and particularly by a lively feeling for poetry and for a further development both of social and political life. It must be remembered that the Danish-Norwegian Normans, who made conquests in England, did not go thither in one great body, but in small divisions, which only by degrees, and in the course of about three centuries, settled themselves in the districts inhabited by the Anglo-Saxons; and that, though far less numerous than the latter, they were not only able firmly to maintain their position among them, but at length even to expel them from a great part of the country north-east of Watling-Stræt. For this proves that the new Scandinavian inhabitants of England, along with greater physical strength and more martial prowess than the Anglo-Saxons possessed, must have been soon able to acquire that skill in the employments of peace, as well as that higher polish and refinement, which in the long run could alone insure them the superiority and preponderance which they enjoyed over the Anglo-Saxons, not only in the rural districts, but in many towns of the north of England; and secure for them such an influence as they obtained in England's best and greatest city, even London itself.

Further, that those Northmen, who by the Danish conquests became the progenitors of a great part, probably as much as half, of the present population of England, were just as brave men, and just as great lovers of liberty, as their Norman brethern, the ancestors of the English nobility; and that they played a part not much inferior to theirs in the development of England's freedom and greatness, a closer examination will probably place in a clearer light.

SECTION X

Commerce and Navigation.

The Northmen, who in ancient times sailed to foreign shores, were far from always being Vikings, bent only on rapine and plunder, and the conquest of new possessions. They were very often peaceful merchants. The remote situation of Scandinavia, and the dangers which the natives of more southern countries pictured to themselves as attendant upon a voyage to that *ultima Thule* and its heathenish inhabitants, must in ancient days, when navigation was very limited, have deterred foreign merchants from visiting it regularly, and bartering their wares. The Scandinavian tribes, on the contrary, were at that time almost the only seamen. From the want of all that belonged to the exterior comforts and conveniences of life in Scandinavia, the business of a merchant who bartered the products of the north and south, and who brought home with him a knowledge of distant and unknown lands, must early have become a profitable, and, from the dangers connected with it, an honourable profession. The trading voyages of the merchant were not, indeed, held in such esteem as those of the Vikings; yet from the most ancient times certain established customs were observed in the north for the protection of merchant vessels; and the merchant who, as was frequently the case, had distinguished himself by warlike qualities and shrewdness of understanding, was neither despised in the company of Vikings, nor in the King's hall. Even chiefs of royal descent did not regard it as anything dishonourable to exercise the mercantile profession. Already, in the most ancient times, a number of trading places were scattered round the north, and large annual fairs were held. Once a year the ships of the merchants assembled together from the whole of Scandinavia, perhaps even from the other nearest situated countries, in the Sound of Haleyri, or, as it is now called, Elsinore. Booths were erected along the shore; foreign wares were bartered for fish, hides, and valuable furs; whilst various games, and all sorts of merry-making, took place.

During the Roman dominion in England, and probably even in far earlier times, a tolerably brisk commerce appears to have been kept up between England and the countries of Scandinavia, especially Jutland, Vendsyssel, and the districts round the Limfjord; where also, as a consequence of this, genuine Roman antiquities have been dug up at various times. After the conquest of England by the Jutes, the Angles, and the Saxons, and still more after the Danes and Norwegians had begun to settle there, this intercourse became still more frequent. We may safely assert that, so early as the close of the ninth and beginning of the tenth century, a very brisk trade must have existed between England and the North. The Scandinavian element was then so well established, that not only did Scandinavian kings reign, and coin money, in the north of England, but even that extremely important old Saxon city, "North-weorig," which lay in the very heart of England, was called by the Saxon kings themselves, on their coins, by the foreign name of "Deorabui" (Deoraby, Dyreby, Derby); although this name, according to the English chroniclers' own statements, was first given to it by the immigrant Danes. Some will even recognise Derby in the name of "Doribi," which stands on a coin of King Ethelwulf of the middle of the ninth century (837-857). At all events it is a certain and remarkable proof of the early and wide-extended influence of the Scandinavian settlers, even in places far in the interior of the country, that "Deorabui" appears repeatedly on coins of King Athelstane (924-940), and of his immediate successors. It was this same Athelstane who is said to have visited Scandinavia, where he learned the language; and who afterwards educated at his court Hagen Adelsteen, the law-giver, who subsequently became the first Christian king in Norway. This fact also indicates the wide-spread and peaceful connection between England and the North, which not long afterwards induced the Norwegian King's son, Olaf Trygvesön, in his treaty of peace with the English king, Ethelred, whose lands he had long harried, expressly to stipulate for certain rights and privileges in favour of the Scandinavian merchant ships in the English harbours.

Even in Alfred the Great's time (A.D. 900) the seas and lands of Scandinavia were but very little known to the Anglo-Saxons; for which reason Alfred, chiefly with a view to trade and commerce, sent Ulfsten and the Norwegian Ottar on voyages of discovery to the

Baltic, and along the coast of Norway to the White Sea. That according to the laws of his country an Anglo-Saxon merchant obtained the rank and title of Thane, or Chief, when he had thrice crossed the sea in his own ship, sufficiently attests how desirous the Anglo-Saxon kings were to awaken among their subjects, by means of large rewards, a desire for such voyages. Subsequently, however, during the expeditions of the Vikings and Normans, when the dangers attending long voyages had become still greater than before for the Anglo-Saxons, owing to the perfectly overwhelming force of the Northmen at sea, the trade, with Scandinavia at least, must have continued to remain in the hands of the Scandinavian merchants; who, as we learn from the Sagas, were continually making voyages, as well from Denmark, Sweden, and Norway, as from the still more distant Iceland, to England, and the other countries of the West. Wherever the Normans had won new settlements, Scandinavian merchants likewise established themselves in order to maintain a steady connection with their ancient home. It is for this simple reason that we find in those times so many Danes and Norwegians settled in the most important trading places, not only in England (in London, Southwark, Derby, Grimsby, York, Whitby, and other towns), but also, as we shall see in the sequel, in Ireland and in Normandy, where the city of "Ruda," or Rouen, is spoken of as an important place of trade often visited by the Northmen.

The Scandinavian merchant vessels brought not only the wares of Scandinavia to the British Islands and other countries of the West; they likewise brought merchandise from the remote East. From the most ancient times, indeed, the Northmen had maintained connections with the eastern countries; which was a natural consequence of their having emigrated thence into the North, and left friends behind them there. By means of these connections, metals otherwise totally unknown in the North, and especially gold, were certainly brought thither at a very early period from the mountains of the East. Subsequently, in the fifth and sixth centuries, when fresh migrations from the East had taken place, a closer connection was opened with the eastern Roman Empire, and particularly with Constantinople, so that coins of that empire, and other valuables, began to be circulated in the North. After the Scandinavian colonists, too, had conquered kingdoms for themselves in the countries which now form modern Russia, and taken possession of the city of Novgorod, a regular commercial *route* appears to have been opened, through Russia, between Constantinople and the North, by which the Varangians passed, who entered as body-guards into the service of the Emperors of the East. But as far as regards trade, Novgorod and the Scandinavian colonists in Russia promoted a connection with Asia, which was of far greater extent and importance.

Before the passage to the East Indies by sea was discovered, and particularly before the Genoese and Venetians began to trade in the Black Sea and on the coasts of Asia, the main road from Arabia and the countries round the Caspian Sea to the Baltic and Scandinavia, lay through Russia, along the great rivers. To judge from the Oriental coins found both in Russia and in the Scandinavian countries, this commercial road must have been used from the eighth until far in the eleventh century, when it was broken up by internal disturbances in Asia, and by contemporary revolutions in Russia and the North. The road ran either from Transoxana (in Turan) to the countries north of the Caspian Sea, whence the merchandise was then brought along the river Volga to the Baltic; or else from Khorasan (in Iran), through Armenia, to the Black Sea; whence the Khazars and other people again conveyed it up the rivers farther towards the North. How considerable this trade must have been may be seen from the numerous hints in the Sagas, as well as from the still-existing Arabian accounts of merchants who in those days visited the coasts of the Baltic for the sake of trade, where considerable trading places, such as Sleswick and many others, are mentioned; but still more than all these, from the very great number of Arabian coins that have been dug up both in Russia and Scandinavia. In Sweden, and particularly in the island of Gothland, such an immense quantity of these has been found at various times, that in Stockholm alone above twenty thousand pieces have been preserved, presenting more than a thousand different dies, and coined in about seventy towns in the eastern and northern districts of the dominions of the Caliphs. Five-sixths of them were coined by Samanidic Caliphs. Together with the coins, a great mass of ornaments has been dug up, consisting of rings and other articles in silver, which are distinguished by a peculiar workmanship. On the whole, it appears that silver first came by this way into the North, where it was not generally circulated before the ninth and tenth centuries, and consequently at the time when the trade with Arabia was in full activity.

These discoveries of Arabian coins in the north of Europe, but which are confined to the shores of the Baltic, the German Ocean, and the Irish Sea, undoubtedly prove that Scandinavia, and particularly the countries on its eastern coasts, together with the islands of Gothland, Öland, and Bornholm, must have been the principal depôt for Arabian merchandise. It was the trade with the East that originally gave considerable importance to the city of Visby in Gothland; and it was subsequently the Russian trade that made Visby, in conjunction with Novgorod, important members of the German Hanseatic league. As long as the Arabian trade flourished, Gothland was the centre of a very animated traffic. Even now an almost incredible number of German, Hungarian, and particularly Anglo-Saxon coins, of the tenth and eleventh centuries, is dug up in the island. The collection of coins in Stockholm comprises an assortment of Anglo-Saxon coins, mostly the product of these discoveries, which, for extent and completeness, surpasses the greatest collections of the sort even in London and England.

The important and extensive commercial intercourse between Scandinavia and England, to which this so decidedly points, can also be traced in England itself. Oriental or Arabian coins, struck in the countries near the Caspian Sea, are dug up both in England and Ireland in conjunction with the very same kind of peculiar silver rings, and other ornaments of the same metal, that are also found with the Arabian coins in Scandinavia and Russia; nay, they are sometimes dug up, as in Cuerdale, in conjunction with coins of Danish-Norwegian kings and jarls; a fact which still further confirms the opinion that they were brought over to the British Isles by the Northmen. This connection with Arabia through the countries of Scandinavia may probably have brought to England, as well as to the North, such a mass of silver as enabled the Anglo-Saxon kings to mint that surprising number of silver coins, which appears at once in such forcible contrast to the want of silver in the preceding centuries. The ancient Britons had little or no silver before the Roman conquest. The Romans, who had large silver mines in Spain, certainly brought silver money with them into the British Islands; but after the overthrow of their dominion, a want of silver again prevailed, and continued, as the coins show, until far into the eighth and ninth centuries. Silver was consequently introduced into England and Scandinavia, generally speaking, about the same time; and there is undoubtedly far greater probability that it was brought into these countries in the same way—that is, from Asia through Russia—than that it should have come into England through the Moors in Spain; of whose caliphs there are very rarely any coins found in England, and between whom and the English the intercourse at that period seems to have been but very limited. In the treasure found at Cuerdale the rings and other silver ornaments were for the most part broken, and twisted, or even melted, together. Something similar has been observed in the treasure trove in the countries round the Baltic, and in Russia. This clearly proves that silver, as an article of commerce, was brought from Asia to the North, where it was melted and converted into ornaments and coins.

As long as the Norman expeditions lasted, and on the whole as long as the Scandinavian supremacy at sea sufficed to protect the Scandinavian merchants and their ships, they continued to make voyages on their own account to the countries colonized by the Northmen. Thus the Anglo-Saxon coins dug up in the island of Gothland indicate a brisk and uninterrupted commerce between Scandinavia and England from the time of the Anglo-Saxon king Edgar (959-975) down to the death of Edward the Confessor and the Norman conquest (1066). But from that time, and particularly after the year 1100, there is a remarkable decrease in the Anglo-Saxon coins found in Gothland; which is a natural result of the interruption of the previous connection, through the hostile relations that ensued between the descendants of William the Conqueror and the Scandinavian kings, who steadily continued to claim the crown of England. Later in the middle ages the countries of Scandinavia fell more and more under the commercial yoke of the German Hanse Towns; whilst in England, on the contrary, a freer and healthier state of commerce was continually developing itself. The Danish king,

Canute the Great, made it a point of the utmost importance to conclude commercial treaties with various foreign nations; and the Scandinavian merchants settled in England essentially contributed to make these leagues profitable. Old authors expressly notice the influence of these merchants on British trade. We also find evidence of it not only in their great number, and the weight they possessed in several English towns,—especially London, where they had their own churches, markets, and courts of law, and where, as before stated, they even at times decided the election of a king, as in the case of Harald Harefoot,—but also in the names of money afterwards retained in the English language, as "March" and "Ora," from the Scandinavian "Mark" and "Ore." It was a natural consequence that commerce should at the same time make great progress, as the numerous Scandinavian settlers in England, and the Danish conquest, had infused a new and hitherto unknown life into everything relating to navigation, without which no animated trade could have flourished in the British Islands.

The ancient Britons were by no means a seafaring people. They appear to have confined themselves to short coasting voyages between the islands, and over the Irish and English channels. They had, therefore, no fleet to protect their coasts from the attacks of the Romans. Their vessels consisted either of the trunks of trees hollowed out, or of small frail boats formed of interwoven branches, or wicker-work, covered with hides. The Celtic nations have, on the whole, never been remarkable for their love of the sea, or of a seafaring life. On the contrary, they seem to have derived from nature a decided antipathy to it; and even to the present day it is very striking to observe how unwillingly their descendants venture out to sea. They prefer, under all circumstances, a landsman's life, even in remote and barren mountain tracts; nay, their disinclination for everything relating to a seaman's life is carried so far that they neglect, in a way almost incredible, the rich fisheries on the western coast of Scotland, and on the greater part of the coasts of Ireland; although, in the last-named country especially, famine carries off the inhabitants in shoals. In those villages where fishing is carried on to any extent, the inhabitants are in general descended from immigrant foreigners. Thus it is said that the fishermen on the west coast of Ireland are descended from Spaniards; and, to judge from their appearance, the assertion finds some confirmation.

Nor were the Anglo-Saxons a seafaring people, in the proper sense of the term. They comprised, it is true, Jutes, Angles, and Frisians; but the Saxons were the most numerous, and the Saxon disposition has always clung to a life ashore. It was natural, however, that the art of navigation should gradually develop itself among the Anglo-Saxons as they advanced in civilization and refinement. But how little they were at home on the sea, even in the time of Alfred the Great, is shown by the feeble resistance they were able to offer to the Danes. It is true that Alfred had large ships of war built in order to protect the coasts; but he was obliged to man them, in part at least, with Frisians. We are further told that these ships were much larger than those of the Danes. Yet the history of the tenth and eleventh centuries affords no proof that these ships were able in the long run to prevent the conquests of the Danes, or that they served to increase the Anglo-Saxon skill in seamanship.

Even the Greeks and Romans, however much they distinguished themselves in other ways, as in literature and art, did not make any remarkable progress in seamanship. Their navigation chiefly consisted of trips along the coast or voyages across the Mediterranean; and if an adventurer was now and then bold enough to pass the Pillars of Hercules, or Straits of Gibraltar, out into the Atlantic Ocean, in order to sail along the west coast of Europe to the British Isles, or countries still farther north, it was regarded as a great exploit. Regular voyages thither were scarcely known; nor do the Greek and Roman ships appear to have been well adapted to keep the sea in the wide and stormy Atlantic.

It was reserved for a land washed by the waters of that ocean—the Scandinavian North—to build the first large "sea-going" ships, capable not merely of successfully conveying, in calm weather, and under favourable circumstances, a solitary daring navigator over the Atlantic, but of affording, in spite of storm and tempest, a secure passage over its enormous waves. It is only by duly considering how much experience and talent must have been exerted, and, above all, how many calculations must have been made previous to the building of such a vessel, and before the art could be acquired of steering it with safety through breakers and in storms, that we shall perceive how much it redounds to the honour of Scandinavia to have made these great and most important advances; which, by founding modern navigation, by extending commercial intercourse to a degree before unknown, and by thus uniting parts of the globe which were previously separated, may be said in a manner to have changed the face of the world.

Even before the time when the Danes conquered England, the Northmen had long possessed large and splendid sea-going ships. The Norwegians, in particular, were then constantly making voyages across the Atlantic, to the Shetland Isles, Iceland, and Greenland; nay, they undoubtedly reached the continent of America several times; of which Scandinavian and German historical traditions, as well as internal probabilities, bear witness. For, first, it was a natural consequence that a people who could navigate the dangerous and ice-bound sea that surrounds the coast of Greenland, and who could establish considerable colonies both in north and south Greenland—traces of which are still preserved by runic inscriptions, ruins of churches, and the foundations of numerous houses—should also be able to sail to the coast of America, the navigation to which was always attended with less danger. And, again, it would have been very strange if the Northmen, who sailed without a compass, should always have succeeded in reaching Greenland, and never have been driven by storms to the neighbouring coast of America. It was, besides, just in this manner, according to the statements of history, that America was first discovered. It is quite another matter whether traces of these early visits of the Scandinavians could really be still found in America, which there is good reason to doubt.

The above-mentioned voyages, in the ninth and tenth centuries, are sufficient proofs of the excellence of the Scandinavian ships. It is not, therefore, to be regarded as pure exaggeration if the Sagas use strong expressions in celebrating the war-ships of that time, particularly the galleys, or, as they were called, long ships; and amongst others that magnificent royal vessel "Ormen hin Lange" (the long snake), which bore the Norwegian king, Olaf Trygvesön, in the celebrated sea-fight of Svöldr (near Greifswald) in the year 1000. These long ships were also called "Dragons," because the stems were frequently ornamented with carved, and even gilded, images of dragons; or else were beheld there figures of vultures, lions, and other animals, ornamented with gold. These long ships had sometimes crews of several hundred men. Other, and partly smaller, ships had different names, such as "snekken," "barden," "skeiden," "karven," "barken," and several others. Both Scandinavian and English chronicles dwell on the description of the splendour with which the fleets of the Danish conquerors, Svend and Canute, were adorned. Magnificent images glittered on the prows; the sails were worked, or embroidered, with gold; the ropes were of a purple colour; and on the top of the gilded masts sat curiously-carved images of birds, which spread out their wings to the breeze.

[+ +] Sailing Ship

With the exception of very imperfect representations carved on rocks and runic stones, there are no images left in the countries of Scandinavia of these ships of the olden time. But the celebrated tapestry at Bayeux, in Normandy, on which the conquest of England by the Normans is depicted, is a contemporary evidence of the appearance of the Normanic ships; and the accompanying woodcut taken from it, representing probably the ship in which William the Conqueror himself sailed, will clearly prove how splendid they really must have been. Both this and the rest of the Norman ships in the tapestry perfectly agree with the contemporary Danish and Norwegian ships, just as we know them from the Sagas, even to the shields hung out along the bulwarks. This, however, is nothing more than what one might naturally expect, since the Normans and Danes, on the conquest of Normandy, must have brought such ships with them, as well as that art of ship-building which they afterwards carried to greater perfection. For this, however, they found no models in the wretched vessels of the Franks and Bretons. But their steady connection with the Scandinavian fatherlands, at all events through the Danes and Norwegians in England, communicated to them those improvements in the form and arrangement of ships which the very extensive ship-building of the Northmen, and their long and uninterrupted voyages to Iceland and Greenland, must gradually have produced. That influence on maritime affairs, which, on the whole, was exercised by the Scandinavian settlers in Normandy, showed itself also in the circumstance that Scandinavian names of ships, together with other maritime terms, passed into the Romance language; as, for instance, *flotte* (*Dan.*, Flaade; *Eng.*, fleet), *verec* (*Dan.*, Vrag; *Eng.*, wreck), *bord* (*Dan.*, Skibsborde, Rand; *Eng.*, ship-board), *windas* (*Dan.*, Vinde, Spil; *Eng.*, windlass) *mast* (*Dan.*, Mast; *Eng.*, mast), *sigler* (*Dan.*, Seile; *Eng.*, sails), *esturman* (*Dan.*, Styrmand; *Eng.*, steersman), *eschiper* (*Old Northern*, skipa; *Eng.*, equip), from which are derived the now commonly used French words, *équiper*, *équipage*, (and with us Danes likewise, eqvipere, Eqvipage-mester; *Eng.*, master of ordnance.)

As a consequence of the Danish-Norwegian immigrations, the art of ship-building must also have necessarily developed itself in a similar manner in England, on whose eastern and north-western coasts the descendants of the Vikings had everywhere spread themselves, both by the sea and on the rivers. Christianity certainly put an end to the life of the Viking. "Söhaner" (sea-cocks) were no longer to be found, who scorned "to sleep by the corner of the hearth, or under sooty beams." But the Vikings' spirit was not therefore dead. The Scandinavian colonists could never entirely degenerate from their fathers, who had joyfully "ridden on the backs of the waves," and who in the icy sea, and on the Atlantic Ocean, had greeted the storm only as a welcome friend, which assisted the oars and speeded the merry passage. Among the Vikings were many like the Danish chief made prisoner by King Athelstane at the siege of York in 927. The King treated him well, and retained him in his hall more as an equal than a prisoner. But in a few days the chief fled and put out again to sea; for it was, the chronicle adds, just as impossible for him to live on land "as for the fish to live out of the water." The immediate descendants of such men, for whom a seafaring life was a necessity of their very nature, must have continued to dash through the water, particularly when, as in England, they were settled near seas and rivers. During all the internal dissensions and foreign wars that occupied England in the first centuries after the conquest by William the Norman—and which ended by binding more firmly together the various Celtic, Teutonic, and Scandinavian races which composed its population—the maritime affairs of the English were no longer confined, as in more ancient times, only to commerce with the nearest neighbouring countries. Through the mother countries of Scandinavia, and especially Norway, they continued during the early part of the middle ages to maintain a lively intercourse with the distant Scandinavian republics in Iceland and Greenland. But when, in the thirteenth century, the independence of these republics was overthrown, and they were placed as tributary countries under the Norwegian crown, the free trade that had previously flourished became much more restricted. The consequence of this was, that the navigation to Greenland from the north decreased more and more, until, in the fifteenth century, when the Scandinavian population of Greenland had been annihilated by sickness and by the assaults of the natives, it entirely ceased. What also much contributed to this was, that the trade which the Northmen themselves carried on with Iceland became gradually, and in the fifteenth century was almost entirely, although illegally, transferred to the English, who under the guidance of their Scandinavian kinsmen had found their way thither. Hull and Bristol—which latter place is named as early as the twelfth century as the port for ships from Norway (and Iceland?)—were the two English harbours whence this trade with Iceland was carried on. There are even some who think that Christopher Columbus during his stay in these harbours, through conversations with Iceland navigators, and possibly by a voyage to Iceland itself, obtained information of the ancient voyages of the Northmen to Greenland and America; and that he was thus first completely confirmed in his opinion, that a large and unknown continent must lie in the far west, across the Atlantic Ocean. But even if this supposition be unfounded, or destitute as yet of certain historical proof, may it not at least be probable that Columbus had heard in some other way of the Northmen's former voyages to Greenland; and that this might have had some influence on the resolution he afterwards formed to set out across the Atlantic on a voyage of discovery towards the west?

But under any circumstances, the regular voyages of the English to Iceland were certainly connected with the subsequent complete discovery of the New World. They had served to make them familiar with more extensive voyages on the open ocean, and thus essentially contributed to foster that daring Viking spirit, which they had inherited from their Scandinavian forefathers, and which in process of time was to become so important in cementing the connection between the Old and the New World. No sooner was the latter a second time discovered than the Vikings' spirit again strongly displayed itself in a renewed form among the English people. There was the same lofty tranquillity, the same daring and contempt of danger, that characterised the Vikings of ancient times. But the English seaman had now more experience and knowledge, and quite other means were at his disposal than had ever before existed. He therefore entered on his first voyage to the New World with undaunted courage, and not only soon became familiar with that ocean which his Scandinavian forefathers had ploughed in the remote days of antiquity, but also opened a way to new lands over seas before unknown. Thus was established that maritime supremacy which has been one of the most important props of the wealth and power of England.

The first accidental discovery of America by bold adventurers from the remote north took place so early, and under such peculiar circumstances, that neither Scandinavia nor the rest of the world derived any use or benefit from it. After a transient glimpse, the golden treasure again sank beneath the waves. It lay, nevertheless, in the dispensations of Providence, that the descendants of those Scandinavian adventurers should bear an essential part in raising the re-discovered treasure, and in making it productive for mankind. And had not the Scandinavians, by their numerous settlements in the British Islands, engrafted on the population a skill in seamanship before unknown, together with a daring spirit of enterprise, England, in spite of its fertility, its wealth, and its favourable maritime situation, would scarcely have succeeded in solving such a problem as that of closely knitting together lands separated from each other by the Atlantic in all its breadth and vastness.

SECTION XI

Art and Literature.

At the period when the Danes were making their conquests in the West, art and literature did not occupy any very high position in Europe. The severe shock which the fall of the Roman Empire had given to all the more elevated pursuits was still far from being overcome. Christian art was in its childhood, and groped its way with weak attempts, and imitations of Roman models; whilst literature, confined for the most part to one-sided theological inquiries, or to the inditing of dry and annalistic chronicles, could scarcely be said to deserve the name.

It was, however, a natural result of the long-continued domiciliation of the Romans in France and England, where they founded so many and such important works, and where Christianity was adopted at a comparatively early period, that a taste for art and literature should develop itself in no mean degree in those countries; particularly in comparison of the far North, where the Romans had never ruled, and where the darkness of heathenism still rested on the people.

Nevertheless we should be grievously mistaken if we imagined that the Scandinavian people was at that time entirely unfitted for the ennobling occupations of art and literature. It has been before stated that the Northmen early distinguished themselves not only by an extraordinary skill, for those times, in the art of ship-building, but that they had also developed, previously to the conquest of England, a taste, in some respects peculiar, in the manufacture of their ornaments, domestic utensils, and weapons, and which had principally sprung from characteristic imitations of the Roman and Arabian articles of commerce brought into the North. The Scandinavian antiquities that are dug up, belonging to the older period, or what is called "the age of bronze," as well as those of the latest times of heathenism, or "the iron age," may on the whole, with regard to form and workmanship, be even ranked with contemporary objects of a similar kind manufactured in England, France, or Germany. The Sagas, moreover, state that the carving of images was sometimes very skilfully practised in the North; and the English chronicles, which depict in such glowing colours the splendidly-carved figures on the prows of the Danish or Scandinavian vessels, confirm the truth of these statements. In Olaf Paas' Hall, at Hjarderholdt, in Iceland, the walls were even adorned with whole rows of carvings, representing the ancient gods, and their exploits. On the other hand there could naturally as yet be no possibility of erecting such buildings in the North as those which the spirit of Christianity had already produced in other countries.

But no sooner were the Normans from Denmark and Norway settled in Normandy, and converted to Christianity, than they began to manifest a lively desire to erect splendid buildings, and particularly churches and monasteries. Scarcely had the first violent revolutions in that country been brought to a close when there sprang up such a number of great architectural works among the Normans, that Normandy can still show more such monuments of art, of the eleventh century, than any other district of France. After William's conquest of England, the Normans also founded there a somewhat peculiar style of building, which, though only a branch of the Byzantine-Gothic, or a further development of the older Saxon, constantly bears in England the name of "Norman."

Previous to the Norman conquest, the Danes settled in England were naturally unable to influence, in a like degree, the style of English architectural works. Their sway there was both too short and too unsettled for such a purpose: not to mention that the Danes had still much to learn from the Anglo-Saxons in the art of building; for the latter had long been Christians, and were besides settled in a country possessing abundant remains of the magnificent architectural works of the Romans. Nevertheless it is not incredible that several of the many churches and convents then and subsequently erected by Danish princes and chiefs, and especially in the northern parts of England, but which are now for the most part either rebuilt, or have entirely disappeared, may have borne the stamp of their Scandinavian origin. We are led to this opinion by the ruling inclination manifested by the ancient Northmen to let their own conceptions pierce through, even in their imitations of foreign objects. Numerous and contemporary evidences in England itself also sufficiently prove to what a remarkable extent the Danes must have devoted themselves to peaceful occupations, long before the Norman conquest. In these, indeed, which relate to only a single branch of art, the Anglo-Saxons were their teachers; still they will show that the Danes were neither wanting in a natural capacity for art, nor in faculty or will for its further development.

It has been stated before that the Danes, previously to the conquest of England, were unacquainted with the art of coining money. At most they only imitated the Byzantine coins by fabricating the (so-called) "*Bracteates*," which, however, were stamped only on one side, and were for the most part used merely as ornaments. But the art of coining was very ancient in England. It was customary among the Anglo-Saxons for the coiners to put their names on the coins struck by them. The quantity of Anglo-Saxon coins that has in the course of time been found and examined, has afforded an opportunity for inspecting and comparing a considerable number of names of coiners in England, especially from the eighth and ninth centuries until far into the thirteenth. About Edward the First's time, the names of the coiners were no longer suffered to occupy so conspicuous a place on the coins as previously.

In the eighth and ninth centuries the names of these coiners are purely Anglo-Saxon. But in the tenth century, and especially after the year 950, pure Danish or Scandinavian names begin to appear; for instance, Thurmod, Grim, under King Edgar (959-975); Rafn, Thurstan, under King Edward (975-978); Ingolf, Hafgrim, and others. These Scandinavian names are more particularly found on coins minted in the northern part of England, or at all events in the districts that were early occupied by the Danes to the north-east of Watlinga Stræt. But under King Ethelred the Second (979-1013), who contended so long with Svend Tveskjæg and Canute the Great (and consequently, therefore, before the conquest of England by the Danes was completed), such a number of Scandinavian coiners arose all at once, in consequence of the rapidly-increasing power of the Danes, that the names of forty or fifty may be pointed out on coins of Ethelred alone that have been found in different parts of England. During the Danish dominion, Scandinavian names naturally appear no less frequently on the coins of Canute the Great and Harald Harefoot; nay, even after the fall of the Danish power, they are to be met with, in almost the same number as before, on coins of the Anglo-Saxon king, Edward the Confessor (+ A.D. 1066).

The following table exhibits, from the coins themselves, a list of fifty names of Danish-Norwegian coiners in England that appear most frequently from 979 to 1066; or in that period which embraced, as well as immediately preceded and followed, the Danish dominion; together with the names of the places in which the respective coins were minted. We must remember, besides, that there must have been several coiners of the same name at one and the same time. Thus, for instance, we find coins of Ethelred bearing the name of "As-" or "Oscytel," though minted in cities so far distant from one another as Exeter, London, Cambridge, Leicester, and York. Again, as it is nowhere stated that "Arncytel," for instance, who was coiner in York under King Ethelred, was the same man as Edward the Confessor's coiner in that city, it is clear that the fifty names here given might very easily have belonged to ninety or a hundred different persons; yet they are but a selection from a greater number. The same difficulty, however, occurs with these names as in the previous consideration of the Scandinavian names of places and of the popular language; namely, that owing to the great similarity between the Saxon and Scandinavian tribes in ancient times, it is often almost impossible to decide with certainty what is exclusively Saxon and what Scandinavian. But at all events, the annexed list contains, at most, hardly more than a couple of names that might have been current in Saxon England before the Danish conquests.

FIFTY NAMES OF DANISH-NORWEGIAN COINERS IN ENGLAND IN THE YEARS 979-1066, CHIEFLY FROM HILDEBRAND'S "ANGLO-SAXON COINS IN THE ROYAL SWEDISH COLLECTION OF COINS FOUND ON SWEDISH GROUND." (Stockholm, 1846. 4to.)

	Ethelred (979-1013)	Canute (+1035)	Harald (+1040)	Edward Confessor (+1066)
Arncytel	York	York		York
Arngrim		York	York	York
Arnkil			York, Stamford	
Arnthor	York			
Ascil	London			
As, or Oscytel	Exeter, London, Cambridge, York, Leicester			
As, or Oslac		London, Lincoln, Norwich		
Auti				London, Lincoln
Beorn (Björn)		York	York	York
Cetel	Exeter, York	Exeter, York		York
Colgrim	Lincoln, York	Lincoln, York	Lincoln, York	Lincoln, York
Dreng	Lincoln	Lincoln		
Eilaf	York			
Eistan			Winchester	
Escer	Stamford		Stamford	
Grim	Lincoln, Thetford	Shrewsbury		
Grimcytel		Lincoln		
Hardacnut			Lincoln	
Huscarl				Leicester
Iric		London		
Jelmer (Hjalmar)				Lincoln
Justan, or Justegen		Lincoln		
Northman			Lewes	
Othgrim	Lincoln, York	York		Lincoln
Othin		York	York	York

Oustman, or Ustman		York	Winchester	
Rœfen (Ravn)		York		
Rœienhold	Lincoln			
Siafuel, Sœfuhel				York
Scula		Exeter, York	York	York
Stgncil (Stekil)	Lincoln			
Styrcar, Stirceir	Lincoln, York	York		
Sumerled	Deptford, Nottingham, York, Lincoln	Lincoln, Norwich	Lincoln	Lincoln
Swan		York		
Swarti		Leicester, Lincoln		
Swartgar	York, Stamford			
Sweartabrand		Lincoln	Lincoln	
Swegen	London, Leicester	Leicester		
Thor				York
Thorald	Leicester			
Thorcetel	Torksey, Lincoln	London, Torksey		
Thorstan	York	York, Stamford		Norwich
Thorulf	Chester, York		Stamford	
Thurcil				Wilton
Thurgrim		York	York	York
Ulfcetel	York, Lincoln, Norwich	London, Lincoln		York
Valrefenn			Lincoln	Lincoln
Widfara			Ipswich	
Winterfugl				York
Wintrieda	York			

Although this list cannot make any pretensions to completeness, still it will prove, even in its present form, that these Scandinavian names exist on coins from places in the most distant parts of England, both south and north of Watlinga-Stræt; as well as from those most essentially Anglo-Saxon cities, Exeter, Winchester, Wilton, Lewes, and London. From this last circumstance, some might, perhaps, contend that Scandinavian names were frequently borne by Anglo-Saxons, who in one way or another were related to the Danes; and in this respect one might cite the instance of the Anglo-Saxon Earl Godwin, whose sons—possibly by a Danish wife—were called Harald and Svend; and it might consequently be argued, that the proof adduced from these Scandinavian names of the Danish capacity for skill in art is not sufficiently conclusive.

It cannot of course be denied that the Anglo-Saxons, in whose veins there was a mixture of Scandinavian blood, sometimes bore Scandinavian names. But as a rule, the names that have been cited must have belonged to Danes or Northmen, and their immediate descendants. It is well known that the Danes were settled everywhere in England, even in the southern cities, particularly those just cited; and that, too, in considerable numbers: as, for instance, in Exeter, where in later times there was a St. Olave's Church; in Winchester, which obtained a Scandinavian "Husting;" not to speak of London. This alone affords a natural explanation why Scandinavian coiners should be found in the south of England; but we should further observe, that those names of coiners about which there might be most doubt are found to the north-east of Watlinga-Stræt. The preceding tabular view will clearly prove that they occur especially in the old Danish part of England, in the five Danish fortified towns, and in York. The two cities, Lincoln and York, which, according to the statements of history, had, in the eleventh century, a very numerous, if not preponderating, Scandinavian population, are remarkable for having the greatest number of coiners with Scandinavian names. Some of these names are so peculiarly Scandinavian, that we cannot without difficulty assume them to have been borne at that time by Anglo-Saxons. Such are "Othin" (*Anglo-Saxon*, Woden) and "Thor;" names that did not sound well in the ears of Christians: also "Northman" and "Ustman," or "Östman," by which the Anglo-Saxons designated the Norwegians and Danes, who came from the North and East. "Östman," especially, was an appellation commonly given by the inhabitants of the British Isles in those times to the Scandinavian tribes that dwelt to the east of them.

Among other names, those of "Colgrim" and "Valrefenn" may be noticed as frequently appearing, and as peculiar to Lincolnshire, a district occupied in such great numbers by the Danes. Names of birds appear on the whole to have been often assumed in the old Danish part of England. Thus in York we find a "Ræfn," or "Ravn" (Raven); "Siafucl," "Sæfuhel," or "Söfugl" (Seafowl); "Swan" or "Svane" (Swan); and "Winterfugl" (Winterfowl). Strangely enough, there also appears a "Sumrfugl" (Summer-fowl) as the name of a coiner, who minted coins for the Danish-Norwegian king Magnus the Good, in Odensee; and as English coiners were at that time employed in Denmark, this Sommerfugl perhaps came over from the north of England. It was, indeed, quite natural that Denmark and the rest of the North should procure their earliest coiners from Danish North England, where there were plenty of them of Scandinavian origin. The English names found on the oldest Scandinavian coins (of the first half of the eleventh century) are consequently by no means universally Anglo-Saxon, but often Scandinavian; as Svein, Thorbaern (Thorbjörn), Ketil, Thorkil, Othin, Thorstein, Thurgod, Thord, and others. It is remarkable, that the names of "Sumerled" and "Winterled," answering to those of Sommerfugl and Winterfugl, were also found at that time in York. Another remarkable name is that of "Widfara" (the far-travelled), which seems to indicate either that its bearer had come from a great distance, or had made long voyages.

These Scandinavian names, which, as I have said, are just as frequent on coins minted immediately after, as on those struck during, or just previously to, the Danish-English kings' dominion, by no means cease with Edward the Confessor (+ 1066). During Harald Godvinsön's short reign, we further meet with Outhgrim, Snaebeorn (or Snéebjörn), Spraceling (Sprakeleg), Thurcil, Ulfcetel, &c.; nay, even after the Norman conquest, and as long as it was customary to place the coiners' names on the coins, Scandinavian names may be recognised. Thus, under William the Conqueror (+ 1087) we find Colsvegen, Thor, Thurgrim, Jestan (Jostein or Eistein, Justan and Justegen), Siword, Thorstan; under Henry the First (1100-1135),

Chitel (Ketil), Runcebi (Rynkeby), Spracheling, Winterled; under Stephen (+ 1154), Ericus, Siward, and Svein; and under Henry the Second (+ 1189), Achetil (Asketil), Colbrand, Elaf, Raven, Svein, Thurstan, and others. A great number of these names appear in connection with towns in the north of England; and we have thus a new and instructive proof that the remarkable influence of the Danish element in England, and especially in the northern part, before the Norman conquest, was not entirely lost *after* that conquest had long been completely effected.

Considering the distant period in which the Danish conquests in England fall, it is fortunate that we can obtain so many palpable evidences of the state of domestic civilization as these coins afford; and more will assuredly follow from the discovery of others hitherto unknown. These coins prove much, and justify us in inferring still more. They place, as it were, before our eyes, the earnestness with which the Danish Vikings, and the rest of the colonists in England, must have applied themselves shortly after their settlement, to rival the Saxons in art, and to retrieve what they had neglected in this respect. In like manner, there is every reason to believe that they must have devoted themselves with no less zeal to other peaceful occupations which they had already cultivated in their own native homes; and that thus they must have also preserved and cherished in England, both in war and peace, that love for poetry and history, which flourished in the homes of their ancient forefathers, and which, on the whole, harmonized so completely with the heroic life of the olden times in the North. It was not natural that the deep desire which filled the Northman to enjoy posthumous fame in chronicles, and in the songs of the poets—which left him no peace at home, but drove him out to sea on daring expeditions—should immediately desert him because he had removed to a foreign soil. It is expressly related of the Normans that they cherished eloquence and poetry in a high degree, and that they were accustomed to entertain their guests with songs and legends. Scandinavian bards, especially from Iceland, continued to visit the Scandinavian colonists in France, as well as in the British Isles. As court-minstrels, they were in constant attendance upon the Scandinavian princes in Scotland, Ireland, and England. Their office partly was, to entertain the warriors with lays of past exploits in the

North; and, partly, to accompany the chiefs on their warlike expeditions; that they might, as eye-witnesses, be able to sing their heroic deeds, and by these lays convey to the North a knowledge of what passed among the Scandinavian colonists in the western regions. When we add that the Scandinavian kings, as, for instance, Canute the Great himself, practised at times the art of poetry, it will be easily perceived in what high honour the bard and his lays must have been held.

But it lay in the nature of things that a pure Scandinavian poetry could not grow up either among the Normans in France, or their Danish kinsmen in England. For the development of such a poetry it was necessary that they should preserve their Scandinavian nationality intact. But it is well known, that a foreign education and refinement soon caused them to abandon their belief in Odin, as well as many of the habits and customs which they had inherited from their forefathers. Of the change that took place in them nothing bears stronger evidence than their mother tongue, which, by degrees, lost more and more of its characteristics, and at length passed entirely into the modern French and English languages.

The old predilection for poetry which the Normans brought with them from the North, was reflected in many ways in their foreign refinement. Of all France, Normandy was the country where most historical and warlike songs were heard. The Normans sang them in battle, and derived from them a sort of inspiration. Before the battle of Hastings, William the Conqueror's bard, Taillefer, recited songs about Charlemagne, Roland, and others, to the Norman host, to cheer and enliven the warriors after the old Scandinavian fashion; just as Thormod Kolbrunaskjald, before the battle of Stiklestad, in Norway, (1030), sang the far-famed Bjarkemaal. When the poetry of the Troubadours of Provence began to spread itself throughout France, it found another home in Normandy; where it so peculiarly developed itself, that the French troubadour poetry is generally divided into two principal kinds, the "Provençal" and the "Norman." Even in Italy, where the Normans conquered fresh kingdoms, their peculiar poetry had a perceptible and important influence on the development of the art.

In England, likewise, there arose, partly as a consequence of the Danish and Norman conquests, a particular kind of composition which, in England, is called Anglo-Danish and Anglo-Norman. That all poems of this sort were written by Danes or Normans, I do not venture to assert. All that is meant is, that they were partly produced by the Danish and Norman wars; and that, partly, they were the expressions of the new adventurous and knightly spirit, which, through the Danish-Normanic conquests, became prevalent in England. Some of the most celebrated of them are romances about "Beowulf," "Havelock the Dane," and "Guy, Earl of Warwick." In the oldest romances, which are composed of the same mythic materials as our Scandinavian Edda songs, and some of the Sagas or legends, adventurous combats against dragons, serpents, and similar plagues, are celebrated; whereas, in the later romances of the age of chivalry, warriors are sung who had fallen in love with beautiful damsels far above them in birth or rank, and whose hand and heart they could acquire only by a series of brilliant adventures and exploits. Valour, which before was exerted for the welfare of all, and for the honour that accompanied it, now obtained a new object and a new reward, and that was—love. The heathen poems of the Scandinavian North are all conceived in the selfsame spirit; and it is therefore not altogether unreasonable, perhaps, to recognise in this striking agreement traces of a Scandinavian influence on English compositions. In later times, and down to the middle ages, this influence is still more clearly apparent in the before-mentioned ballads, or popular songs (p. 89), which are only to be found in the northern, or old Danish, part of England, and which betray such a striking likeness to our Scandinavian national ballads.

The Danes in England do not appear to have occupied themselves with any compositions that can be properly called historical; at all events all remains of such composition have disappeared. It is related of the contemporary Normans in France, that, down to the days of William the Conqueror, they devoted themselves more to war than to reading and writing. This, however, is not surprising, since even the Anglo-Saxon clergy in Alfred the Great's time, according to that monarch's own statement, were so ignorant and so unaccustomed to literary occupations, that exceedingly few of them could read the daily prayers in English, much less translate a Latin letter. Even if we should admit that the Danes in England, by reason of their earlier and more extended settlements there, had somewhat better opportunities for study than the Normans in Normandy, still there is not sufficient ground to suppose that they wrote any other chronicles than such dry annals as some few monks, and other learned men of that time, composed. The reason of this seems partly to have been because they preferred preserving the remembrance of important events in historical lays; and partly, because neither their national nor political development could proceed in a foreign land with such freedom from all admixture, and in such tranquillity, as to allow of more important historical works, and especially in their mother tongue, being produced among them.

In Iceland, on the contrary, where a great number of the most powerful and shrewdest of the heathens of Norway sought, after the year 870, a refuge against spiritual and political oppression, and where they founded a republic which retained its independence for centuries, the Scandinavian spirit obtained a free field. Not only did the old bardic lays, and the remembrance of the deeds of former times, continue to live among the Icelandic people, but new bards arose in numbers, who, spreading themselves over the whole north of Europe, returned "with their breasts full of Sagas." There also speedily arose in Iceland, immediately after the Viking expeditions, and altogether independently of any external influence, an historical Saga literature in the old Scandinavian tongue, which, viewed by itself, is, from its simplicity and elevation, extremely remarkable, but which, when compared with the contemporary dry Latin monkish chronicles and annals in the rest of Europe, is truly astonishing. The Edda songs, the purely historical Sagas, the historical novels, and other peculiarly bold and original productions of the Icelandic literature, in an age when the European mind was singularly contracted, form, in the intellectual world, manifestations of the same thorough individual freedom, which stamped itself on the arms, endeavours, and whole life of the heathen Northman.

Section XII.

Ecclesiastical and Secular Aristocracy.

The supposition that the Danes in England devoted themselves to study both earlier, and to a greater extent, than the Normans in France, is not founded only on loose conjectures. The English chronicles of the earlier middle ages contain traces of the Danes having not unfrequently entered into the English Church, in which they sometimes obtained the highest preferment. On this point we still possess an important source of information, which has, besides, the advantage of being for the most part contemporary with the events and circumstances which it elucidates. This consists of a considerable quantity of letters and diplomas issued by kings, bishops, and other leading men in England, from about the year 600 to 1066. These documents, which have lately been collected and published by a gentleman celebrated for historical research, Mr. J. M. Kemble, (under the title of "Codex Diplomaticus Ævi Saxonici," vol. i.-vi., London, 1839-1848, 8vo.) more especially regard the southern and midland parts of England, as unfortunately the greater part of the letters relating to the north of England are lost. Nevertheless, those that remain, taken in conjunction with the chronicles, afford valuable information, both respecting the Danish clergy in the south-east of England, and their diffusion throughout that country.

In the centre of the east coast of England, in Lincolnshire, and near the Wash, stood in the Anglo-Saxon times the large and famous convent of Croyland, or Crowland, dedicated to St. Guthlac. It was built upon an island, and so protected on the land side by the vast morasses which in those times covered the districts nearest the Wash, that it was a sort of natural fortress. According to the chronicles of the convent, compiled by one of the abbots in the eleventh century, it was governed, shortly after the year 800, by an abbot of the name of Sivard; in whose time there is also mentioned in the convent a priest (presbyter) named "Turstan," and a monk "Eskil" (Askillus monachus). In the same ancient chronicle are also recorded several deeds of gift, which possibly, with regard to the rights conveyed to the convent, may have been forgeries of the times, but which, at all events, so far as regards the names of persons and places mentioned in them, must be perfectly correct and trustworthy; since incorrectness in these particulars would have easily led to the discovery of the intended frauds. These deeds mention, between the years 800 and 868, amongst the benefactors of the convent, three viscounts in Lincolnshire, “Thorold” (or Thurold), “Norman,” and “Sivard;” and also “Grymketil” and “Asketellus” (or Asketil), who was cook to the Mercian king Viglaf. Lastly there appear (particularly in the year 833) the following names of places:—Langtoft, Asuuiktoft, Gernthorp, Holbeck, Pyncebek, Laithorp, Badby, and Kyrkeby.

The names of persons in the convent, and of places about it, here cited are all, perhaps, or at most with a single exception, of undoubted Danish or Scandinavian origin. They not only prove that, even long before the treaty between Alfred the Great and the Viking King Gudrum or Gorm, which in the year 879 secured to the Danes their conquests on the south-east coast of England, and therefore, more than one hundred and fifty years before Canute the Great’s time, the Danes really had such a footing round the Wash that they could give their villages Danish names, and were governed by their own chiefs; but they likewise indicate the remarkable fact, that at least a great number of these Danes must have been already Christians, since they had villages with churches (Kyrkeby) and gave landed property to a convent, in which we find both Danish monks (Eskil and Thurstan), and a Danish abbot (Sivard.) It was about the same time that the Jutland king, Harald Klag, was baptized, together with his whole suite, during a sojourn with the Emperor Ludvig, at Ingelheim, near Mayence, in the year 826. This christening of Danish men abroad, in Germany and England, was the beginning of the subsequent introduction of Christianity into the Scandinavian North.

The genuineness of the above-mentioned Scandinavian names is placed beyond all doubt by the circumstance that similar names appear in other documents connected with the history of Croyland at the same period, or the ninth century. In the year 867, swarms of Danish-Norwegian Vikings landed on the east coast of England, and the Christians who then lived there, whether Danes or Anglo-Saxons, as well as their churches and convents, suffered from the ferocity of these heathens. After a great battle in Lincolnshire, in which, however, the heathens lost three of their kings, whom they buried in a place afterwards called "Trekyngham" (the three kings' home), they marched against Croyland. In vain did the Christians seek to arrest their progress. In a battle near the convent many of the Christians fell, and amongst them "Toli" or "Tule," who had previously been a knight, but who had now entered the cloisters of Croyland. The Vikings stormed the convent, and committed a terrible massacre. Their king, "Oskytyl," cut down the abbot before the altar; after which the convent was plundered and destroyed. The Danish Viking Jarl Sidroc, or Sigtryg, saved a boy called Turgar (Thorgeir) from this massacre, who afterwards escaped to the neighbouring convent of Ely, and gave an account, which is still preserved, of this terrible devastation. Meanwhile, however, the convent of Ely, as well as that of Medehamstede (Peterborough), was plundered and destroyed by the Vikings.

Amongst the monks then killed in Croyland, we may cite from the chronicle, the prior, Asker, and the friars Grimketulus (Grimketil) and Agamundus (Amund); and among the few saved, Sveinus or Svend:—names which, not less than Tule and Thorgeir, indicate a Danish origin. Men of Danish extraction continued in the following centuries to play a considerable part in the history of this and of the neighbouring convents. A Dane named "Thurstan" is said to have rebuilt that of Ely; and another man of Danish family, "Turketul" (Thorketil), certainly rebuilt Croyland. Thorketil, who (it is stated) was nearly related to the royal Saxon family, had previously distinguished himself both as a warrior and statesman. In the battle of Brunanborg he commanded the citizens of London who were in Athelstane's army, and during a long series of years was chancellor to several kings. Subsequently, however, he took the vows of the convent, and governed Croyland with honour, as abbot, till his death in the year 975.

It is, indeed, very striking to observe how many abbots of Danish origin governed the convent of Croyland from the ninth to the twelfth century. Sivard and Thorketil have been already mentioned. Thorketil was succeeded by two of his relations, both named Egelrik; and after the death of the last of these in 992, followed an abbot with the pure Danish or Scandinavian name of "Oscytel." This Asketil had long been prior of Croyland before he became its abbot, which he continued to be till his death in the year 1005. To what extent Asketil's immediate successors were Danes is at least very uncertain, as they have Anglo-Saxon names. During the invasions of the Danish kings, however, the convent was at times suspected of being in league with the Danes. Canute the Great is said to have presented a chalice, and his son Hardicanute his coronation mantle, to Croyland. Other Danes also made similar gifts to that convent. In the year 1053 it again had an abbot with the Danish name of Ulfketil (Wulketulus); and, what is very significant, after the Norman conquest, the swampy districts round it became places of refuge for the Danes and Anglo-Saxons who had in vain fought the last battle for freedom against the victorious and advancing Norman conquerors. One of the chief leaders in this battle was the Jarl Valthiof, a son of the far-famed Danish Jarl, Sivard Digre (*Eng.* Sivard the Stout) of Northumberland. Valthiof, it is expressly stated, was one of Croyland's best benefactors and protectors. Subsequently he made his peace with William, but was at last executed by that monarch's directions, and immediately buried at Winchester. Nevertheless the abbot Ulfketil, together with his monks, obtained permission to convey Valthiof's body to Croyland, where many miracles were soon performed at the shrine of the innocent and murdered martyr of freedom. Exasperated probably by this, as well as by the refuge which their opponents found in and about Croyland, the Normans inflicted many calamities on it, and at length deposed the abbot Ulfketil. He was succeeded by an Englishman with the Scandinavian name of "Ingulf," to whom we are indebted for having indited the ancient chronicles of the convent.

The close connection of Croyland with the Danes, as well as its Danish monks and abbots, was a natural consequence of the convent's being situated in Lincolnshire, a part of England which was pretty nearly the earliest and most numerously occupied by them. Satisfactory reasons certainly exist even to justify us in calling this convent peculiarly a Danish one. In consequence of its size and importance, it is highly probable that it was one of the principal places whence the Danish settlers in England derived their civilization. In this manner Croyland answers in England to the convent of Bec in

Normandy (from the Danish Bæk, a small rivulet), founded by the Northmen, and afterwards very celebrated; which also seems to have been one of the most important nurseries for the diffusion of a higher Christian and intellectual cultivation among the Scandinavian colonists in Normandy.

The very remarkable evidence which the history of Croyland affords of the Christianity of the Danes in England so early as the ninth century, is, however, by no means solitary. Before the treaty concluded between Gorm (Gudrum) and Alfred in the year 879, the former had already been converted, and received at his baptism the name of Athelstane. In a somewhat later treaty concluded by the same King Gorm with Alfred's successor Edward, it is assumed that there must long have been Christians among the Danes settled in East Anglia, and that they had at all events allowed the ecclesiastical institutions to exist unmolested among them. In the year 890 there was in Northumberland a king called Guthred (Gutfred, Godfred?), a son of the Danish king Hardicanute, of whom it is stated that he extended the bishopric of Durham, and conferred on it considerable rights and privileges, which even at the present day distinguish that see above all others in England. The coins of Danish-Norwegian kings minted in the north of England in the ninth and first half of the tenth century (as mentioned at p. 49), also indicate an early conversion to Christianity; as they show both the cross, and frequently also parts of the Christian legend: "Dominus, dominus, omnipotens rex mirabilia fecit;" or, "The Lord, the Lord, the Almighty King, hath performed wonderful things."

About the year 940, Christianity must, on the whole, have had a firm footing among the Northumbrian Danes. It would otherwise be inexplicable how, in the wars which Edmund waged at that time with the Danish king Anlaf, or Olaf, in Northumberland, even the Archbishop of York, "Wulfstan," should have sided with the Danes against the Anglo-Saxons. Wulfstan subsequently, in the year 943, negotiated a peace between Olaf and Edmund, whereby the latter ceded the country east of Watlinga-Stræt to Olaf. In this treaty a great man, of Danish extraction, took part on the Anglo-Saxon side; namely, Odo, Archbishop of Canterbury, whose father was a Dane who had fought in the host of the Vikings against Alfred the Great. One might almost be led to believe that Wulfstan himself was of Danish origin, and that his name was only the Anglo-Saxon form of the Scandinavian "Ulfsteen." For under King Edmund's successor, Edred, we again find the Archbishop, together with his clergy, paying homage to the Danish king's son, Erik (son of Harald Blaatand?), although he had shortly before, in common with the Northumbrians, taken an oath of fidelity to the Anglo-Saxon king. After the murder of Erik, King Edred caused the Archbishop to be deposed and thrown into prison; but afterwards gave him the bishopric of Dorchester, though far removed from the Danish possessions.

Another argument in favour of the Danish extraction of Bishop Wulfstan (or Ulfsteen) is, that several of his successors in the archbishopric were undoubtedly Danish; which shows that in those days such men were chiefly elevated to that dignity, as, through their common descent and kinsmanship, possessed an influence over the Danish population in Northumberland; where, also, there was doubtless a great body of Danish clergy. Contemporary with Abbot Thorketil, a certain "Oscetel," or Osketil, is also named as churchwarden (circeværd) in the King's letters-patent in the year 949; probably the same Osketil who, between the years 955 and 970, constantly signed the King's letters as Archbishop of York. As Odo, the Danish Archbishop of Canterbury, lived long after Osketil had become Archbishop of York, we are thus presented, half a century before the reign of Canute the Great, with the singular spectacle of the two chief ecclesiastics of England, the Archbishops of Canterbury and York, being both of Danish extraction. Oscytel's successor in the archiepiscopal see of York was also a Danish man, although he bore the Anglo-Saxon name of Oswald. He was both nearly related to Oscytel (his "nepos"), and, moreover, a brother's son of Archbishop Odo; consequently descended in a direct line from the Danish Viking, Odo's father. This Archbishop Oswald published some laws for the Northumbrian clergy which are still extant, and in which, according to Danish custom, fines are computed in *marks* and *öre*; whilst in the rest of England they were reckoned in pounds and shillings.

As these facts lead us to suppose that, at that time, a great part of the inferior clergy in England must have been of Danish extraction, and particularly in Danish North and East England; it thus becomes still clearer that the English priests or missionaries, with Scandinavian names—as, for instance, Eskild, Grimkild, and Sigurd—who went over to Scandinavia in the tenth century for the purpose of converting the heathens, were, as their names show, of Danish origin, and undoubtedly natives of the Danish part of England. Sprung from Scandinavian families, which, though settled in a foreign land, could scarcely have so soon forgotten their mother tongue, or the customs which they had inherited, they could enter with greater safety than other priests on their dangerous proselytizing travels in the heathen North; where, also, from their familiarity with the Scandinavian language, they were manifestly best suited successfully to prepare the entrance of Christianity.

The rapid accession of the Danes to the highest ecclesiastical offices in England must satisfactorily convince every impartial person how carefully we should discriminate between the Danish or Scandinavian Vikings, who, only for a certain period, robbed and plundered, and the Danish colonists, who, from the beginning of the ninth century were settled down—particularly in the east and north of England—as peaceful Christian citizens; and whose sons soon became sufficiently accomplished and respected to fill the highest places among the already powerful ecclesiastical aristocracy of England. Nor should it be forgotten, that the Danes in England, who, though fewer in number than the natives, yet aimed at the supreme authority, were early obliged to apply themselves to study, and to permit their sons to enter the clerical order; for, the greater the influence they could acquire among the clergy, who at that time held a very large share of power, the stronger and more secure would their position become in the land of their adoption.

After having had, at least, three archbishops of Danish family during the tenth century, it is not surprising that in the following one the English clergy had lost a great deal of their horror for the Danes, and were so willing to do homage to the Danish conqueror, Canute the Great, in preference to any prince of Anglo-Saxon descent. Nor did

Canute betray their confidence. He conformed to their manners, and built churches and convents, whilst his followers imitated his example. Under such a state of things the English clergy must have become still more mixed with Danes. In Canute's time the royal letters are signed by the abbots "Oscytel" (1020-1023) and "Siuuard" (in Abingdon, Berkshire); as also by "Grimkytel," bishop in Essex; and under Hardicanute, by "Sivard" and "Grimkytel" as well as by the *diaconus* Thurkil. Even long after the fall of the Danish power, as, for instance, in Edward the Confessor's time, we still meet with many high dignitaries of the church, with Scandinavian names; such as the abbots Sivard, Sihtric, Uvi or Ove, abbot of St. Edmundsbury, in East Anglia, and Brand; who was also abbot of a convent on the east coast, namely Peterborough, close to Croyland. We further have Sitric, chaplain to the Bishop of Dorchester, and lastly the Kentish bishop, Siward. William the Conqueror's Doomsday Book likewise mentions several such Danish clergymen; for instance, in the old Danish city of Lincoln, the priests "Siuuard" and Aldene or Haldan. In St. Edmundsbury there was still later (1157) a Danish abbot named Hugo.

The secular nobility, or chiefs, were closely connected with the high church dignitaries of that time. The royal letters before mentioned also show, that whilst the Danes succeeded in placing men of their own race amongst the highest clergy in England, they likewise procured admittance into the ranks of the nobility, and even into the suite that surrounded the Anglo-Saxon kings themselves. This happened not only from the Danish chiefs frequently entering the service of the Anglo-Saxon kings, and often marrying among the Anglo-Saxon aristocracy; but still more from the circumstance, that certain districts became in time so strongly occupied by the Danes, as to fall under Danish chieftains; and consequently the Anglo-Saxon kings, inasmuch as they held dominion over such districts, were compelled to take these chiefs into their court and councils. History informs us that the Danish kings Halvdan and Gudrum divided the districts they had conquered in Northumbria and East Anglia among their followers, and thus formed there, at an early period, a resident and wealthy Danish aristocracy.

It has been before shown that, so early as the ninth century, Lincolnshire had had at least three Shire-greves (Sheriffs), or earls of the shire, of Danish or Scandinavian extraction; viz., Thurold, Norman, and Sivard. In the ninth century, indeed, as well as in the first part of the tenth, the Danish possessions in England were almost entirely independent of the Anglo-Saxon kings. It was at this period that the Danish-Norwegian kings in the districts north-east of Watlinga-Stræt minted, as independent sovereigns, the many coins before described. There could not, consequently, have then existed in the courts of the Anglo-Saxon monarchs so many Danish chiefs, or vassals, as when those monarchs subsequently began to acquire dominion over the previously more independent Danish kingdoms. Thus, among the regular followers of King Athelstane (925-941), who subdued the Danish kingdoms in England, we find, even before his successful expeditions into the North, not a few Danish-Norwegian chiefs, who signed diplomas in conjunction with him, and particularly during the years 929 to 931; namely, besides the Thane "Syeweard" (his minister), the Jarls Urm, Gudrum, Healden or Halfdene, Inhwær (Ingvard), Rengwald, Hadder, Haward, Scule, and Gunner. This may, perhaps, partly confirm the statement of the chronicles, that Athelstane availed himself of Danish warriors to suppress rebellion in his kingdom. It is expressly stated that, at the battle of Brunanborg (treated of at p. 34), there were Scandinavian warriors in his army; and, among the rest, two Iceland brothers, namely, Thorolf, who fell in the battle, and the bard, or scald, Egil Skallegrimsen, who stayed for some time with King Athelstane, by whom he was presented with rich gifts for his lays. It is by no means improbable that Egil entertained, with his songs, the Scandinavian chiefs then at King Athelstane's court.

Between the years 940 and 960, several of the above-named Jarls, as Gunner, Scule, Haldan, and Urm, together with Grim and the chiefs, or ministers, Thurkytel and Thurmod, continued to sign the Anglo-Saxon letters-patent, in conjunction with their countrymen or relatives, the Abbot Thurcytel, and Oscytel, Archbishop of York. At this time the Latin title "dux" varies alternately with the Scandinavian title of Jarl, which the Anglo-Saxons called "Eorl."

With King Edgar's reign (959-975) began a fortunate epoch for the Danish dominion in England. Edgar himself was educated among the Danes in East Anglia, under the care of his relative, Alfwena, dowager queen of the converted Viking king, Gudrum, or Gorm. Hence he had early conceived such a partiality for the Danes, that during his reign he was accused of showing too much favour to those foreigners at the expense of the natives. It was in his time that the two highest ecclesiastics in England, the archbishops of Canterbury and York, were men of Danish extraction; and to judge from the diplomas issued by him, he must certainly have been served by several Scandinavians; for instance (959), by the Jarl Oscytel, and by the Thanes (or ministers) Ulfkytel, Rold, and Thurkytel. Thored, or Thured, a son of the before-mentioned Danish jarl, Gunner, is likewise named in the chronicles as one of Edgar's most trusted chiefs.

The Scandinavian, or Danish aristocracy had now gradually taken such deep root in England, that Ethelred the Second, who can scarcely have favoured the Danes, since he was repeatedly forced by their kings, Svend and Canute, to fly his kingdom, was even unable to remove the Danish chiefs from about his person, and to put in their places Anglo-Saxons of unmixed descent. In the first years of his reign there were in his suite, as the letters-patent show, several chiefs with Scandinavian names; as the Jarl Nordman, and the thanes Ulfkytel, Siweard, Wolfeby, and Styr, as well as the knights (milites) Ulfkytel and Thurcytel; whence it is clear that there must have been several chiefs of the same name at one and the same time in his court, and particularly of the names of Ulfkytel and Siweard. Nay, Ethelred himself was united, in first marriage, with a queen of Danish descent; namely, Elfleda, a daughter of the Danish chief Thured, Jarl Gunner's son. By this at least semi-Danish queen, he had several children, and amongst them a son, who afterwards became the renowned Edmund Ironsides. According to the chronicles, many powerful Danes had now obtained large fiefs even in the southern and western parts of England; as, for instance, the Jarl Paling, who was married to Gunhilde, a sister of the Danish king, Svend Tveskjæg, and who had extensive fiefs in Devonshire. This Paling, or Palne, however, to judge from the name, was probably the celebrated Scandinavian hero Palnetoke, whose possessions are said to have lain in that district.

The Danes were now so spread over the whole of England, that the Danish invaders were sure of finding support in almost every corner of it; and Ethelred consequently saw that, if their power was not crushed at once, the Anglo-Saxon dominion was threatened with imminent ruin. But it was too late. The secret massacre planned by him in the year 1002 was far from sufficing to annihilate, even in South England, the numerous traces of Danish influence; and to North England, as is well known, it did not extend. Even after the slaughter, we continue to find in the royal letters-patent nearly the same Scandinavian names of chiefs as before: such as Siward, Styr, Ulfkytel, Nordman, and the knights Ulfkytel and Thurkytel. The Icelandic scald, or bard, Gunlaug Ormstunge, also remained some time afterwards with Ethelred, just as Egil Skallegrimsen had before resided at the court of King Edgar, a monarch favourably disposed towards the Danes. The old chronicles also mention a powerful chief of Danish extraction who was in Ethelred's army after the massacre. This was Thorketil, surnamed Myrehoved (Ant-head); and, according to the same chronicles, a Dane named Ulfketil Snilling, sheriff or earl in East Anglia, was even married to Ethelred's own daughter Ulfhilde!

Thus, even before the conquest by Canute the Great, Danish families had frequently ingrafted themselves on the families of the Anglo-Saxon nobility; nay, even on the royal family itself. After that conquest the line of demarcation between the Danes and Anglo-Saxons cannot have been so strongly drawn as is generally imagined. Thus the descriptions given in the Sagas of the bold chiefs of the heathen North, as being also shrewd, amiable, and eloquent men, gain more and more credibility; and we cannot help admiring the ability and manliness which enabled the heathen Danish chiefs, and their immediate Christian successors, to maintain their difficult position against a hostile aristocracy, and, in spite of it, gradually to extend their power in the very midst of Anglo-Saxon England. Nay, they not only maintained their ground as the equals of the Anglo-Saxons, but soon became their superiors. The weakness and depravity of the Anglo-Saxon nobles under the reign of Ethelred were the best proof that their day was past. Faintheartedness, bordering very closely on cowardice, want of union, treachery, and every other vice, reigned no less among the chiefs than among their dependents. Luxury and effeminacy had usurped the place of the old Anglo-Saxon simplicity and vigour. Scarcely any great men appeared among them, notwithstanding the urgent need that there was for such characters. Even the greatest of their few warriors, Edmund Ironsides, was, as we have seen, of Danish descent on the mother's side.

We may almost say that England was the spoil of the Danes before Canute came over and seized the sceptre. What a contrast does Canute the Great, with his proud jarls and chiefs, present to the weak Anglo-Saxons! What vigour was at once developed in the government! What bravery was displayed in the field!

Canute the conqueror must, from motives of gratitude alone, if not for other reasons, have rewarded his Danes, and especially his chiefs, with landed estates, large fiefs, and lucrative posts of honour. He divided all England into four earldoms (Jarledömmer):—Wessex, the most Saxon part of England, he himself took, as being the most dangerous and hostile district. Mercia, or the middle part of England, which was half Saxon and half Danish, he gave to Edrik Streon, who was in favour with the mixed population there, possibly because, as the proverb runs, he wore his cloak on both shoulders. The Danish districts of Northumbria and East Anglia he assigned to his companion in arms, the Norwegian jarl, Erik, and the Danish jarl, Thorkil the Tall. Thorkil, meanwhile, had married King Ethelred's daughter, Ulfhilde, after her first husband, Ulfkytel, had fallen in the battle of Ashingdon. A number of smaller fiefs in different parts of England were made over, in a similar way, to Danish warriors of lower rank. Canute increased, moreover, the number of his guards of Scandinavian Huskarle, or *Thingmen*, of whom his forefathers had already availed themselves; and drew up for them a special code of laws, of such severity, that even the king himself could not infringe them with impunity. These Huskarle, or body-guards, being thus totally separated from the English by a peculiar system of law, became, in consequence, a really firm support for the kings. This Huskarle law, called Witherlagsretten, remained in force in the Danish court long after Canute's time.

The letters-patent issued by Canute show him surrounded by a great number of Danish or Norwegian chieftains. Among the signatures we find the names of men celebrated in history, such as "Thurkil hoga," "Yric," or "Iric," jarls in East Anglia and Northumberland; Ulf, Canute's brother-in-law, and father of King Svend Estridsen of Denmark; and also Hacun, a sister's son of Canute, and for a long time jarl in Worcestershire. All of these met a tragical fate. Thorkil and Erik had to wander in exile; Ulf was killed by Canute's order in Roeskilde; and Hagen, after many vicissitudes of fortune, perished on a voyage to Norway, where Canute had appointed him Stadtholder. Besides these we find named the jarl Eglaf or Ælaf (probably the leader of the *Thingmen*), Eilif Thorgilson, the jarls Haldenne ("princeps regis"), Ranig (Rane), Thrym, Siuard, Suuegen, Svend (1026), Tosti (1026), Sihtric, and others. Among the Thanes (ministri), appear Aslac, Tobi, Acun (Hagen), Boui (Bue), Toui, Siward, Haldan, Thurstan, Thord, Hastin(g), Broðor, Tofig, and several others; and among the knights (milites), Thord, Thirkil, Thrim, Broðor, Tokig, Ulf, and Siward. Several of Canute's chieftains, according to the genuine old Scandinavian custom, had surnames, mostly taken from their personal appearance; as, besides "Thurcyl hoga," we find Thurcyl hwita (white), Thurcyl blaca (black), Thoui hwita, Toui reada (red), and Haldan scarpæ (Halfdan the Sharp). A letter dated in the year 1033, is signed among others, by the chiefs: Jarl Siward, Osgod Clapa, Toui Pruda, Thurcyl, Harald, Thord, Halfden, Rold, Swane, Orm, Ulfkitel, Ketel, Gamal, and Orm; and as the document relates to some land in Yorkshire, it is probable that many of these Danish chieftains dwelt in that old Danish district. A powerful Dane, named Ulf, a son of Thorald, is named as of York in Canute's time. He gave many estates to the cathedral there, together with a carved horn, by way of conveyance or title-deed, which is still preserved in the cathedral under the name of "Ulph's horn," or "the Danish horn." This Ulf is possibly the knight of that name before mentioned. A similar horn is said to have been given by Canute the Great, with some landed property, to the family of Pusey, of Berkshire.

Under Canute's immediate successor, Harald Harefoot, as well as under Hardicanute, the power and grandeur of the Danish chieftains continued steadily to increase. Many besides those just mentioned are spoken of in letters of Hardicanute's reign; and above all the celebrated Danish jarl Siward, surnamed Digre, who in the year 1040 became jarl in Northumberland. We also meet with the jarl Thuri; the thanes Urki, Atsere (Adzer), and Thurgils; the knight Ækig (Aage); and, in the chronicles, Styr and Thrand. Lastly, Osgod Clapa, and Toui Pruda are mentioned in the history of Hardicanute, but on a mournful occasion. It was at the marriage festival which Osgod Clapa made for his daughter and Toui Pruda, that Hardicanute had a stroke of apoplexy, from which he never recovered. Some, therefore, are of opinion that the marriage did not take place at Lambeth (see p. 20,) but at Clapham (Clapa-ham, or Clapa's home), in Surrey, to the south of Kennington, which now forms part of London.

As long as their supremacy lasted, the Danes must naturally have behaved as conquerors in the land which they had subdued. Their innate love of splendour and profusion found ample nourishment, whilst at the same time their pride was flattered, by the subjugation of the Anglo-Saxons. The old English chroniclers complain bitterly of the severe humiliations which the natives were compelled to endure. If, for instance, Anglo-Saxons met a Dane upon a bridge, they were obliged to stand still, and make low bows; nay, even if they were on horseback, they must dismount, and wait till the Dane had passed. At the same time the Anglo-Saxon nobility gradually lost the many fiefs and lucrative posts of honour which had formerly been in their possession, but which were now transferred to their powerful conquerors. But what really injured the Anglo-Saxon aristocracy more than anything else, was the wise and conciliatory policy of Canute the Great, which, by extinguishing the hatred between the Anglo-Saxons and Danes, amalgamated the aristocracy of the two nations to such a degree that the Anglo-Saxon nobility at length existed only in name, having become by imperceptible degrees more than half Danish. A contrary method of proceeding, a violent and sanguinary oppression of the Anglo-Saxon aristocracy, would, perhaps, in some respects, have been more serviceable to them, as it would have inflamed their hatred, and provoked them to a desperate resistance; and would thus have incited them to keep themselves free from the intrusion of all foreign admixture.

As the matter stood, the Danish power apparently gave way to the Anglo-Saxon dominion; but, in reality, it was little more than the name that was changed. It is said, indeed, that the new Anglo-Saxon king, Edward the Confessor, some years after his accession (in 1048), expelled the great Danish chiefs and their descendants from his court, and drove them into exile; as, for instance, Osgod Clapa, sheriff of Middlesex, and Asbjörn, a brother of King Svend Estridsen of Denmark, whose second brother Björn, a jarl in the west of England, had shortly before been killed by the jarl Svend Godvinsön. He also banished Canute the Great's niece, Gunhilde. By her first marriage with her cousin, Hagen Jarl, Stadtholder of Norway, Gunhilde had a daughter named Bothilde; by her second with Harald, a son of Thorkil the Tall, who also succeeded to the Stadtholdership, she had two sons, Hemming and Thorkil. Gunhilde went into exile with her sons by way of Bruges in Flanders, and thence to her relatives in Denmark.

Nevertheless the signatures to Edward's letters-patent prove that this king, alleged to have been so favourably disposed towards the Anglo-Saxons, must have had many chiefs of Danish extraction about his person, even after this expulsion of the Danes; nay, even to the day of his death. We need not look for them among the "Huskarle," or body-guards, alone, amongst whom are named Thurstan and Urk; for Huskarle with Scandinavian names are mentioned at a still later period in England; and we find, under William the Conqueror (1071), Eylif Huscarl, and, even in 1230, Roger Huscarl. Even in King Edward's suite, and occupying considerable offices, were such men as "Atsere Swerte (Adser the black), Atsur röda (Adser the red), Eiglaf (Eylif), Guðmund, Ulfketil, Thord, Siward, Thurstan, Harold, Turi, Yrc (Erik), Anschitil (Osketil), Tofi, Neuetofig, Esgar, Ingold, Tosti, Thorgils, Wagen, Ulf Tofis sune, Askyl Toke's sune, Jaulf Malte's sune." Also the knights Esbern (Asbjörn) and Siward, together with several others, the greater part of whose names appear in letters that were issued after the expulsion of the Danes in 1048. Many of the royal fiefs were still in the hands of Danes. Jarl Siward Digre governed the extensive district of Northumberland with the same power and influence as before, till his death in the year 1055. Somersetshire, lying far towards the west in the Saxon part of England, had a sheriff (vice-comes) named "Touid," or "Tofig," who can scarcely have been an Anglo-Saxon. We find a person named "Toli" filling the same high office in East Anglia; as well as in Huntingdonshire a "Tuli;" in

Hamptonshire, a "Norman;" in Lincolnshire a "Marlesuuein." Northmen, or at least chiefs of Scandinavian origin, filled the highest posts at Edward's court. Between the years 1060 and 1066, a letter mentions the following royal chiefs, or "Hofsinder:" "Jaulf, Agamund, Ulf, Wegga (Viggo), Locar (Loke), and Hacun." In one of Edward's letters, dated 1062, the following names appear:—"Esgarus, regiæ procurator aulæ;" "Bundinus, regis palatinus;" "Adzurus, regis dapifer;" "Esbernus princeps;" "Siwardus princeps;" "Hesbernus regis consanguineus." These are all pure Danish names, viz., Esgar, or Asgier, Bonde, Adser, Asbjörn, and Sivard. The different Latin titles here given to Esgar, Bonde, and Adser, are translated in contemporary letters by one and the same word, "steallere" or "stalre." The dignity of "Staller" was also, as is well known, an established one in the courts of the Scandinavian kings, at all events after the time of Canute the Great. The Staller was superintendent of the court, or a sort of High Steward, and attended the "Thing" meetings for the king, but more particularly in cases which concerned the court. From an English diploma, dated 1060-1066, and signed by "Esegar steallere," "Bondig steallere," and "Roulf steallere," we see that there were several "Stallers" at the same time in England; which certainly arose from the Stallers being also the king's commissaries.

The last-named, "Roulf steallere," is probably the Ralph so much in favour with King Edward, and who was a son of Edward's sister and a Norman nobleman. Another Staller of Norman descent is mentioned in letters of the years 1044 and 1065, namely, Roldburtus, or Rodbertus, son of Winwarc. Indeed Norman names begin to be frequent in Edward's letters-patent; for, as a consequence of the favour which he bore towards the Normans, many of whom he gradually placed in the highest posts of honour in England, there quickly grew up by the side of the pure Danish elements, what may be called a half-Danish or half-Scandinavian influence from Normandy, which was soon to supplant the Danish power, as well as annihilate once for all the apparent dominion of the Anglo-Saxons in England. Thus Edward's reign was clearly only a state of transition from the Danish to the Norman dominion; a national Anglo-Saxon reign it could not well be called.

How, indeed, should Edward have been able to maintain, or rather to reinstate upon the throne of England a purely national Anglo-Saxon line, after it had long been broken by the Danes? Edward's own race may, in a manner, be said to show how weak and irretrievably declining was the Anglo-Saxon element. Edward himself was a son of the Norman princess, Emma, and thus brother-in-law to the Danish jarl, Thorkil the Tall, who had married his sister Ulfhilde, widow of the Danish jarl Ulfketil Snilling; he was half-brother to his predecessor on the throne, the Danish king Hardicanute; and he was married to Editha, daughter of Jarl Godwin, by his second wife, Gyda, who, being a daughter of the Jarl Thorkil Sprakaleg, nephew of the Danish king Harald Blaatand, was of Danish descent. Godwin, moreover, in his first marriage, is said to have espoused a Danish woman, a daughter of Svend Tveskjsæg, and sister to Canute the Great. Thus Edward the Confessor's queen, Editha, and her well-known brothers Svend, Harald, Gurth, and Toste, who, both during and after Edward's reign, played a highly remarkable part in English history, were on the mother's side of Danish extraction, of which the Scandinavian names of Godwin's sons bear sufficient evidence. It was partly also in consideration of this Scandinavian kinsmanship that Toste sought assistance in Denmark and Norway against his brother, King Harald; and that afterwards (in the year 1066), both Toste's son, Skule, and Harald's son, Edmund, fled to Scandinavia—the former through Orkney to Norway, the latter straight to Denmark—after their fathers had fallen, within a short period, in the battles of Stamford Bridge and Hastings. It is remarkable enough that Godwin's race should return to, and even flourish in, that same Scandinavian North whence, on the mother's side, it had sprung. Toste's son, Skule, married in Norway Gudrun, a daughter of Harald Haardraade's sister, and became by her the progenitor of so mighty a race, both of jarls and kings, that their branches extended over the whole of Scandinavia.

[+ +] Gravestone: Magnus

During the last period of the declining house of the Anglo-Saxon kings, we further meet with the Scandinavian names of Guttorm, Hagen, and Magnus. The name of Magnus, borne by King Harald Godvinsön's youngest son, was introduced into Norway through a mistake. It is related that a son having been born one night to King Olaf (Saint Olaf), no one dared to awake the King and inform him of it. The child, however, being very weakly, the priest Sighvat Skjaldt took upon himself to baptize it, and called it Magnus, after "the best man in the world," Karl Magnus, or Charlemagne; probably in the belief that the Latin word *magnus*, which was only the Emperor Charles' surname, was a real name. The boy grew up, and afterwards became king of Norway, where he was usually called "Magnus the Good." Magnus's grave is said to have been discovered in St. John's Church, in the town of Lewes, in Sussex. In the new church, which has lately been built on the site of the old one, has been preserved, and built into the wall, the monumental stone, which bears the following inscription:—

"Clauditur hic miles Danorum regia proles; Mangnus nome(n) ei Mangne nota progeniei. Deponens Mangnum, se moribus induit agnum P(re)pete p(ro) vita fit parvulus arnacorita."

Or, "Here lies a warrior (or knight) of the royal Danish race; his name, Mangnus, is the mark of his great descent. Laying aside his greatness he adopted the habits of a lamb, and exchanged his busy life for that of a simple hermit."

That this Magnus, "of the royal Danish race," was the son of the Harald Godvinsön lately mentioned (whose mother Gyda, it is true, was of the Danish royal family) is, however, a mere conjecture. An older legend states that he was a Danish chief, or commander, taken prisoner by the English in a sanguinary battle near Lewes, and who, being well treated, afterwards laid aside his sword, and became a hermit at that place. (See Lower, in "Transactions of the British Archæological Association at its second Congress at Winchester," pp. 307-310.) It may, perhaps, be most probable that he was one of those scions of the Danish aristocracy that remained in the south of England after the Norman conquest had overthrown the supremacy of the Danish chiefs in that part.

It was in the south of England, where William the Conqueror first established his power, that the Norman nobility obtained their earliest possessions. In the midland and northern districts, on the contrary, it was neither easy to subdue the country, nor to annihilate entirely the

Danish aristocracy, which had completely coalesced with the essentially Danish population. Long after the conquest, therefore, the Danish chiefs continued to preserve their independence, or at least their influence, in those parts. A remarkable instance of this, though taken only from a single district, is afforded by William's own "Domesday-Book," drawn up about twenty years after the conquest. In this, under the head of Lincolnshire, are mentioned the great persons who possessed the right of administering justice on their estates, together with other privileges belonging to noblemen, such as sacam and socam, and Tol and Thiam; and among them are found "Harald Jarl; the Jarl Waltef (Valthjof); Radulf Jarl; Merlesuen; Turgot; Tochi, son of Outi; Stori (Styr); Radulf "stalre;" Rolf, son of Sceldeware; Harold "stalre;" "Siuuard barn;" Achi (Aage), son of Sivard; Azer, son of Sualena; Outi, son of Azer; Tori, son of Rold; Toli, son of Alsi; Azer, son of Burg; "Uluuard uuite;" Ulf; Haminc (Hemming); Bardt; Suan, son of Suane." Now even if it be certain that several of these chiefs were Normans, particularly since the Norman names at that time still preserved their primitive Scandinavian form, yet it is clear that most of them were Danish-English. It is to be regretted that Domesday-Book does not comprise the ancient Northumberland, as that district would certainly have afforded more names of Danish chieftains than even the old Danish Lincolnshire; for the Danish aristocracy were never driven out or entirely subdued in those parts; but rather must have amalgamated in the course of time with their countrymen, the Norman nobility, until the latter by degrees gained the ascendancy. This is at once shown by the notorious fact that neither William the Conqueror, nor his immediate successors, obtained such mastery over the north of England and its Danish population, as over the rest of that country; since the inhabitants of the north fought, with the bravery inherited from their forefathers, for their Danish chiefs, and for their peculiar, and partly Danish, institutions, manners, and customs.

SECTION XIII

The Danelag.—Holmgang, or Duel.—Jury.—The Feeling of

Freedom.

The Anglo-Saxons were the teachers of the Danes in several ways; above all they made them Christians, and thus communicated to them a new and higher civilization. The Danes in England reaped advantage from the civilization of the Anglo-Saxons, just as the Anglo-Saxons themselves had once begun their own, by building on that refinement which their predecessors, the Romans, had disseminated in England.

But as the Anglo-Saxons did not become Romans, because they adopted and remodelled the Roman civilization; nor the Normans in Normandy Frenchmen, because after their settlement in France they soon assumed many of the French manners and customs; so neither did the Danes in England become Anglo-Saxons, however much they might have been indebted to them for their civilization. The Normans in France retained, in spite of their Christianity and French refinement, the characteristic stamp of their Scandinavian origin, which afterwards caused them to play quite a peculiar part in history. In like manner the Danes in England, amidst the refinements of the Anglo-Saxons, undoubtedly preserved many of their Scandinavian characteristics, which did not disappear without leaving visible and very remarkable traces. But the Scandinavian spirit stamped itself, though perhaps only apparently, in a somewhat different manner on the Norman race in Normandy, and on the Danes in England.

Among the Normans in France the Scandinavian spirit worked, so to speak, only outwardly, in magnificent conquests, of which the chief theatres were England, Italy, and Sicily. Chivalry and feudalism, with their crusades, communicated a new impulse to it; but, internally, it effected comparatively little for France. It did not manifest itself in Normandy by forming political institutions capable of supplanting the oldest and most essential French laws and constitutions; nor, indeed, are we able to point out with exactness what really Scandinavian customs the Normans established in that country. Yet it can scarcely be doubted that they introduced there trial by jury, as well as trial by battle, and other Scandinavian legal institutions.

In England, on the other hand, the northern character showed itself so far outwardly active as to exercise a vast and unmistakable influence on her commerce and navigation, and on the bold and adventurous spirit of enterprise among her people; which, though at a much later period than the conquests of the Normans, has nevertheless extended her dominion over every sea. But in England it has also been internally a living and guiding spirit, in the formation of her judicial and political institutions. It is an incontrovertible and notorious fact, which has, however, hardly been sufficiently insisted upon, that about half of England—the so-called "Danelag," or community of the Danes—was for centuries subject to Danish laws; that these laws existed even after the Norman conquest; and that they did not pass into the general or common law of England, till the successors of William the Conqueror at last united into a whole the various discordant parts into which England had been previously divided. When we remember that the Normans long retained a predilection for old Scandinavian institutions and forms of judicature, it seems highly probable that the Danish laws, which had for so long a period prevailed in England, did not disappear under their sway without the new laws, which they established, deriving from the old a particular colour, and certain Scandinavian stamp. A further examination of this point will scarcely be superfluous, as it will enable us to judge how far those are right who, in company with one of England's most celebrated statesmen (Sir R. Peel, in a speech in Parliament), are proud that "the Danes tried in vain to overthrow the institutions of England, instead of securing them;" and then reproach the Danes that, on the whole, they did not, after all their devastating expeditions, establish anything new, great, and durable.

The population of the heathen North, as was the case everywhere else at that period, was divided into serfs and freemen. Even after the introduction of Christianity, many centuries elapsed in all countries before thraldom was abolished, and the worth of man, as man, generally recognised. The serf was always regarded more as an animal than as a human being. The freeman, on the contrary, enjoyed a high degree of civil liberty. He was not only uncontrolled master in his own house, and among his nearest dependents, but likewise exercised an important influence on the management of the public concerns of his own district and of his country. He took part in the decision of law cases in the "Thing," and gave his vote at the great "Thing," where the election of a monarch, war, treaties of peace, and other important matters, came under consideration. Scandinavia was, besides, in ancient times, divided into a number of small kingdoms; and the smaller these were, so much the greater was the individual freeman's power and importance.

The old inhabitants of the North entertained, therefore, a sincere affection for those institutions which gratified their proud feeling of freedom. Personal participation in the administration of justice, at a time when written laws did not exist, must have made every freeman a lawyer and a zealous defender of existing institutions, especially so far as regarded the main point, namely, the freedom they ensured. A general knowledge of the laws was still further promoted by the innate love of the Northmen for disputes and law-suits. Respect for the law was speedily carried to such an extent, and in the administration of justice at the *Things* old established customs and usages were so strictly observed, that the slightest formal flaw was sufficient to ensure the rejection even of the most important cause. How deeply rooted the old national law was, is best shown by the fact that the Roman law, which had been adopted in the greater part of Europe, could never gain the supremacy in the countries of Scandinavia. The present Scandinavian law is by no means the offspring of any foreign code, but is founded on, and independently developed from, the law which already existed in the North in the days of heathenism.

The powerful warriors, who in those remote times emigrated from the North, were, for the most part, men no less high-spirited and fond of freedom than their fathers before them. The old chronicles state, that among the warriors who came over to England with the conquerors Svend and Canute, there was not a single serf. The history of Iceland shows, even at an earlier period, that most of the colonists who went thither were descendants of kings, jarls, and other of the most powerful freemen of the North. These emigrants did not leave their paternal home because they were dissatisfied with their ancient hereditary rights and liberties, but because those rights and liberties were gradually threatened with restriction, and even annihilation, by ambitious and absolute monarchs. It was this that led them to undertake the conquest of foreign lands, and thus to acquire a freedom which might indemnify them for what they had been compelled to relinquish.

It is therefore no wonder that the Scandinavian colonists introduced their national laws, which had always proved the surest defence of their liberties, at once and completely both into countries previously uninhabited, and into those from which the ancient inhabitants were expelled by their invasions. This was the case, for instance, in Greenland, the Faroe Isles, the Shetland Isles, and the Orkneys. But with regard to freedom they even went still further than in Scandinavia, and sometimes abolished the regal power, whose caprices and dangers they had learned to appreciate and fear, and founded republics in its place. Even in countries like France and England, where a large and civilized population, possessing a complete system of national law, previously existed—and where the Scandinavian colonists, till they became strong enough to assume the authority of masters, were for a long time inferior both in numbers and power—they adhered immovably to their ancient legal customs, and caused them to be observed, in spite of Christianity, and of that foreign civilization which they themselves soon adopted. But it was at the same time a natural result of this state of things, that they were neither able to introduce into such countries *all* the ancient legal usages of Scandinavia, nor, generally speaking, *any* law of a comprehensive character, without adapting it to the peculiar situation which they, as conquerors and strangers, now occupied in regard to the natives and their existing institutions.

A strong proof, not only of the affection of the Danes for their Scandinavian institutions, but of the complete settlement of that people in England at a very early period, is, that in the beginning of the tenth century, and consequently more than a hundred years before the time of Canute the Great, they had already established their own laws on the east coast of England, notwithstanding that Christianity, as before stated, had gained a footing amongst them. It appears, from the remarkable treaty concluded at that time between Kings Edward and Gudrum, that the Danes settled in East Anglia, and on the eastern coast of England, were not only placed on an equal footing with the English with regard to legal rights, but that it was also determined how disputes between the English and Danes should be decided, and what fine each people should pay for certain crimes. Thus the English were to pay "*wite*," or fines, according to the English law, in pounds and shillings; whilst the Danes were to make compensation for "*lah-slit*" (i. e., *infraction of the law*, from the old Norsk, *lög*, law, and *slita*, to rend in two, break), according to the Danish law, in "marks" and "ores."

About the same time the chronicles testify that the "five burghs" occupied by the Danes in the heart of England, together with large districts both in the east and north, were subject to Danish laws. The Anglo-Saxon king Edgar (959-975) says, in a passage of his laws (cap. 12), which shows his partiality for the Danes, "Then will I that with the Danes such good laws stand as they may best choose, and as I have ever permitted to them, and will permit so long as life shall last me, for their fidelity, which they have ever shown me." He likewise says in the next chapter, where mention is made of a fixed punishment: "Let the Danes chuse, according to their laws, what punishment they will adopt."

From this state of things, it happened that four different sorts of law were in force in four different parts of the kingdom. Farthest towards the west, where the remnant of the ancient Britons dwelt, the Welsh law was in force; among the West Saxons, the West-Saxon law; in Mercia, the Mercian law; and in the so-called Danelag, or country to the north-east of Watlinga-Stræt, the Danish law. Of these four systems of law, the Danish, beyond comparison, most prevailed. Its decrees were in later times constantly recognised, not only by Ethelred (not to speak of the Danish kings), but by Edward the Confessor and William the Conqueror, whose laws usually treat of the "Danes-law" (Dene-lahe), with its fines, or "*lah-slit*," in marks and ores. Even in the laws promulgated by Henry the First (1100-1135), it is stated (vi. § 1), that England is divided into three parts, Wessex, Mercia, and the province of the Danes. ("Regnum Anglie trifariam dividitur in regno Britannie, in Westsexiam, et Mircenos, et Danorum provinciam.") And it is further said (§ 2), that the law of England falls into three parts, according to the above division, viz., the West Saxon, the Mercian, and the Danish law, or Denelaga. ("Legis eciam Anglice trina est particio, ad superiorem modum; alia enim Westsexie, alia Mircena, alia Denelaga est.")

A cursory view of these different laws will soon show, both that Scandinavian words and juridical terms were employed in the *Danelag*, and that by degrees, but mostly in the time of Canute the Great and William the Conqueror, they were introduced into the common laws of England: as, for instance, "hor-qwene" (Hoerquinde; *Eng.*, adultress), "nam," "halsfang," "heimillborch," (Hjemmelborg), "husting," and others. For the rest, it is natural that most traces of the old Scandinavian institutions should be found in the districts to the north-east of Watlinga-Stræt.

The Danes settled there had from the beginning several chiefs with the title of king, who were for the most part independent of the Anglo-Saxon kings, and reigned by means of their jarls and the chiefs to whom they had portioned out the conquered land. These numerous small kingdoms were afterwards subdued by the Anglo-Saxons, and converted into Earldoms. A peculiar sort of Danish chiefs or Udallers ("*holdas*," from the old Norsk *hölldr*), is mentioned in East Anglia, who, like the Norwegian "Höldar," or "Odelsmænd," held their properties by a perfectly free tenure. It is probable that the original Udallers were the chief leaders, or generals, of the Danish conquerors settled in East Anglia. From the fines fixed for the murder of such "holdas," it is plain that they held a very high rank. The old Scandinavian name for a peasant, "*Bonda*," was also disseminated in the north of England. There, as in Scandinavia, the peasants undoubtedly constituted the pith of the landed proprietary. The names of places in the north of England beginning or ending with *garth* (or *Gaard*), such as Watgarth (*Vadegaard*, on the river Tees), Grassgarth, Hall Garth, Garthorpe, Garthwaite, and others, show that the peasants, as in Scandinavia, were settled in *Gaarde*, or farms, which belonged indeed to the before-mentioned "*holdas*" ("*Odelsmænd*"), or other feudal lords; but which nevertheless seem, in some degree, to have been the property of the peasants, on condition of their paying certain rents to their feudal lords, and binding themselves to contribute to the defence of the country. Other landed proprietors, or agriculturists, with pure Scandinavian names, appear in Cheshire under the appellation of "*drenghs*" or *Drenge*.

The Danes and Norwegians in North England settled their disputes and arranged their public affairs at the *Things*, according to Scandinavian custom. The present village of Thingwall (or the *Thing-fields*), in Cheshire, was a place of meeting for the *Thing*; and not only bore the same name as the old chief *Thing* place in Iceland, but also as the old Scandinavian *Thing* places, "Dingwall," in the north of Scotland; "Tingwall," in the Shetland Isles; and "Tynewald," or "Tingwall," in the Isle of Man. There were incontestably in the Danish parts of England certain larger or common Thing-meetings for the several districts, which were superior to the *Things* of separate ones; and it may even be a question whether traces of them are not to be found in the division into *Ridings*, at present used only in Yorkshire, but which formerly prevailed also in Lincolnshire. Originally these divisions had not the name of *reding* or *riding*, which they did not obtain till later, and undoubtedly through a misconception. Yorkshire is at the present time divided into the North, East, and West Ridings; and, according to Domesday-Book, Lincolnshire also was (about the year 1080) divided into Nort-treding, Westreding, and Sudtreding; consequently, like Yorkshire, into three parts. These divisions were called by the Anglo-Saxons "Þriding," or "Thriting." Now, as they were foreign to the Anglo-Saxons, whose historians did not even know how to explain their origin, and as they also appear exclusively in the two most Danish districts in England, it is surely not unreasonable to seek their origin in Scandinavian institutions, in which a simple and natural explanation of them may certainly be found. In Scandinavia, and particularly in the south of Norway, provinces or Fylker (petty kingdoms), were not only divided into halves (hálfur) and fourths (fjórðjungar), but also into thirds, or *Tredinger* (Þriðjungar), which completely answer to the North-English "thrithing." It was, moreover, precisely to the *Tredings-things* that all disputed causes were referred from the smaller district *Things*.

It is more doubtful whether we may ascribe to the Danes alone the introduction of the word "Wapentake" (*Vaabentag*), as the peculiar designation for a district. In the northern counties of England, viz., Northamptonshire, Nottinghamshire, Lincolnshire, and Yorkshire, this term is still used instead of the customary one of "Hundred." Yet there is some probability that it may have been derived from the circumstance that the Danes, like the ancient inhabitants of the North in general, elected their chiefs, and signified their assent to any proposition at the *Things*, by Vaabentag, or Vaabenlarm (sound, or clang of arms). Vaabentag (Wapentake) might thus have become the name of a small district, having its own chief and its own *Thing*. A law of King Ethelred's (see Thorpe, *Leges et Instit. Anglo-Sax.*, Glossary, *Lahman*), which seems to have been promulgated only for the five Danish burghs, and the rest of the Danish part of England, orders that there shall be in every Wapentake a *Gemot* or *Thing*. It is at all events very remarkable, that the division into Wapentakes should exist only in old Danish North England.

In the towns occupied by the Danes, as in the five burghs—or, if Chester and York be included, in the "seven cities"—there was certainly a Danish *Thing*, as well as in the rural districts. The English word *by-law*—still used to denote municipal or corporate law, which is neither more nor less than the Danish "*By-Lov*," and which, consequently, must have retained its name ever since the times of the Danes—shows at once that they must at least have had some share in developing the system of judicature in the English cities. It is, besides, well known that there was in remote times a Scandinavian "husting" in Sheppey, London, and Winchester, as well as York and Lincoln, and consequently in places south of Watlinga-Stræt. Of the seven cities before mentioned, only York and Lincoln are with certainty known to have had "hustings;" but nevertheless, it can scarcely be doubted that there must have been similar *Things* in the other five cities. I may add, that the tribunals existing in them are called, in the Anglo-Saxon text of Ethelred's laws for the five burghs just alluded to, "*Gethingd*"—a word which bears an undeniable resemblance to the Scandinavian *Thing*; whilst in Anglo-Saxon such courts were called "*Gemot*."

According to old English records, the Danish laws in force in the Danish part of England, though in several respects strikingly similar to the Anglo-Saxon laws, differed from them in many points. It is not, indeed, clearly determined in what these differences and resemblances consisted; but it is at all events certain that the dissimilarity cannot have been confined merely to the difference before mentioned in the amount of the fines, nor to the mode of calculating them; which, as previously stated, was in marks and ores in the Danish part of England, and in pounds and shillings in the Anglo-Saxon districts.

In law-suits among the Anglo-Saxons, the usual kinds of proof were by oath, by witnesses, by cojurors, and by the ordeal of hot iron, or the judgment of God. It was at an early period also customary, in the heathen North, to use by way of proof oaths, cojurors, and witnesses; but instead of the ordeal by hot iron, which was first introduced under Christianity, the old Northmen had quite a different way of deciding their legal disputes, and one which agreed better with their martial spirit, namely, by duel. By some this method was also considered a peculiar kind of God's judgment; but it should rather, perhaps, be regarded as the subjecting of the original feud, or quarrel, to certain settled forms. This sort of combat was called "*holmgang*," because the duel generally took place on a small island, or *holm*, where it was conducted according to fixed laws. Both plaintiff and defendant had the right of challenging their adversary. Although this mode of deciding legal disputes might easily be, and indeed sometimes was, abused by evildoers—who did not scruple to take advantage of the weakness and want of warlike skill in others, in order to obtain possession of their estates—still it was far more in favour in the North than the proofs by oath and cojurors. The Normans carried it with them into Normandy; and there can scarcely be a doubt that the Danes and Normans, long before the Norman conquest of England—nay, long before Canute the Great's time—introduced it into the *Danelag* in the north of England; where, at least, the word "*Holmgang*," in its pure Scandinavian meaning, was in use for many generations.

But a peculiar, and in its results highly important, judicial institution prevailed in the North, namely "*Næfn*," "*Næfninger*" (Nævninger); or, as it has been called in later times in English, "Jury." According to the most ancient Danish laws the accuser had a right, particularly in important criminal causes, to select from among the people a certain number of jurors (Nævninger), who, after taking an oath, were to condemn or acquit the accused; and judgment was not pronounced till they had given their verdict. The accuser's choice of jurors was limited by law to owners of landed property who were not related to him; neither were they to be inimically disposed towards the accused, who had the right of challenging any of them. The decision of the jury was declared according to the majority of votes. In some districts at least, as for instance in Scania (Skaane), the accused was allowed, if the decision of the jury was against him, to appeal to the ordeal by red-hot iron, which, after the introduction of Christianity, became an important mode of proof in the North. But after the abolition of that ordeal in Denmark (in 1218), and after the heathen mode of duelling, or *holmgang*, had been abolished by Christianity, and superseded by the institution of juries, this last method of trial played an important part, and became popular with the people because it afforded them a participation in the administration of justice, and at the same time secured their civil liberties. Nevertheless trial by jury was at length obliged to yield to newer forms of law in Scandinavia; and just in proportion as the ancient freedom of the people was lost, the political institutions which had originated from it also disappeared.

England, as is well known, is the only country that, in spite of all commotions, has preserved trial by jury down to modern times. But it is a matter of much dispute to what people may be more particularly ascribed the honour of introducing an institution which has not only for many centuries been of much service to freedom in England, but which has also been transplanted in later times into many other countries, and is now on the point of being disseminated over all that part of Europe which may be called free. Many learned men assert that trial by jury was unknown to the Anglo-Saxons, and maintain that its proper home was the Scandinavian North, whence it was carried by the Northmen into Normandy, and from that country into England by means of the conquest. Others again assert almost the direct contrary; maintaining, that the tradition which ascribes the introduction of juries to the Anglo-Saxon king, Alfred the Great, though it does not speak the literal truth in deriving the institution merely from that monarch, is still thus far deserving of credence, that trial by jury was known and used by the Anglo-Saxons long before the Norman conquest. These persons are of opinion, that the Danes and Normans even set aside the jury for the barbarous *Holmgang*, or duel, until in the course of time that venerable relic of ancient Saxon freedom again obtained the ascendancy. In order to prove this, they point especially to a passage in one of Ethelred's laws (Ethelred, iii. § 3), which ordains "that every Wapentake shall have its *Thing*;" and "that a 'Gemot' be held in every Wapentake, and the XII senior Thanes go out, and the reeve with them, and swear on the relic that is given to them in hand, that they will accuse no innocent man, nor conceal any guilty one." Further (§ 13): "And let doom stand where Thanes are of one voice; if they disagree let that stand which VIII of them say; and let those who are outvoted pay, each of them, VI half-marks." To these passages may be added another, also of Ethelred's time (Ordinance respecting the Dun-Setas, § 3), wherein it is ordered that: "XII lahmen shall explain the law to the Wealas and English, VI English, and VI Wealas. Let them forfeit all they possess if they explain it wrongly; or clear themselves that they knew no better."

That a jury is here spoken of is beyond all doubt. But a highly remarkable circumstance has been too much overlooked, namely, that Ethelred's above-mentioned regulation as to the composition of the jury is contained only in the law just cited; which, according to the opinion of its latest English editor, was intended only for the Five Burghs and the surrounding Danish districts. ("*The document of Ethelred, above referred to, seems, in a great measure, to have been published for the sake of the Five Burgs.*"—Thorpe.) That it cannot have been intended for the Anglo-Saxon part of England may be immediately seen from the circumstance that all the fines mentioned in it are, without exception, fixed, according to Danish custom, in *marks* and *ores*, or *öre*, and not, after the Anglo-Saxon custom, in pounds and shillings. In this concise law, moreover, we find several Danish legal terms which were not in use in the south of England; for instance, "lahcop" (Old Norsk, "lögkaup"); "wit-word" (Old N., "vitorð"); and "thrinna XII," or "trende Tylvter Eed" (i. e. three twelves oath). With respect also to the "XII lahmen," or, as they are called in Latin, "lagemanni" (Old Norsk, lögmaðr), mentioned in Ethelred's time, it has long been agreed in

England that they must have been originally instituted by the Danes. (Thorpe says: "*The institution was most probably of Danish origin, as we generally meet with them in the Danish portion of the country.*") They were constantly twelve in number, and it can scarcely admit of a doubt that their functions were the same as those of "the twelve eldest Thanes" before mentioned, and that consequently they were regular jurymen. We see, moreover, from Domesday-Book, which mentions "Lagemanni" only in the Danish portion of North England, viz., in Cambridge, Stamford, Lincoln, and Chester, that they were Thanes, or at least equal to Thanes in rank and privileges. Among other things, jurisdiction (sacam and socam) was conceded to them over their inferiors, or subjects. In the old Danish city of Lincoln the names are recited of those who were previously Lahmen, and of those who remained so when Domesday-Book was compiled. These names, which are partly pure Danish—as, for instance, Hardecnut, Ulf, son of Suertebrand, Walrauen, Siuuard, Aldene (Haldan), and others—prove that sons frequently succeeded their fathers in the office of Lah-man (for instance, "Suardinc loco Hardecnut patris sui. Sortebrand loco Ulf patris sui. Agemund loco Walrauen patris sui. Godvinus fil. Brictric").

For the rest, since we might search the old Saxon laws in vain for any other certain traces of jurymen besides these, and as special care must be taken not to confound jurymen with cojurors, it becomes quite clear, first, that those authors who conclude, from the above often-quoted passages of Ethelred's law, that the English jury is of Anglo-Saxon origin, are in error; and secondly, that their opponents have not taken a quite impartial view of the matter when they ascribe the introduction of the jury into England to the conquest by William of Normandy. For it must now be regarded as a point quite decided that THE EARLIEST POSITIVE TRACES OF A JURY IN ENGLAND APPEAR IN THE DANELAG, AMONG THE DANES ESTABLISHED THERE, and that, long before William the Conqueror's time, they had brought over from their old home the Scandinavian *Nævn*, or jury, into the districts north-east of Watlinga-Stræt, colonized by them, just as their kinsmen and brothers introduced that powerful safeguard of popular freedom into Iceland and Normandy. It would, indeed, have been quite inexplicable that the Danes should have given up their peculiar Scandinavian *Nævn* in a country like England, where the Danish law obtained by degrees so extensive a footing that, during the reign of the first Norman kings, it was still in force in one-half of the kingdom.

The provisions in Ethelred's law, so frequently cited, respecting the force of the majority of votes in the verdict of the jury, also betray a likeness, which can scarcely have been accidental, to the regulations of the *Nævn*, or jury, at that time observed in Denmark. According to the most ancient Danish laws, the outvoted jurymen were also to pay fines. For the rest, there is this peculiarity in the jury of the Danish part of England, that from the time of Ethelred it was no longer chosen by the complainant, as was originally the case in Denmark, but by the court, or by the sheriff of the district ("gerefa"); which was a considerable step gained towards security against partiality. The choice of jurymen was, besides, still more limited in England than in Denmark. Instead of landed proprietors in general, the twelve eldest Thanes alone were eligible; whence it followed that the jurymen were not only fixed, but also obtained, as a reward for their labour, a certain rank, with the rights and income attached to it. This more aristocratical form of the jury undoubtedly sprang from the circumstance that the Danes had entered the northern and eastern districts of England as lords and conquerors. They could not, consequently, appoint as jurors native Anglo-Saxons, unacquainted with the customs of the Danish law courts; nor would they, assuredly, have permitted a conquered people to take a part in verdicts affecting themselves and their Scandinavian brethern. The consequence was, that they chose from among themselves men of consideration, and acquainted with the law, to conduct the administration of justice. It is very remarkable that a later development of the law in Denmark produced a similar change in the jury, the jurors not being chosen for a single cause, but for a period. In Jutland even "*Sandemænd*," or jurors appointed by the crown, were instituted, who seem to have answered to the before-mentioned *Lag-men*, or *Lahmen*, in the north of England. Eight landed proprietors were selected in every district by the king, and discharged the office of jurymen for life, unless they forfeited it by some misdemeanour.

Not the least trace is to be found in the old English laws and chronicles that the Danish laws in force in the Danelag were more barbarous than the contemporary Anglo-Saxon ones in the south of England. On the contrary, the fact lately mentioned, that the beneficial change in the composition and working powers of the jury, which had long been in force in Danish North England, was in far later times adopted in Norman England, seems rather to attest, in no slight degree, the superiority of the laws of the Danelag. On the whole, the Danish kings in England, and particularly Canute the Great, seem to have been excellent lawgivers. Canute's laws respecting the limitation of capital punishment, the right of every man to hunt on his own land, and others, evince a mildness and humanity scarcely to be expected in those rude times.

From what has been said, it appears that the Danish part of England must, in William the Conqueror's time, have had just as many old Danish popular institutions as Normandy, nay, doubtless still more. It is, therefore, no wonder that William and his Normans were highly partial to the Danish laws then in force in England. Immediately after he assumed the reins of government, he commanded that these laws should be in force throughout the kingdom, and consequently even in the purely Anglo-Saxon districts, as both his own forefathers, and those of almost all his barons, had been Northmen, who had formerly emigrated from Norway. But in an assembly held at London in the fourth year of his reign, he suffered himself to be persuaded, by the urgent entreaties of the leading men among the Anglo-Saxons, to restore the laws of Edward the Confessor in the districts in which they had before prevailed. Nevertheless, the Anglo-Saxon laws gradually gave place to the Scandinavian institutions in force in the north of England. Thus duel, under the name of "trial by battle," came to be considered throughout England as lawful proof in judicial suits; an evident result of the bold and chivalrous spirit of the new Norman lords. This kind of proof caused, however, much disturbance in England, and at length, though tardily, grew out of use. It was not formally abolished by law till the year 1818, after a prosecutor had challenged his adversary to trial by battle; a proceeding which even the legal tribunals were obliged to acknowledge that the law, taken in its strictest sense, fully authorised him in adopting. It is, however, remarkable enough that the proof by duel, which in Scandinavia itself was abolished on the introduction of Christianity, should have maintained its ground for several centuries in England, which had long been Christianized. We might even say that down to the present times it has everywhere left perceptible traces in Europe. For what are duels but trials by battle, or sort of judgment of God? They were, however, much disseminated by chivalry, in the development of which the warlike Normans took so considerable a part. The ancient *holmgang* was, as we have seen, called, both in Normandy and England, "duel."

The institution of the jury ("Nævninger," or "Nævn"), before mentioned as originally Scandinavian, was established throughout England by the Normans in such a manner that it has maintained its place to our times. Under the first Norman kings we find traces of a more general employment of the jury, which was previously confined to the Danish part of England, where it continued to exist after the conquest by William. When, in the following century, *holmgang* or trial by battle, began, in spite of the limitations it had undergone, to become too grievous in England, a law was published in 1164, that a jury of twelve knights, chosen by four knights of the district, should be substituted in its place. Thus at its first general establishment in England the jury had much the same form as it possessed in earlier times in the Danish part of the kingdom. The provision that the jury should be composed of knights soon fell to the ground. Subsequently, after the ordeal by red-hot iron, or the judgment of God, had been abolished (in the year 1219), it was appointed, in the reign of Henry the Third, that the accused, who might previously have liberated himself by that ordeal, should submit his case to the decision of twelve *Nævninger*, or jurymen. In this manner an influence was secured to the jury in England, which has since been continually increasing; trial by jury having become, as it were, the central point of the judicial system in that country. The English themselves, with just reason, regard the jury as a wise and happy institution, which has much contributed to develope the excellence of the national character, and to maintain the free constitution of their country. What is more, foreigners pass the same judgment on it; and it especially deserves to be remembered, that at the present moment, after the introduction of popular freedom into the Scandinavian North, its people are seeking to re-establish the native *Nævn*, or jury, which formerly crossed the seas with the conquerors of England and Normandy, and which has victoriously stood the trial of centuries in those countries.

We have already seen it proved, from contemporary laws, that the germ of at least one of England's freest and most important institutions was to be found, as early as the ninth century, among the numerous Danes and Norwegians settled in that country, to whose successors and kinsmen may be justly ascribed the honour of further developing the institution of trial by jury. In like manner contemporary chronicles bear witness that these Danish and Norwegian settlements in many ways essentially contributed to promote political liberty and the spirit of freedom. According to that remarkable document, Domesday-Book, there was, about twenty years after the Norman conquest, a greater number of independent landed proprietors, if not, in the strictest sense of the word, freeholders, in the districts occupied by the Danes, and under the Danelag, than in the other, or Anglo-Saxon, part of England. The smaller Anglo-Saxon agriculturists were frequently serfs, though, for the most part, perhaps, leaseholders, or holding other subordinate situations; whilst the Danish settlers, being conquerors, were mostly freemen, and, in general, proprietors of the soil. Domesday-Book mentions, under the name of "Sochmanni," a numerous class of landowners, or peasants, in the Danish districts north-east of Watlinga-Stræt, who, to the south of that line, and even then only just upon the borders of it, are rarely to be found, (viz., in Buckinghamshire, 19, and in Surrey, 9). It also mentions a great number of freemen in those districts, or, as they are called in Latin, "liberi homines." Neither *Sochmanni* nor *liberi homines* seem, however, to have been freeholders, in the present sense of that term. They certainly stood in a sort of feudal relation to a superior lord; but in such a manner that the "Sochmanni" may be best compared with our present hereditary lessees. Their farms passed by inheritance to their sons, they paying certain rents, and performing certain feudal duties; but the feudal lord had no power to dispose of the property as he pleased.

The counties occupied by the Danes and Norwegians, viz., Northumberland, Durham, Westmoreland, Cumberland, and Lancashire, are not mentioned in Domesday-Book. In the other fifteen counties to the north and east of Watlinga-Stræt, the "Sochmanni" and "liberi homines" are summed up as follows (see Turner's "History of the Anglo-Saxons"):—

Essex	Sochmanni	343
	liberi homines	306
Suffolk	Sochmanni	1,014
	liberi homines	8,012
Norfolk	Sochmanni	5,521
	liberi homines	4,981
Cambridge	Sochmanni	245
Hertford	"	57
Bedford	"	88
Northampton	"	915
Huntingdon	"	23
Rutland	"	2
Leicester	"	1,716
Derby	"	127
Nottingham	"	1,565
Lincoln	"	11,322
Yorkshire	"	438
Cheshire, drenches	"	54
	Total	36,729

The so-called "freemen" (liberi homines), who, it may be assumed, most resembled our freeholders, seem from this to have been principally confined to Essex (306) and the ancient East Anglia, or Norfolk and Suffolk (together, 12,993). "Sochmanni" were also very numerous in these three counties (together, 6878); yet they appear in the greatest numbers in the old Danish Lincolnshire, which alone had 11,322. In the other districts round the Danish five burghs, they were also pretty numerous: in Leicestershire, 1716; and in Nottinghamshire, 1565. The number of these independent landowners was consequently greatest in the districts earliest occupied by the Danes, where they naturally sprung up from the Danish chiefs' parcelling out the soil to their victorious warriors. That the large county of York had not more than about 440 *Sochmanni* can hardly be used by way of counter-proof; partly because Yorkshire had been terribly exhausted in the wars of William the Conqueror, which took place before Domesday-Book was compiled; and partly because it is clear that Yorkshire is not so fully described in that document as the more southern counties. Lastly, it is remarkable that extremely few serfs are mentioned in the districts north-east of Watlinga-Stræt, in comparison of the many that are recorded in the south and south-west of England.

English authors admit that the Danish settlers in England bestowed a great benefit on the country, in a political point of view, by the introduction of a numerous class of independent peasantry, who formed a striking contrast to the oppressed Anglo-Saxon commonalty. ("The Danes seem to have planted in the colonies they occupied a numerous race of freemen, and their counties seem to have been well peopled."—Turner.) But unfortunately the number of Danish-Norwegian freeholders and freemen at that time in England cannot now be given more closely than by the above sum of 36,729, which is evidently too low, and in every respect highly inaccurate.

It is, however, large enough to strengthen and throw light upon the statements of the chronicles, that the descendants of the Danes and Norwegians in the country to the north-east of Watlinga-Stræt, especially distinguished themselves by a lively feeling of freedom and independence. From the time of their very first settlement, they desperately resisted every chief who attempted to deprive them of their rights as free and independent men. It was, indeed, but reasonable that they should, with persevering boldness, defend in a foreign land that freedom for the sake of which they had abandoned their Scandinavian homes. Their severest and most perilous struggle for liberty naturally took place after the destruction of the Danish power under Hardicanute (1042): although the extensive Danish tract north of the Humber still retained its Danish jarl, Siward.

But on Siward's death (1055), his son, Valthjof (Waltheof), was too young to govern that important district, which was therefore made over to Toste Godvinsön, who afterwards fell at Stamford Bridge. Toste ruled with despotic power, set aside the laws of Canute the Great, and levied taxes which were contrary to the people's ancient rights. The Northumbrians therefore deposed him at a *Thing*, and expelled him in 1064. When Toste's brother, Harald, afterwards endeavoured to effect a reconciliation, on the condition that Toste should be reinstated in the earldom, the Northumbrians unanimously rejected the proposal. "We were born and bred up in freedom," they exclaimed; "a proud and ambitious chief we will not endure, for we have learnt from our fathers either to live like freemen or to die."

When, two years afterwards, William began to conquer England, and to parcel it out among his warriors, it was chiefly the inhabitants of the old Danish districts who opposed him with all the energy of despair. The successors of the Danes and Norwegians, under ordinary circumstances, would have joined their kinsmen the Normans; especially as they gave out that one of their objects in coming to England was to avenge their Danish and Norwegian relatives, secretly massacred by Ethelred. But the Normans aimed at nothing less than the abolition of the free tenure of estates, and the complete establishment of a feudal constitution; a mode of proceeding which, by depriving the previously independent man of his right to house and land, and transferring it to powerful nobles, shook the very foundation of freedom. The descendants of the Danes turned from them, therefore, with disgust, and now no longer hesitated to enter into an alliance with the equally oppressed Anglo-Saxons; for the common danger made both races forget their ancient animosities. Many of the Anglo-Saxon chiefs and warriors who had been defeated by William in the west and south-west of England, fled towards the north, and prepared, in conjunction with the inhabitants of that district, to venture everything in self-defence.

It was not till the year 1068 that the Normans succeeded, after a severe contest, in taking Oxford, Warwick, and the old Danish burghs Leicester, Derby, Nottingham, Lincoln, and York. In these places, but especially in Lincoln and York, the Normans were obliged to build strong fortifications, for fear of the people of Scandinavian descent, who abounded both in the towns and in the adjacent rural districts. But what the Normans chiefly apprehended was, attacks from the Danes who, there was good reason to suppose, might come over with their fleets to the assistance of their countrymen in the north of England.

Meantime, whilst the remains of the united Anglo-Saxon and Danish-Norwegian armies had withdrawn to the mountains of Northumberland, where they often surprised and killed whole detachments of Norman troops, numerous fugitives and messengers repaired to King Svend in Denmark, to implore him, in the name of his English friends, and in that of freedom, to assist them against William the Conqueror. Svend sent his brother Asbjörn, and his sons Harald and Canute, over with a fleet, who, after a vain attempt to land at Sandwich, entered the Humber, in the year 1069. The Northumbrians, and the rest of the aggrieved inhabitants, both Northmen and Anglo-Saxons, flocked gladly together under the Danish banner. Edgar, who had been chosen king by the Anglo-Saxons, Valthjof (Waltheof), a son of the old Northumbrian jarl Siward, and many other fugitives, joined the Danish host. York was taken, the Normans put to flight, and their fortifications levelled with the ground. In these encounters Waltheof gained great honour for courage and bravery.

But the joy of victory was only of short duration. William, who had sworn in his anger to lay all Northumberland waste, knew how to avert by persuasion, cunning, and bribery, the danger that threatened him from Denmark. The Danish fleet went home in the spring; and William retook York, and extended his dominion in Northumberland; where his progress was marked by slaughter, incendiarism, and rapine. The unfortunate inhabitants fled to the forests and morasses; their last place of refuge was the marshes near the Wash. Moved by the cries of complaint which continually reached him from England, the Danish king Svend again sent a number of vessels, which appeared in the Humber in the year 1074. But these were not able to render any effectual assistance. Waltheof, whom William, in order to conciliate the Northumbrians, had appointed Jarl in his father's earldom, fell under the axe of the executioner on suspicion of being concerned in this naval expedition; and fresh devastations promoted William's dominion over Northumberland, which was so terribly harassed that large districts were left without houses or human inhabitants.

The forests of the north of England now became the last refuge of numberless outlaws, who would not submit to the ferocious conqueror, preferring a free and merry life in the green woods; where they united together, and defied William's powerful armies and severe laws. They had secret connections among the people, who saw in them the last defenders of their ancient freedom. Among the leaders of these outlaws, who, long after William's time, continued to wander about in the English forests, but who were most numerous in the north of England, we meet with Scandinavian names, such as Sweyn, and Sihtrik; and in the legends and songs which have preserved the remembrance of them, are found Scandinavian traits of character, such as the story of William of Cloudesley, who shot the apple from his son's head. It is the identical legend related in our old Sagas of the Scandinavian hero, Palnatoke.

The last gleam of any well-founded hope of deliverance shone upon the successors of the Anglo-Saxons and Danish-Norwegians in the north of England, when, in the year 1085, the Danish king Canute, afterwards called the Saint, assembled a powerful fleet in the Liimfjord, in order to release England from the Conqueror's yoke, and if possible to seat himself on the throne. Sixty Norwegian vessels had joined Canute's fleet. William, on his side, made great preparations in order to resist the expected attack. Danegelt was again collected for the defence of the kingdom against the Danes. The inhabitants of Scandinavian descent in the north of England were compelled to alter their dress, and to cut off their long beards, that the Danes might not thereby recognise their kinsmen. The coasts were occupied by soldiers, who erected strong defences; whilst William at the same time endeavoured, by means of secret envoys and bribery, to sow disunion in the Danish fleet. Canute's progress was impeded by unfortunate circumstances; the fleet separated, and a mutiny broke out, which ended in the murder of Canute at Odensee, in the year 1086. No further attempt was made by Denmark to conquer England; for the expedition said to have been prepared by King Erik Lam in the year 1138 was, at all events, a very poor and unsuccessful one. Thus the Northmen in England, being no longer able to obtain support from Denmark or Norway, were forced to submit to the Norman dominion.

Nevertheless, in spite of the terrible devastations by which William coerced the north of England, "the half-Saxon half-Danish population of these districts" (says the French historian, Thierry) "long continued to preserve their old feeling of independence and their ancient indomitable pride. The Norman kings who succeeded the Conqueror dwelt with perfect safety in the southern districts, but did not venture north of the Humber without some fear; and a chronicler, who lived at the close of the twelfth century, assures us that they never visited that part of the kingdom without being accompanied by a strong army."

Although no very great number of Northmen, or men of Scandinavian extraction, could have remained in Normandy after William's conquest of England, and after the Norman expeditions into Italy, yet even these few, as we have before stated, were subsequently able to impart to the popular spirit in Normandy a peculiar Scandinavian colouring. The Norman knights distinguished themselves from the effeminate, dreaming, and excitable knights of the south of France, not only by a greater inclination for adventures and a bolder martial spirit, but also by a genuine Scandinavian sedateness and an all-subduing perseverance. The old Scandinavian feeling of freedom revealed itself, even in the middle ages, in the cities of Normandy, which were long the seats of a democratic spirit and of republican movements. According to William the Conqueror's own statement, the ancient Normans, and, above all, their Scandinavian forefathers, were, in a high degree, quarrelsome and litigious; and, even to this day, Normandy is remarkable, above all other provinces of France, for the great number of law-suits which annually take place in it. Frenchmen themselves have remarked that their most skilful and persevering seamen are to be found among the inhabitants of Dieppe, and that the most celebrated admirals of France have been natives of Normandy.

If such was the influence of the Normans in France, were not the Danes and Norwegians, who had been settled for centuries in England, in a still better position to fix a lasting stamp upon the life and character of the people; more particularly as the Danish-Norwegian elements continued, long after the Norman conquest, to exercise a very considerable influence in England? We may truly assert that the Scandinavian spirit is still clearly to be discerned, not merely in separate districts, but throughout England. The love of the English for bold adventures, especially at sea, their unshaken calmness in the greatest dangers, their apparent coolness during the most violent emotions, and their proud feeling of freedom, are surely not to be ascribed exclusively to the Normans. These qualities must, in a great degree, be attributed to the English, as the descendants of those Danish and Norwegian warriors who sought dangers on unknown seas; who looked death steadily in the face, come in whatever shape it might; who gloried in the feeling that their countenances should not betray the passions which fermented in their breasts; and who prized liberty far more than life.

It deserves at least to be mentioned, as affording a remarkable analogy to Normandy, that England's most celebrated and successful admiral, Nelson, bore a genuine Scandinavian name (Nielsen, with the characteristic Scandinavian termination of *son*, or *sön*). He was, besides, a native of one of the districts early colonized by the Danes, having been born in the town of Burnham-thorpe, in Norfolk, or East Anglia. In fact, the perceptible difference of character still actually found between the people in old Saxon South England and in the more northern old Danish districts, is very remarkable. The southern Englishman is softer and more compliant. The northern Englishman is of a firmness of character, bordering on the hard and severe, and possesses an unusually strong feeling of freedom. The Yorkshireman is well known in England as a hasty and touchy, but determined and independent, character. Great political movements have therefore not only found reception and encouragement among the population of the north of England; but this population, from the interest it takes in the progress of public affairs, and from its love of freedom, has played a leading part in the great internal revolutions which mark the recent political history of England. Public men regard it as a great honour to represent the northern districts of England in Parliament (for instance, the West Riding of Yorkshire), merely from the intelligent political character of the voters; and it is certainly through the adherence of the lovers of freedom in the north, that Cobden has been able to struggle so successfully for the promotion of free trade, for financial reform, and for similar liberal measures. That this spirit of liberty in the north of England is chiefly derived from the old Scandinavian colonists is by no means merely the partial assertion of a Dane. The celebrated English writer, Sir E. Bulwer Lytton, who, in his "Harold," has successfully begun to awaken the attention of his countrymen to a juster view of the Danish conquest, says in a note appended to that work: "It might be easy to show, were this the place, that though the Anglo-Saxons never lost their love of liberty, yet that the victories which gradually regained liberty from the gripe of the Anglo-Norman kings were achieved by the Anglo-Norman aristocracy. And even to this day, the few rare descendants of that race (whatever their political faction) will generally exhibit that impatience of despotic influence, and that disdain of corruption, which characterize the homely bonders of Norway, in whom we may still recognise the sturdy likeness of our fathers; while it is also remarkable that the modern inhabitants of those portions of the kingdom originally peopled by the Danes, are, irrespectively of mere party divisions, noted for their intolerance of all oppression, and their resolute independence of character; to wit, Yorkshire, Norfolk, Cumberland, and large districts in the Scottish lowlands."

It would be impossible to deny that the Danes and Norwegians settled in England before the arrival of the Normans not only essentially contributed to the preservation of popular liberty—which, through the weakness and effeminacy of the Anglo-Saxons, was threatened with destruction—but that they also laid the foundation of its further development, and powerfully contributed to its complete establishment. We need, therefore, be no longer surprised that memorials of the Danes are mixed up with England's freest and most liberal institutions; and that to the present day, for instance, the place whence the candidates for a seat in Parliament address the electors, bears, throughout England, the pure Danish name "*husting*."

SECTION XIV

General View.—Anglo-Saxon and Danish-Norman England.—Sympathies

for Denmark.—The Dane in England.

The various kinds of Danish and Danish-Norwegian memorials which I have alluded to, such as names of places, coins, and peculiarities of language (not to mention contemporary letters-patent and laws), afford so many incontrovertible proofs that the Danish influence in England was neither of short duration, nor, on the whole, of a transient nature. Future and more successful investigations and comparisons, more particularly in England itself, will undoubtedly much extend the circle of known Danish memorials existing there. So much, however, is already placed beyond all doubt, that in no country out of the present homes of the Scandinavian race have its colonists left such various, such considerable, and such clear traces of their existence, as the Danes, especially, have left in England. The Scandinavian spirit has not ruled with so much power in any other, still less in any greater, European kingdom; nor been able to retain so powerful a dominion for such a length of time.

The Danes, and their successors the Normans, did not content themselves with the temporary overthrow of the Anglo-Saxon dominion; they annihilated it for ever. In this the Danes may be said to have been more active than the Normans. They not only gradually settled themselves under their own laws and their own chiefs, in half of England, but spread themselves over the whole of it. In the time of Alfred the Great, they once held all England in subjection; and at an early period obtained places amongst the highest ecclesiastical and secular aristocracy of the country. In the tenth century, the Anglo-Saxon king Edgar favoured the Danes so much, that during his reign the Danish power had an opportunity to consolidate and extend itself. Even the Anglo-Saxon royal family became mixed with Danish blood. Among the Anglo-Saxons, both high and low, weakness and proneness to vice went on continually increasing; whilst the Danish dominion, prepared by two centuries of independent Viking expeditions, and by the subsequent settlements of the Northmen, established itself completely, as soon as the sea kings and wandering Vikings were succeeded by Danish monarchs with considerable fleets at their command.

All England yielded to the conqueror Canute, and under his wise, powerful, and just administration, enjoyed that tranquillity and happiness of which it had long felt the want. The Anglo-Saxons and Danes now became more amalgamated. But Canute's sons wanted their father's ability and strength of purpose. The old dissensions and quarrels broke out afresh; whilst violent internal disturbances in the newly Christianized Scandinavian North, where the Viking spirit became extinguished, deprived the Danes in England of the succour necessary in their contests with the natives. The Danish power in England fell, but left the population completely mixed and saturated with Danish elements. The Anglo-Saxon royal race, as it was called, was now half Danish. The higher clergy and nobility were connected by the closest ties of relationship with the Danes and their chiefs, in whose hands several of the most important fiefs remained. The Danes had acquired considerable influence in many of the largest cities; and in about half of England the majority of the population was of Danish extraction, and possessed Danish laws and other Danish characteristics. The Danes who, naturally enough, could not forget that they had been absolute masters in that conquered land, obeyed unwillingly a king of another race, though they had not the power to place one of their own race upon the throne. The unmixed Saxon population, on the other hand, could not endure that the royal sceptre should continue to be borne, in the once independent country of their forefathers, by foreign conquerors from Denmark, whose power, besides, seemed at that time on the wane. Inward dissensions increased; the kings were too feeble to maintain efficiently their difficult position; and the power falling more and more out of the hands of the degenerate Anglo-Saxons, passed over to the stronger Danes and their Norman kinsmen.

With an unmixed population, England would have been able to maintain herself united and powerful in the hour of danger, and when threatened by foreign conquerors. But split and divided as she now was among different races contending for the mastery, real unanimity was impossible; and, in case of a powerful attack from without, dissolution was inevitable. Through the Danish expeditions, the Danish colonizations, and finally through the fall of the Danish supremacy, it became practicable for William of Normandy to conquer England with an army of only 60,000 men. Had not those events prepared the way, it would be inconceivable that with such a force a foreign conqueror should have been able to subdue a country so extensive, so well peopled, and so favoured by nature; still less that he should have succeeded in retaining such a conquest for any length of time. William won the battle of Hastings, which decided the fate of England, only because Harald Godvinsön's Anglo-Saxon army entered the field weakened and exhausted by the sanguinary battle of Stamford Bridge. This was fought against the Norwegian king, Harald Haardraade, and the discontented Scandinavians in the north of England, who wanted to re-establish a king of their own race on the English throne.

The Danish-Norwegian settlements, and the Danish dominion in England, by subduing for a time the political power of the Anglo-Saxons, had not only prepared the way for the first victory of the Normans, but also for the future progress and establishment of the Norman power in England, and especially for the ultimate triumph of the Norman popular spirit over the remains of the ancient Saxon nationality. The Danes, by expelling the Anglo-Saxons from the northern and eastern parts of England, as well as by mixing with them in the south, had by degrees undermined their national independence and their popular characteristics, and had thus prepared an entrance for the Scandinavian spirit, which was so nearly allied to the Norman, into a great, if not the greater, portion of the English population. The bold and chivalrous spirit of the Norman aristocracy, their love of daring adventures, and their lofty feeling of freedom, completely agreed with the characteristics of the Scandinavians settled in England at an earlier period. The Normans found among the Scandinavian population of England, and particularly the Danish portion of it, several of those free institutions already in full force which they themselves, with much advantage to liberty, afterwards extended to the whole country.

Thus the conquest of England by Danish Normans, undoubtedly prepared, or, more properly speaking, was the indispensable and necessary foundation of the subsequent French-Norman conquest; and it may therefore be justly called the first act of that great historical drama, "The Norman Conquest," of which William of Normandy's conquest is only the concluding act.

But many will undoubtedly ask, was the Norman conquest, on the whole, beneficial to England? Would it not have been better had the Anglo-Saxon nationality been permitted to develope itself, instead of being arrested by such violent devastations and by such bloodshed as the Danish-Norman expeditions occasioned? And is it not a proof of the nobleness of the Anglo-Saxon nationality, that it has since prevailed so preponderantly in England?

On this point let us hear a learned and impartial Englishman. The latest and most celebrated Anglo-Saxon historian, Mr. Kemble, says, in his preface to the before-mentioned Collection of Anglo-Saxon Diplomas:—"With the close of the fourth volume of this work we arrive at the reign of Harald, and the Norman conquest of England; an event which our contemporary forefathers could only regard as deplorable, but which we must look back upon with gratitude and pride, as the remote origin of our own peculiar character and power. It is hardly possible to compare the signatures to the charters contained respectively in this and in the previous volumes, without seeing how widely a foreign element had become predominant. The Scandinavians of Ingwar, Guðorm, Swegen, and Cnut, successively prepared the way for the descendants of other Scandinavians under William; and the Saxon national character, like the national dynasty, was too weak to offer a successful resistance. Defeated, yet still holding a portion of its domain with unabated perseverance, yielding somewhat in one place, to break out with unshaken obstinacy at another, it accommodated itself partially to the peculiar habits of each successive invader; till, after the closing scene of the great drama commenced at Hastings, it ceased to exist as a national character, and the beaten, ruined, and demoralized Anglo-Saxon, found himself launched in a new career of honour, and rising into all the might and dignity of an Englishman. Let us reflect that defeats upon the Thames and Avon were probably necessary preliminaries to victories upon the Sutlej."

The weakness and degeneracy of the Anglo-Saxon national character contained the seeds of its decay. It has long since been agreed that, in an historical view, we ought not to complain that the degenerate, though highly-civilized, Romans in Britain were compelled to make way for the rude Anglo-Saxons, since the latter brought with them the germ of a new and higher development. In like manner we can hardly regret that the degenerate, but to a certain degree civilized, Anglo-Saxons, were in turn expelled by the more powerful, but ruder Danes; since these also were to prepare, and lay the foundation of a new and more flourishing state of society. Under the reign of Ethelred the Second, the supremacy of the Anglo-Saxons had already passed away. As a people, they sank entirely, and left only a part of their civilization and of their institutions to their successors in dominion, the Danes and Normans. The transition took place amidst the same shocks and the same bloodshed which still mark every important and radical revolution in the history of nations. The Danish-Norman, or perhaps more properly, the Scandinavian national character, usurped the place of the Anglo-Saxon. It was certainly built upon the foundation laid by the Anglo-Saxons, but it must be observed that it has made greater progress in all respects. To it especially is owing the development in England of a maritime skill before unknown, of a bold and manly spirit of enterprise, and of a political liberty, which, by preserving a balance between the freedom of the nobles and of the rest of the people, has long ensured to England a powerful and comparatively peaceful and fortunate existence.

The Englishman is justly proud of his native land, of its internal freedom, and external greatness. But when he extols his country in respect only of its being "Anglo-Saxon," or praises the merits of the Anglo-Saxons and Norman-French, whilst he unconditionally condemns the Danish expeditions and settlements, as having been merely devastating and destructive, he commits both an historical error and an evident injustice. The Anglo-Saxons performed their share in the civilization of England, and the Norman-French did still more; but it ought not to be forgotten—and least of all by Englishmen, who are so nearly related to the Danes—that the latter also very essentially contributed to win freedom and greatness for England, and that this freedom, and this greatness, are in no slight degree sealed with Danish blood. From at least the Danish-Normanic conquest (about the year 1000), the Danish-Normanic, or Scandinavian, national character has been the prevailing and leading one in England's history, and so it certainly continues to be at the present day.

A perceptible and very remarkable evidence of this is the sympathy which the English people in general feel for the North, the ancient home of their fathers, and particularly for Denmark. The Englishman himself will generally aver, with a sort of pride, that he derives his descent from the North. A Dane travelling in England will everywhere find an unusually cordial reception. He will in general be regarded more as a countryman than as a foreigner, merely because he is a Dane. He will discover that the English, instead of having forgotten their kinsmen beyond the sea, with whom they were formerly united, feel themselves attracted to them by the ties of blood and friendship. He will continually hear complaints of the deplorable attitude which the policy of England assumed with regard to Denmark at the commencement of the present century; and he will adopt the conviction that in this mistaken policy, the people themselves, at least, were not to blame. He will at times be induced to forget that he is at a distance from his native land and from his nearest relatives; for the highly-striking agreement between the character of the English and that of their Scandinavian kinsmen causes a Dane to imagine that he is still among his own friends, in the home which he has long since left. It was certainly also something more than mere accident that, during the last war in Denmark, the Danish cause nowhere, out of the North itself, awakened such general sympathy among the people, nor found so many bold champions, both in speeches and publications, as in England. May we not in these facts trace the effects of near relationship, and perceive the ties of blood?

It should not pass altogether unnoticed that the sympathies of the English for Denmark, and their fraternal feeling towards the Danish people, have increased in proportion as they have been obliged to acknowledge that the Danes of modern times still know how to defend their independence, liberty, and honour, with the bravery inherited from their forefathers. Not to speak of the last contest, so glorious for Denmark, it is particularly the battle in Copenhagen Roads, the 2nd of April, 1801, which has maintained in England the ancient fame of Danish valour. The English regard this action not only as one of Nelson's greatest triumphs, but as one of their most glorious naval battles, particularly on account of the sturdy resistance which they encountered. On Nelson's monument in Westminster Abbey, on which his most glorious battles are recorded, that of Copenhagen is named first. Nelson himself describes the action as the bloodiest and most desperate he had ever beheld. That he is correct in this respect, and that he has not extolled the bravery of our nation merely to enhance his own, we Danes, at least, cannot doubt, since we cannot even admit that the battle must be unconditionally regarded as lost by us.

For the rest, it is remarkable how frequently the English confound the battle in Copenhagen Roads in 1801 with the carrying off of our fleet in 1807, and place these two entirely distinct events under one and the same head. The English historians have endeavoured gradually to conceal the dishonour attaching to the robbery of our fleet in 1807; and this has even been carried to such an extent, that the rising generation but too often reckons that ignominious act amongst Nelson's triumphs. They imagine that the surrender of our fleet was the result of the battle in the Roads; and yet Nelson had fallen two years before, at the battle of Trafalgar, in 1805. Fortunately for his honour, he was thus spared from partaking in the robbery of the fleet of a nearly-related people, with whom England was at peace.

But this is not the only error which the Dane must correct when he hears in England the name of Nelson extolled at the expense of Denmark and of historical truth. Yet he will find it difficult to refute another similar mistake, namely, a firm belief in Nelson's "*complete victory*" in the battle of 1801. It is just as unshaken an article of faith among the British people that Nelson then gained a brilliant victory, as it is an acknowledged certainty, founded on fact, that at all events the battle was neither won by the English nor lost by the Danes. Nay it is certain that almost the whole of Nelson's fleet would have been destroyed, or taken, if the Crown Prince of Denmark—for fear of engaging in a lengthened war with England, and from other purely political reasons, as well as, it must also be observed, at Nelson's own request—had not put a stop to the battle. Curiously enough, in two of the finest poems which the English and Danish people can produce, Campbell's "Battle of the Baltic," and Hertz's "Slaget paa Rheden," the combat is represented in each as honourable to the respective nations.

Not long since, a Dane in England was led into a warm argument respecting the disputed result of this battle; when the master of the house suddenly recollected that an old invalid, who looked after the boats on the canals in the garden, had served under Nelson. He called out to him that "here was a Dane, and that he had certainly seen that sort of folks before." "Yes, master," answered the honest tar, "but on that day the Danes made it much hotter than we liked."

This terminated the dispute. The time, however, in the order of Nature, cannot be far distant when the Dane in England may look in vain for such support from men who were present at the battle. He must then be contented to state his opinion, without the least hope of its carrying any weight; though he can, at all events, console himself with the reflection that, when the conversation turns on the mutual relations between England and Denmark, the latter may point to conquests of a very different, as well as far more important and altogether undisputed kind.

In the long series of brilliant victories, won not only by the Danish sword, but by the Danish national character in England, and which, by the conquest of that country, essentially contributed to found there a greatness and a power before unknown, the Danish people possess memorials so proud and brilliant, that they may be reckoned among the most beautiful ornaments in that glorious wreath which from time immemorial encircles the Danish name. We may safely leave them by the side of the best and most imposing memorials of most other nations.

THE
NORWEGIANS IN SCOTLAND.

SECTION I

Nature of Scotland.—The Highlands and Lowlands.—Population.—Original

Inhabitants.

None of the seas of Europe are so rough and stormy as that which washes its northern and north-western coasts. Even in Jutland the effects of the cold north-west wind which sweeps down from the icy sea between Norway, Iceland, and Scotland, are severely felt. Along its west coast, for a distance of several miles inland, there are no woods, but only low stunted oak bushes, which in many places scarcely rise above the tall heather. Still farther eastward, and even in Funen and Zealand, which the north-west wind does not reach till it has passed over considerable tracts of land, it has such an influence on the woods, that in their western outskirts the trees are bent, and as it were scorched or blighted at the top. The North Sea, whose surges, breaking on the coast of Jutland, are heard even in calm weather far in the interior, rises to a fearful height during a storm. It would long since have washed over Jutland, and perhaps the whole of Denmark, if Nature had not placed sand-banks or shoals along the coast, as a sort of bulwark, against which the highest waves break harmlessly.

The North Sea is, however, an enclosed one, and little more than a bay of the Atlantic. Its swell is not so great, nor its storms so violent, as those of the open sea beyond, towards the north and west; where the Atlantic breaks on one side against Greenland and North America, and on the other against Norway, Scotland, and Ireland. The sand-banks and shoals which form a sufficient defence for Jutland against the North Sea, would there scarcely be able to resist the open and agitated ocean. On the extreme north-western coasts of Europe, the Atlantic has completely washed away the earth and sand; the bare cliffs, which often rise to a considerable height, alone remain, and still defy the fury of the waves. These rocky coasts, with their numerous towering and ragged crags, with their many and deeply-indented fiords, convey an idea of the power and greatness of the sea as striking as it is true. Everywhere outside lie rocky islands, which, like outposts, stop the advancing waves, and only allow them, if with increased speed, yet with diminished power, to approach the land through narrow channels, or sounds. During violent storms some of the islands are flooded by the sea, which, as it rolls forwards, strives to overtop the cliffs; whence it glides back, again to repeat the same vain attempt. The firm, rocky, isle-bound coasts of Norway, Scotland, and Ireland, are evidently for Europe what the sand-banks and shoals of Jutland are for Denmark.

It is natural, therefore, that those countries which in the north-westernmost part of Europe lie farthest out towards the Atlantic Ocean—such as the Scandinavian Peninsula, Scotland, Ireland, and part of England—should have their highest and wildest mountains and cliffs towards the west, and in the neighbourhood of the sea. This is more clearly seen the farther we proceed northwards: namely, in the Scandinavian Peninsula and in Scotland.

In Norway the rocks often rise almost perpendicularly out of the sea. In the neighbourhood of the coast they reach a considerable height, and then sink gradually towards the east, until they lose themselves in the broad and comparatively low valleys of Sweden. Whole rows of islands lie scattered along the west coast of Norway, round which the sea often whirls in impetuous eddies. On the coast itself, where the land is most exposed to the bleak sea winds, such extensive forests are not to be seen as in the interior of the country; nor do any fertilizing streams wind their way through the short and narrow valleys. It is only here and there that the water from the rocky springs or melted snows, leaps, after a short course, over the edge of the cliff into the open sea, or into the deep fiords with which the coasts are everywhere indented. The greatest rivers in Norway take a more eastern course, and often make their way from the Norwegian highlands through the richly-wooded lowlands of Sweden to the Baltic. In Sweden the coasts are neither so steep nor so indented as in Norway. The waves of the enclosed and comparatively quiet Baltic do not require to be resisted like those of the Atlantic Ocean.

Very similar features are found in Scotland. The whole of the northern and western coast lying towards the Atlantic is wild and rocky, with numerous islands, deep firths, and steep shores; behind which, rock towers upon rock, as if to form an impenetrable barrier against the sea. The country is almost without forests, the streams and the valleys are of small extent, and fertility consequently very limited. But by degrees the rocks sink down towards the south-east and east, till they terminate in the broad, well-watered, and fertile coast districts along the North Sea; which, on account of their inconsiderable elevation, are called the Lowlands of Scotland. Thus the Highlands answer very nearly to Norway, and the Lowlands to Sweden. But as the Scandinavian Peninsula is larger than Scotland, so also are its natural features on the whole on a grander scale. The rocks of Norway are mountains of primitive granite, which in some places rise to a height of 8000 feet, and of which large ranges are covered with eternal snow and ice. Scotland, on the contrary, has transition rocks, whose highest peak, Ben Nevis, which is only somewhat more than 4300 feet above the sea, is not even always covered with snow. Nor can the Scottish Lowlands be compared as to extent to the Swedish valleys, with their immense forests and their large rivers and lakes. Nevertheless the natural features of Scotland are in their way no less beautiful than those of the Scandinavian Peninsula. The sea, which indents the coasts on all sides; the well-cultivated, and partly also well-wooded plains, which, particularly towards the mountain districts, undulate in hill and dale; and lastly the Highland itself, with its many streams, waterfalls, firths, and lakes, afford the richest and most magnificent variety. To these features may be added a milder climate, and in the Lowlands a far richer fertility, than in Norway and Sweden; which have considerably contributed to give the landscapes of Scotland, even in the wildest districts of the Highlands, a somewhat softer tinge than is found in the high Scandinavian North.

A very marked difference exists between the Scottish Highlands and Lowlands, not only with regard to the nature of the country, but also to the original descent and the characteristics of the present population. The Lowlands, which are the seat of a highly-developed agricultural, domestic, and manufacturing industry, are inhabited by a strong and laborious people, speaking a peculiar dialect of the English language, and descended partly from the Celtic Scots, but more particularly from immigrant Anglo-Saxons, Danes, Norwegians,

Normans, and Flemings. Commerce and trade, carried on by means of canals, railways, steamships, and similar easy means of communication, thrive vigorously in large and wealthy cities.

The Highlands, on the contrary, which only a century ago were almost inaccessible from the land side, have scarcely a large town. Rocks and heaths are found instead of the fruitful fields of the Lowlands. With the exception of a few districts farthest towards the north-east, where the soil is more fertile, there are only seen in the valleys, along the firths, and by the sea, small fields of barley and oats, which would not yield the most scanty subsistence to the poor inhabitants if the rocks did not afford pasture for cattle and numerous flocks of sheep; and if the sea, the firths, which abound with fish, as well as the rivers and lakes, did not contribute some part of their riches. The hardy Highland Scots, a great part of whom do not understand, or at all events do not speak English, but still commonly use the Celtic or Gaelic tongue, live here thinly scattered in poor and low peat cabins, which it is often difficult to distinguish from the surrounding rocks. The Highlanders in the districts farthest towards the west and north have preserved their language and other national characteristics purest; for farther towards the Lowlands, a more modern civilization has gradually forced its way forwards, in spite of the mountains. The old warlike dress which formerly distinguished the Highlander, particularly so long as clanship was in full vigour, has, since the annihilation of that system, become every day more rare. The kilt, or short skirt, has almost entirely given place to more modern clothing; the tartan plaid alone is still seen wrapped in the old fashion round the shoulders of the Highlander.

In our days the various tribes of the Highland and Lowland populations live in peaceful union under one and the same government. But during several centuries Scotland was the theatre of the most sanguinary contests between the Celtic Highlanders and the Teutonic Lowlanders. The former, who were animated with an inveterate hatred of the Lowlanders, continually made hostile incursions into the Lowlands, and, after burning and ravaging the country, retired with cattle and other booty to their mountains, whither they knew well the Lowlanders durst not follow them. The exasperation and hatred of the Highlanders were not entirely without foundation. In ancient times they had been sole masters in Scotland, from the Cheviot Hills to the Orkneys and the Shetland Isles, and from the Atlantic Ocean to the North Sea; and they had retained this mastery even long after their kinsmen, the Britons in England, had been compelled to yield to the Romans and Anglo-Saxons.

The celebrated Roman commander, Agricola, had, it is true, in the first century after the birth of Christ, made his way so far into the Lowlands that, as a defence against the Highlanders—the much-dreaded Caledonians, or Picts—he constructed a wall with a deep ditch before it, from the Firth of Forth to that of Clyde, in the low tract through which the Glasgow Canal has since been conducted. The Romans even extended their conquests farther northwards, as far as Burghead on the Moray Firth, to which place they formed regular high roads. But they were not able to defend themselves against the persevering attacks of the Caledonians, or Picts, and were soon obliged to retreat to the south of the Cheviot Hills; where the great wall, with its many towers and deep ditches, which they had built from the Solway Firth to the River Tyne, became their chief defence against the harassing inroads of the Highland warriors. But this wall also was surmounted by the Picts, whose courage and daring increased in proportion as the power of the Romans, both at home and abroad, was rapidly waning. At last the Picts destroyed the wall, and after the fall of the Roman dominion, made incursions into England, where neither the descendants of the Romans, nor the Britons, found any means to repel them. It was not till the Anglo-Saxon conquest of England that the Picts were again compelled to fly towards the north over the Cheviot Hills, where they found sufficient employment in defending their own homes.

For, whilst they were spreading themselves over the rich plains of the north of England, a foreign, though nearly related, Celtic people, the Scots from Ireland, had taken possession of their south-western frontier districts. Hence they spread themselves to such a degree over the Lowlands that both these and the Highlands, though the latter were almost entirely independent of the Scottish sovereigns, were called by one name, Scotland. After many battles the older Pictish inhabitants were, about the year 900, entirely amalgamated with the Scots in the Lowlands. Meanwhile a storm had gathered which threatened no less danger to the Scots in the Lowlands, than to their kinsmen, the Picts, in the Highlands. The dominion of the Celts, which had long before ceased in other and more accessible lands, was no longer to find a sure place of refuge even in Scotland, though its coasts were protected by the stormy Northern Sea, and its interior filled with rocks and warlike men.

SECTION II

The Anglo-Saxons.—The Danes and Norwegians.—Effects of their Expeditions.

The same want of unity and the same internal disputes which had brought ruin on the Celts in other places, prepared the way for foreign conquerors in Scotland. An indomitable fate decreed that the newer and higher civilization of Christianity should here, as in the rest of Europe, be founded and promoted by a Teutonic people. But though the Anglo-Saxons had conquered almost all England, they were not able, by their own power, to subdue the Celts in Scotland. The Anglo-Saxon kings undertook, indeed, several expeditions against that country, in which they were at times pretty successful; but they were not able to hold steady possession even of the Lowlands. Subsequently, however, the Anglo-Saxons wandered by degrees, and in a more peaceful manner, from the northernmost parts of England over the Scottish border, and established themselves both in the towns and in the rural districts. The number of these emigrants appears to have increased very considerably after the conquests of the Danes and Norwegians in the midland and northern districts of England in the ninth and tenth centuries, when a great part of the Anglo-Saxons were driven from their old dwellings, and obliged to fly towards the north. Saxon institutions may even have been introduced into the Lowlands in the tenth and eleventh centuries, after an expedition of the Anglo-Saxon king Edgar. But the rocky highlands of the interior constantly defied all conquest; and the northern and western coasts, together with the surrounding islands, could be subdued only by considerable fleets, which the Anglo-Saxons did not possess.

But what in this respect the Anglo-Saxons were obliged to leave undone, was for the most part accomplished by the warlike and shrewd men of the Scandinavian North, who were then masters of the sea. Even from the oldest times, connections, both of a warlike and peaceful nature, had existed between Scotland and the opposite shores of Scandinavia. The old Sagas, for instance, bear witness that the

Danish king Frode's daughter, Ulfhilde, was married to "the founder of the Scottish kingdom;" and that the Danish prince Amleth (Hamlet) married the Scotch queen, Hermuntrude. From Denmark, moreover, and particularly from Jutland, many colonists afterwards emigrated to the Scotch Lowlands, whose coasts were, besides, plundered by the Danish Vikings.

The Danish colonists, even in the north of England, were much mixed with Norwegians, and this was still more the case in the Scottish Lowlands. The more north the districts lay, the farther were they removed from Denmark, and the nearer did they approach Norway; whilst the features of the country much more resembled the Norwegian fiords, valleys, and rocks. Whilst, therefore, the Scandinavian colonists in the Lowlands were of Norwegian-Danish descent, the Highlands and islands farthest towards the north and west, were conquered, and in part peopled, by Norwegians only. This happened about the same time as the Danish conquests and settlements in England. The Norwegians founded kingdoms on the northern and western coasts of Scotland, which existed for centuries after the destruction of the Danish power in England. They introduced their own manners, customs, and laws, and gave Norwegian names to the places colonized by them. They appear not unfrequently to have married native Celts; at least it is often stated that Norwegian chiefs married daughters of the Celtic, or Pictish, and Scotch aristocracy, whose pure nationality and power were thus gradually broken down. The unfortunate Celts were now in a painful position. The Celtic Scots in the Lowlands were pressed upon by the Anglo-Saxons and Northmen, whilst the Pictish Highlanders were assailed both from the Lowlands and from the Norwegian kingdoms in the west and north. The most essential result of the Norwegian conquests and settlements in the Scotch Highlands was, that the Northmen, in conjunction with the Norwegian-Danish colonists in the Lowlands, and with the Anglo-Saxons who dwelt there, overthrew the Celtic dominion, and, like the Danes in England, prepared the way for the eventual triumph of the Norman spirit and Norman institutions. In the Lowlands this took place in the twelfth century, but much later in the Highlands and surrounding islands.

As a close union was thus effected between the long-separated Highlands and Lowlands, and a higher and more widely-diffused civilization introduced among the people in both, it may justly be asserted that the Norwegian conquests in the Highlands, and the Norwegian-Danish settlements in the Lowlands, were particularly fortunate for Scotland. It must always, indeed, be a subject of regret that so brave, and in many respects so noble, a people as the Caledonians and their descendants, should be exterminated. Who can observe without a feeling of sadness how the last feeble remnants of Scotland's ancient masters, after having been expelled from the glorious Lowlands, cannot even now find rest among the barren rocks, and in the few arable valleys of the Highlands, but are obliged, year after year, in increasing numbers, to seek another home farther west, in the new world beyond the Atlantic? But, viewing the matter as it regards enlightenment and civilization, no charge can be reasonably brought against the Norwegians or Northmen, for having co-operated in Scotland to expel a people whose brethern and kinsmen had in every country which they occupied shown themselves incapable of adopting the new and milder manners of Christianity; and who, once before subdued by the Romans, had been compelled to yield to the fresher and more powerful Teutonic tribes of the Franks and Anglo-Saxon.

No small portion of the present population of Scotland, both in the Lowlands and on the remotest coasts and isles of the Highlands, is undoubtedly descended from the Northmen, and particularly from the Norwegians. Both the Norwegians and Danes, wherever they established themselves, introduced their Scandinavian customs, and preserved, in all circumstances, the fundamental traits of their national character. It becomes, therefore, probable that the Norwegian settlers in Scotland must, in certain districts at least, have exercised a vast influence on the development of the more modern life of the Scotch people, and on their national character. This is indeed actually and visibly the case. Yet, although the Norwegian kingdoms on the coasts of Scotland subsisted long after the downfall of the Danish power in England, still the effects of the Norwegian conquests in Scotland were far from being so great, or so universally felt there, as the results of the Danish conquests were in England. The Norwegian language was completely supplanted in the Hebrides by old Celtic or Gaelic; and on the Shetland Isles, the Orkneys, and the north coast of Scotland, by English. The Norwegian laws and institutions either entirely disappeared in these parts, or were formed anew after quite different models. Not even in the purely Norwegian Orkneys and Shetland Isles, though they remained united with Norway and Denmark until far in the fifteenth century, could the inhabitants maintain the ancient freedom which they had inherited from their forefathers. The free tenure of land, or right of "Udal," was, for the most part, annihilated by the most shameful oppression. Established on many small, poor, and widely-separated islands, the Norwegians in Scotland could neither obtain such influence for their laws and institutions, nor concert so united and powerful a resistance against oppression, as their more fortunate Danish kinsmen in the open, rich, and densely-peopled plains of northern England.

In spite of the acknowledged fact that the Norwegians were the most numerous of all the Scandinavian colonists in Scotland, we constantly hear Norwegian achievements and Norwegian memorials referred to "the Danes." Under this common appellation are also generally included, as in England, Norwegians and Swedes. The causes of this must probably be sought in the long dominion of Denmark over Norway, in the brisker and more uninterrupted communication which Scotland maintained with Denmark, in comparison with any other part of the North, and lastly, in the reciprocal marriages between the ancient Scotch and Danish royal families, which in former times contributed, in no small degree, to bind the Scotch and Danish people together. But the preponderance of the Danish name must also be attributed to the pre-eminent power of the Danes in ancient times, and in the early middle ages; and, of course, more particularly to that supreme dominion which they had so gloriously won for themselves in the neighbouring country—England.

SECTION III

The Lowlands.—Population.—Language.—Norwegian-Danish

Names of Places.

The boundaries between Scotland and England were anciently very unsettled. After the time of the Romans, the Anglo-Saxon and Danish kings speedily extended their dominion over the Cheviot Hills, and frequently to the Firths of Clyde and Forth; whilst considerable tracts of the north of England, particularly in the north-western districts, were sometimes united with the Scotch Lowlands, or with kingdoms which existed there. Until England and Scotland were at length united under one crown, the north of England was almost uninterruptedly the theatre of the bitterest border warfare. The blood of many thousands of bold warriors has been spilt on that land which now teems with the blessings of wealth and peace.

Part of this old border land, or the most southern part of the present Scotland, from the Cheviot Hills to the narrow neck of land between the Firths of Clyde and Forth,—a tract of about sixty English miles—has not a much more mountainous character than the north of England. The hills undulate in the same gentle forms; and it is only here and there that a single rugged mountain shows its heath-covered or bare and peaked top. Large and well-cultivated plains alternate with charming valleys, which are frequently narrow, and so fertile that in some places creeping plants, bushes, and trees, almost entirely conceal the rivulets that wind through them.

The Highlands extend themselves from the Firth of Clyde to the north-west and north; whilst the Lowlands take a direction from the Firth of Forth along the eastern border of the Highlands, and by the coasts of the North Sea. To the Firth of Tay, and northwards to the Grampian Hills, the Lowlands are not very broad or extensive, whilst the Highland mountains nearly approach the seashore. It is not till we have crossed the Grampian Hills that those large level plains open upon us which comprehend the north-easternmost part of Scotland, particularly the present Aberdeenshire. From these less-wooded plains we turn towards the north-west into the fertile and well-wooded Moray; whence a transition again takes place to the Highlands, which begin in the adjoining shire of Inverness. At this extreme point the Lowlands have, as it were, exhausted all their splendour and abundance. Down towards the coast the land is filled with gently-sloping hills, and intersected by rivers, whose rapid currents remind one of the neighbourhood of the mountains. At a distance from the coast the land rises, the tops of the mountains become barer and sharper, the valleys have a greater depth, and the roaring of the streams over fragments of rock is heard more distinctly. The mountains, as they rise from the Lowlands to the Highlands, afford in a still higher degree than the more southern border mountains, the most enchanting prospects over the coasts and sea. It is with difficulty that the spectator tears himself from the view of the charms of the Lowlands, to bury himself in the dark mountains that rise so solemn and menacing before him.

Throughout the Lowlands, the people, both in personal appearance and character, very much resemble the inhabitants of the north of England. This is particularly the case with the inhabitants of the southern borders, between the Cheviot Hills and the Firths of Clyde and Forth. The same light-coloured hair and the same frame of body, which, in the north of England, remind us of the people's descent from the Scandinavians, indicate here also considerable immigrations of that people into the southern part of Scotland, and thence farther up along the east coast. According to a very common saying here, even the language of the Lowlands is so much like that of Scandinavia, that Lowland seamen wrecked on the coasts of Jutland and Norway have been able to converse without difficulty in their mother tongue with the common people there. This is undoubtedly a great exaggeration; but this much is certain, that the popular language in the Lowlands contains a still greater number of Scandinavian words and phrases than even the dialect of the north of England. We must not unhesitatingly believe that the Saxon language did not extend itself from the north of England to the Scotch Lowlands till after it had been mixed with Danish; although the remote situation of the latter, so high towards the north, was certainly far more adapted to preserve the old Danish forms of words than that of north England, which was more exposed to the operation of newer fashions. But the Danish or Scandinavian elements in the popular language of the Lowlands are too considerable to admit of such a supposition, not to speak of the Scandinavian appearance of the inhabitants. These necessarily indicate Scandinavian immigrations; and, to judge from the present popular language, we might be easily tempted to believe that a far greater number of Northmen had settled in the Scottish Lowlands than in the middle and northern districts of England. We might, consequently, also expect to meet with a proportionately greater number of Scandinavian names of places in the Lowlands than in England.

But this is very far from being the case. Extremely few places with Scandinavian names are to be found in the Scotch Lowlands; and even those few are confined, almost without exception, to the old border land between the Cheviot Hills and the Firths of Clyde and Forth, and to the counties nearest the English border. Dumfriesshire, lying directly north of Cumberland and the Solway Firth, forms the central point of such places. Northumberland and Durham, the two north-easternmost counties of England, contain but a scanty number of them; and consequently must have possessed, in early times at least, no very numerous Scandinavian population. Cumberland, on the contrary, was early remarkable for such a population; whence it will appear natural enough that the first Scandinavian colonists in the Scotch border lands preferred to settle in the neighbourhood of that county. On the south-easternmost coast of Scotland, they would not only have been separated from their countrymen in the north of England by two intervening counties, but also divided by a broad sea from their kinsmen in Denmark and Norway. Such a situation would have been much more exposed and dangerous for them than the opposite coast, where they had in their neighbourhood the counties of Cumberland and Westmoreland, inhabited by the Northmen, as well as the Scandinavian colonies in Ireland and the Isle of Man.

The Scandinavian population in Dumfriesshire evidently appears to have emigrated from Cumberland over the Liddle and Esk into the plains which spread themselves westward of those rivers; at least the names of places there have the very same character as in Cumberland. Not only are the mountains called "fell" (Fjeld) and "rigg" (Ryg), as is also the case in the other border lands, but, what is more peculiar to Dumfriesshire, the terminations of "thwaite," "beck," and "garth," not to mention "by," or "bie," are transplanted hither from Cumberland: as, Thornythwaite, Twathwaites, Robiethwaite, Murraythwaite, Helbeck, Greenbeck, Botchbeck, Torbeck, Stonybeck, Waterbeck, Hartsgarth, Tundergarth, Applegarth, Locherby, Alby, Middlebie, Dunnaby, Wysebie, Perceby, Denbie, Newby, Milby, Warmanbie, Sorbie, Canoby, and others.

These Scandinavian names of places are chiefly met with between the rivers Esk and Nith. Various authors have also endeavoured to show that the fishermen on the Nith have to the present day characteristic and original Scandinavian terms for their tackle and modes of fishing:—for instance, "pocknet," Icelandic *pokanet*; "leister," or "lister," Icelandic *ljóstr*, Danish *Lyster*; "haaving," Norwegian *haave*, *i.e.*, to draw small nets in the water, &c., &c. Somewhat east of the river, and north of the town of Dumfries, lies the parish of "Tinwald," a name undoubtedly identical with Thingvall, or Tingvold; which, as the appropriate Scandinavian term for places where the *Thing* was held, is found in other districts of the British Isles colonized by the Northmen. And it was, indeed, natural that the Scandinavian colonists in the south-east of Scotland should fix their chief *Thing* place in the district most peopled by them.

From Dumfriesshire the Scandinavian names of places branch off as it were in an arch towards the west and east. Some few appear at intervals towards the west, as in Kircudbright (Begbie, Cogarth), in Wigton (Sorby, Killiness), in Ayr (opposite little Cumbray, Crosby, Sterby, Bushby, and Magby), and also in Lanark (Bushby, close to the south-west of Glasgow). Towards the east, some few are met with in Roxburgh, as, for instance, on the borders of Cumberland, "Corby," and "Stonegarthside," and on the frontier of Northumberland several in *haugh* (Höi, a hill) and *holm*. But on the whole only a few in *by* are still to be found on the borders between Berwick and Haddington (such as Humbie, Blegbie, and Pockbie). Towards Glasgow and Edinburgh the mountains are no longer called "fell" and "rigg." The Scandinavian names of places cease entirely in these districts; and only the Scandinavian word "fjörðr," or Fjord, is heard here, as well as farther towards the north in the names of fiords (or firths) namely: Firth of Forth, Firth of Clyde, Firth of Tay, Moray Firth, and Dornoch Firth.

In the Lowlands, the number of Scandinavian names of places is quite insignificant when compared with the original Celtic, or even with the Anglo-Saxon names. Whence we may conclude that though a considerable immigration of Northmen into the Lowlands undoubtedly took place, it must have occurred under circumstances which prevented them from being sufficiently powerful to change the original names of places. We must, in particular, assume that the immigration took place much later than the Danish conquests in England; and on the whole we shall not be far from the truth in asserting, that as the Danish conquests in England must have driven many Anglo-Saxons into Scotland, so also the subsequent Norman conquest must have compelled many Danes and Norwegians, settled in the north of England, to cross the Scottish border.

According to this view, most of the Scandinavian settlements in the middle and northern parts of the Lowlands are to be referred at the earliest to the close of the eleventh century; and at so late a period an entire change of the ancient names of places then existing there, could not, of course, be effected.

SECTION IV

Traditions concerning "the Danes."—The Southern and Northern Lowlands.—Danish Memorials.—Burghead.

We cannot venture to conclude, from the few Scandinavian names of places found in the Lowlands, that the immigrant Scandinavian population was but inconsiderable; nor can we presume to infer either the extent or the period of the immigration from the numberless traditions respecting the Danes preserved throughout that district. For, although the Lowlands were far from being conquered by the Danes and Norwegians so early as England was, still the number of alleged Danish memorials, even of a remote age, is proportionately as great in the former as in the latter country. Tradition has gradually ascribed almost all the memorials existing in the Lowlands which are of any importance to "the Danes;" nay, even the learned have, down to the present day, been too much inclined to recognise traces of the *bloody Danes* in the much more ancient Pictish, Roman, and Scottish monuments.

The traditions about the Danes have much the same character in the Lowlands as in England. They depict in vivid and touching traits the misery of the people and of the country under the repeated attacks of the wild sons of the sea, whose arrival, departure, and whole conduct, were as variable as the wind. When large bands of Vikings had landed, and the Scots had assembled an army to oppose them, it would sometimes happen that in the morning, when all was ready for the attack, the foreign ravagers were sought for in vain. In the darkness of the night they had taken the opportunity secretly to re-embark, and rumour soon announced to the army that the Vikings had again landed in quite a different part of the country, where they were spreading death and desolation. The Lowlander tells with horror of the many innocent women and children, not to speak of the numbers of brave men, who were slaughtered; of the churches, convents, and towns, that were destroyed by fire; and of the numerous herds of cattle and flocks of sheep, which, together with valuables of all sorts were carried off to the ships of the Vikings.

Although the Vikings are renowned in England for drunkenness and other kinds of dissipation, yet in Scotland tradition still more highly magnifies the inclination of the Danes for intoxicating liquors, and particularly for ale. It is also a general belief among the common people throughout Scotland and Ireland that the Danes brewed their strong ale from heather; a tradition which probably arose from the circumstance that in ancient times the Northmen spiced their ale with herbs; as, for instance, in Denmark with Dutch myrtle, or sweet willow (*Dan.*, Porse), which grows in marshy heaths.

For the rest, there can be no doubt that the Scotch stories about the drunkenness of the Danes were a good deal multiplied in far later times, at the period, namely, when the Princess Anne, a sister of Christian the Fourth, was married to the Scotch king James the Sixth, or James the First of England. Queen Anne was accompanied to Scotland by several Danish noblemen, who introduced at court, and among its hangers-on, the same carousing and revelling which at that time prevailed in far too high a degree at the court of Denmark. Burns, in his poem of "*The Whistle*," celebrates an ebony whistle still preserved in the family of Ferguson of Craigdarrock, which is said to have originally belonged to one of Queen Anne's Danish courtiers.

This Dane, who, even among his own countrymen, had the reputation of a great drinker, challenged the Scotch to drink with him for a wager, and promised the whistle to him who could drink him under the table. At the same time he produced evidence to show that in all his many drinking bouts at various northern courts in Russia and Germany, he had never been vanquished. However, after drinking three consecutive days and nights with Sir Robert Lawrie of Maxwelton, the Dane fell under the table, and Sir Robert gained the whistle. Sir Robert's son afterwards lost it again at a similar drinking bout with Walter Riddel of Glenriddel, from whose descendants it passed in the same way into the family which now possesses it.

But as a contrast to the many naturally exaggerated tales about the excesses committed by the Danes both in earlier and later times, it is refreshing to meet with romantic traditions about Danish warriors, whose bravery and comeliness could win the hearts of Scottish maidens, even whilst the curses of the Scots were heaped on "the Danish Vikings." A Danish warrior had been carried off by the Scots during an expedition into Morayshire, and imprisoned in a strong tower, where a speedy death awaited him. But the daughter of the lord of the castle, who had fallen in love with him, and found a requital of her affection, opened his prison door one night, and fled with him. When morning came the lord of the castle set off in pursuit of the fugitives, and overtook them on the banks of the river Findhorn, which runs through Morayshire. The lovers, who were both on one horse, attempted to swim the river; but the jaded animal could not make head against the stream, and the fugitive couple found a watery grave in the depths of the Findhorn. Near Dalsie, in Nairnshire, is a small sequestered valley on the banks of the Findhorn, inclosed by smooth sloping banks, overgrown with weeping birches. In the midst of this charming spot is seen a grave composed of stones heaped up, at one end of which stands a tall monumental slab, ornamented with carvings of a cross and other antique figures. This slab, the people say, is a monument to the unfortunate lady.

There is nothing intrinsically improbable in this tradition, since history testifies that the daughters of Scottish kings married Norwegian-Danish kings; whilst they, or at all events their countrymen, were making war in Scotland. In the beginning of the tenth century, the Scotch king, Constantine the Third, in conjunction with the more northern Anglo-Saxons, beat the Danes, who had passed over from Dublin under Reginald and Godfrey O'Ivar (Godfred Ivarsön), in a great battle near the Clyde. Although Constantine, during nearly the whole of his reign, had to fight against Danish and Norwegian Vikings, yet he gave his daughter in marriage to Anlaf, or Olaf, king of the Danes in Dublin and Northumberland; nay, he even fought with Olaf and his Danish-Norwegian army against the Anglo-Saxons at the battle of Brunanborg. Sigurd, Jarl of the Orkneys, was also married to a daughter of the Scotch king, Malcolm the Second (1003-1033), although he had made devastating incursions and conquests in Malcolm's lands.

The attacks of the Norwegians and Danes on the Scottish Lowlands were so continuous that out of seven monarchs who reigned over the Scots from 863 to 961, or about a century, three are related to have fallen whilst fighting against the Danes. These monarchs are, however, said to have purchased decisive victories with their blood. If we compare the unsuccessful expeditions of the Northmen into the Scottish Lowlands with the great conquests made by the Danes in England, we shall not wonder that the inhabitants of the former country relate with a sort of pride the many victories of their forefathers over "the Danes;" nor shall we be surprised that the popular traditions, which point out the ancient battle fields, scarcely admit even the possibility of the Danes having been victorious.

In the southern and middle Lowlands (to the south of the Grampian Hills) the Firths of Forth and Tay afforded excellent landing places for the ancient Vikings. Many battles, therefore, were fought in their neighbourhood. In the vicinity of a rampart called "*the Danes' dyke*," in the parish of Crail, close to Fifeness, and between the firths just mentioned, the Scotch king Constantine, Kenneth's son, is said to have fallen in a battle against the Danes in 881. Forteviot, or Abernethy, the ancient capital of the Picts, which the Vikings often tried to plunder, lay in the innermost part of the Firth of Tay. The defence of this place, by King Donald the Fourth, in 961, cost him his life. Near Redgorton, in Perthshire, is a farm called "Denmark;" close to which are to be seen remains of intrenchments, besides tumuli, and monumental stones, said to originate from a defeat suffered by the Danes at this spot.

The most famous battle in these parts is, however, related to have taken place on the northern shore of the mouth of the Firth of Tay. In the reign of Malcolm the Second, after the Danes had already made themselves masters of England, the attacks of the Vikings began to assume a more dangerous character. A number of them landed in the Bay of Lunan, in Forfarshire, whence they plundered and laid waste the country for many miles around. But to the east of Dundee, near Barry, they encountered a Scotch army, which defeated them, and compelled them to make a retreat, during which they were again repeatedly beaten. Even to the present day tradition points out a line of Danish monuments extending from Barry to Aberlemno, in the neighbourhood of which place the last battles were fought, and where human bones of a remarkable size are said to have been often found in the tumuli. At Camuston, not far from Barry, stands a stone cross called "Camus Cross," on which are carved various kneeling figures in an attitude of prayer. According to the statements of the common people the cross was erected in memory of the Danish general Camus, who fell at this spot. At Kirkbuddo were formerly seen the remains of a Danish camp called "*Norway dikes.*" In the parish of Inverkeilor, and near the farm called "Denmark," traces of Danish ramparts are also to be found; and at Aberlemno, Murphy, and many other places, are seen sculptured monuments, said to have been erected in commemoration of the before-mentioned fortunate victories over the Danes.

It is of course by no means incredible that a great battle may have been fought between the Scots and the Scandinavian Vikings in this district, and at about the time mentioned. But it is perfectly clear that most of the Danish monuments before noticed have no connection whatever with this frequently-mentioned battle. The name *Camus* is not at all a Scandinavian one; and it is, besides, not only certain that the village of Camuston was, in more ancient times, called "Cambestowne," but also that there are several similar names of places in the Lowlands, which are most correctly derived from the old Celtic language. The sculptured monuments in question have not, in fact, the least appearance of having been erected after any battle. In a splendid work lately published (P. Chalmers, "The Ancient Sculptured Monuments of the County of Angus," Edinburgh, 1848, folio), are to be found correct delineations of a number of stones of the same kind, which are spread over Perthshire, Forfarshire, Kincardineshire, and Aberdeenshire; and still more are to be met with along the coasts of the northern Lowlands and north-eastern Highlands. One, near St. Vigean, in Forfarshire, has an ancient Celtic inscription; but, with this exception, no inscriptions are found upon them. They are usually ornamented on one side with a cross and various fantastic scrolls and ornaments, and on the other with biblical representations, such as Adam and Eve at the tree of knowledge, Daniel in the lion's den, Samson with the jawbone of an ass, &c. Sometimes all sorts of strange figures are found on them, such as crescents, sceptres, mirrors, combs, and other articles; as well as serpents, lions, elephants, horses, dogs, stags, elks, sphinxes, &c. On some stones we find representations of the chase, with huntsmen, hornblowers, stags, and hounds. The carving is for the most part executed with much skill, and the whole style of the work seems referrible to the tenth or eleventh century. It is beyond all doubt that these stones cannot be ascribed to the Danish or Norwegian settlers, though several authors have asserted the contrary. They are evidently Christian-Scotch monuments, and have been erected with a very different aim from that ascribed to them: some, probably, as boundary stones of landed possessions and hunting-grounds; others as monumental stones to deceased persons.

One of the Aberlemno stones—a rare exception to the rest—which stands close by the church, represents on one side a battle, in which both foot and horse are engaged, and in which a bird attacks a man wearing a helmet, who tries in vain to cover himself with his shield. (See the annexed woodcut.) Above is seen a mirror, and one of those inexplicable figures which appear so frequently on stones of this kind. But in this there is the peculiarity, that the figure intersected by the cross-bar with the sceptres (?) at each end, is square, whilst in other instances it is generally in the form of a crescent. On the back of the stone is carved a cross covered with the finest scrolls and ornaments, and surrounded by fantastic figures of animals interlaced together. The height of the stone is about six feet. This monument might possibly have been erected after a victory; but it still remains uncertain, whether after a victory over the Danes. At all events, the stone is Scotch, and not Scandinavian.

[+ +] Aberlemno Stone

[+ +] Aberlemno Stone: Reverse

The case is much the same with most of the so-called "Danish" forts, camps, stone circles, and bauta stones; which are in general of Pictish or Celtic origin. Had they really been erected by the Danes and Norwegians, those nations must evidently have held confirmed dominion in these parts for a length of time; but it is well known that, in the early period in which these monuments were raised, they can be regarded as masters, in the south and central Lowlands, only at very short and far-distant intervals.

North of the Grampian Hills, and particularly in the district of Moray (the "Mærhæfi" of the Sagas), the Norwegians and the Danes, it is true, firmly established themselves for a somewhat longer period of time. In the beginning of the eleventh century, for instance, they defeated the Scots in a great battle near Kinloss, took the towns of Elgin and Nairn, whose garrisons they put to the sword, and afterwards settled themselves on the sea coast. But the kingdoms which they founded were speedily destroyed without leaving any remarkable traces behind them; so that, even in this district, we cannot place implicit reliance upon the many different stories about the Danish monuments. According to a common and not improbable tradition, the district of Moray, and the present Aberdeenshire, were the theatres on which the last battles between the Danish Vikings and the Scots were fought. Thus it is said that, in the reign of Malcolm the Second, the Danes, after the battle of Kinloss, suffered a great defeat at Mortlach in Banffshire, where Malcolm, as a thank-offering to God, caused a convent to be built. This, again, was partly the cause of Mortlach's becoming the seat of a bishop. Popular tradition states that the Scottish leader vowed during the battle to add to the church in Mortlach as much as the length of his spear if he succeeded in driving away the Danes. An ancient sculptured stone near the church is mentioned as pointing out the Danish leader's grave; and the skulls of three Danish chiefs are still shown, built into the north wall of the church, as a perpetual memorial. A similar tradition is preserved about the church of Gamrie, also in Banffshire. The Earl of Buchan vowed, in the heat of the battle, to build a church to St. John, to replace that which the Danes had destroyed, if he gained the victory over them. Three of the sacrilegious Danish chiefs, by whose command the church had been desecrated, were found upon the field of battle, and in a description of the church lately published we read as follows:—"I have seen their skulls grinning horrid and hollow in the wall where they had been fixed, inside the church, directly east of the pulpit, and where they have remained in their prison house 800 years!"

It is further stated that, on account of the repeated defeats which the Danes and Norwegians had suffered in the Scotch Lowlands, King Svend Tvskjæg sent, in the year 1012, his son Canute, who afterwards became king of England, with a large fleet and army to the northern part of the Lowlands. Canute landed on the coast of Buchan (Aberdeenshire), near the Castle of Slaines, in the parish of Cruden (or Crudane). Here a very fierce battle was fought, which can scarcely have been favourable to the Danes, since a treaty was afterwards concluded between them and the Scotch, according to which the Danes were to evacuate the fortress called "Burghead," in Moray, then occupied by them, as well as the rest of their possessions in the kingdom of Scotland. According to the same treaty the field of battle was to be consecrated by a bishop as a burial-place for the Danes who had fallen on it, and a chapel was to be built there in which masses should be continually sung for their souls. In this neighbourhood also there was certainly, at one time, a chapel dedicated to the Norwegian saint, Olave; but the ruins of this chapel, as well as the old churchyard, have since been destroyed by quicksands. The wind, however, by blowing away the sand, still brings, at times, the fragile bones of the Danes to the light of day.

Straight out of the town of Forres, in Nairnshire, stands a stone nearly twenty feet high, on one side of which is seen a large and handsome cross, and under it some indistinct human figures. On the other side is carved a number of horsemen and people on foot, evidently representing an execution on a great scale; several bodies are seen, and by the side of them the dissevered heads. The sculpture is executed with the greatest care, and displays some very tasteful ornaments, which, however, are now partly effaced through the action of time on the soft stone. The pillar is commonly called "Svenós stone," and tradition relates that it was erected to commemorate the treaty of peace concluded between Svend Tveskjæg and King Malcolm, and the expulsion of the Danes from the coasts of Moray. But the sculptures at present existing on the stone do not in the slightest degree represent anything of the kind. The stone belongs to the same class of monuments as the sculptured Scotch stones before described, which are so numerous in the Lowlands, and in the north-eastern Highlands, particularly Inverness-shire, Ross-shire, and Cromartyshire.

One of the few places in the Lowlands, which may with reason be assumed to have preserved considerable traces of the Danish expeditions, lies in the neighbourhood of the towns of Forres and Elgin. It is a promontory which projects in a north-western direction almost a mile into the sea. Towards its head its steep craggy shores are from eighty to a hundred feet high. This extreme point, which incloses a small harbour, and which presents a level surface on its top, where the fishing village of "Burghead" is situated, was formerly separated from the main land by three immense parallel ramparts, fifteen to twenty feet high, with cross ramparts lying between, as well as deep and broad ditches, of which there are still considerable remains. That the Romans had a fortress here (said to have been named "Ultima Ptoroton") was clearly proved several years ago, when a Roman well, which is still used, was discovered cut in the rock. But for Vikings, like the Norwegians and Danes, this place afforded a still better refuge than for the Romans. Towards the land side, which is in some degree barren and uninhabited, they could easily defend themselves; and from the sea, the Scots could attack them only by entering the harbour, where the well-equipped vessels of the Northmen of course prevented their landing. In all probability, therefore, the Norwegians and Danes still further fortified this important point, and gave it, perhaps, its present name. Tradition, at least, relates that the Danes, after taking Nairn, isolated the town or fortress, and called it "Borgen" (the castle); in which account it is very probable that the names of Nairn and the neighbouring Burghead have been confounded. The latter place gradually gained such importance that it was the last stronghold the Danes possessed in the Lowlands.

It is therefore clear that the Danes, or rather the Norwegians and Danes, have scarcely a right to claim many of the numberless monuments in the Lowlands which both the learned and unlearned ascribe to them. In fact, the whole eastern coast of Scotland, from the Cheviot Hills to Moray Firth, is entirely destitute of characteristic and undoubted Scandinavian monuments. It must, however, be remembered, that the actual Scandinavian immigrations into the Lowlands certainly took place after the Norman conquest of England; or, at all events, at so late a period that the Northmen could not remould the Scotch names of places into Scandinavian forms. Nor is it strange that the Scandinavian colonists in the Lowlands, who at the close of the eleventh century had long been Christians, and influenced by the civilization prevailing in England, should neither have erected such monuments as stone circles, bauta stones, cairns, and barrows, which presuppose a state of heathenism among a people, nor have impressed their characteristics generally on that district by means of peculiar memorials. For at that time they played a subordinate part there, and afterwards gradually became very much mixed with Celts, Anglo-Saxons, and subsequently even with Normans.

The very circumstance, however, that so large a tract of land as the Scottish Lowlands lay out of the path of the Scandinavian conquerors during the ninth, tenth, and first half of the eleventh centuries, was the cause not only that the Danes were able to direct all their power with more effect against England, but also that the Norwegians could more easily subdue the Orkneys and the Shetland Isles, the Hebrides, and various tracts in the northern and western Highlands. In these districts much more perceptible traces of the Norwegian settlers, and of the results which they produced, are still preserved, than in the Lowlands of the in general transient devastations of the Danes and Norwegians.

SECTION V

The Orkneys and Shetland Isles—Natural

Features.—Population.—Oppression.

We might expect that the most northern isles of Scotland, which lie exposed in a stormy sea, should possess the same wild and mountainous character as the Faroe Isles and Iceland. Such a belief gains strength when, for the first time, in passing from Scotland, we obtain a view of the southern Orkneys, especially the considerable mountain heights of the Isle of Hay. Indeed Hay obtained its name (originally "Haey," or the high island) from the old Northmen, on account of the mountains which distinguish it from the rest of the Orkneys; for on sailing farther northwards, past Hay and the adjacent South Ronaldshay (formerly "Rögnvaldsey"), we soon discover that the Orkneys are in general flat and sandy, although with cliff-bound coasts. Their heath-covered hills scarce deserve the name of mountains, though here and there called by the inhabitants "fjolds," or Fjelde (mountain rocks). The islands are destitute of wood, and exhibit frequent ling moors and desert tracts of heath. But there is also much, and by no means unfertile, cornland to be found; and an improved system of agriculture has made such advances, that the stranger is sometimes surprised, in these distant isles, by the sight of luxuriant fields of wheat.

The waves of the sea, and the powerful currents, have intersected the Orkneys with innumerable winding bays, or sounds. Besides Mainland, the chief island (first called by the Norwegians "Hrossey," and afterwards "Meginland," or the continent), the archipelago includes a great number of islands of different sizes, which spread themselves in a north-east direction from the north coast of Scotland. The farthest of the Orkneys is Fairhill, or Fair Isle (formerly "Friðarey"). It lies almost midway between the Orkneys and the Shetland Islands, in the midst of the rapid current now called Sumburg Roost, but which the Norwegians in former times called Dynröst (from "röst," a maelstrom, or whirlpool); whence, again, the most southern promontory of the Shetland Islands has obtained the name of Dunrossness (Dynrasternes). The Shetland archipelago (the old Northern "Hjaltland," "Hjatland," or "Hetland"), like that of the Orkneys, forms a long-extended line, but differs from it in consisting principally of one large island, Mainland ("Meginland"), surrounded by a great number of proportionately small and insignificant ones.

The most southern point of Dunrossness, on Mainland, forms the promontory of Sumburg Head ("Sunnbœjar-höfði"), which, however, is of no very great height; indeed the highest mountain in Shetland is only about fifteen hundred feet above the sea. Although the Shetland Islands, with regard to mountains, are not to be compared with the Faroe Isles, still they exhibit a sort of transition from the flatter Orkneys to the mountainous character of the Faroe group. Before the coasts of Shetland stand many high and ragged rocks, called "stacks" (old Norsk, "stackr"). The coasts themselves are steeper, and the mountains larger than in the Orkneys. On the other hand, however, the valleys are both longer and broader than the mountain valleys of the Faroe Islands. Heath and moorland abound, whilst the corn-fields are small, and the corn harvest in general very uncertain and difficult to gather. Fishing is the most important source of profit for the inhabitants.

The Orkneys and the Shetland Isles were, as is well known, completely colonized by Norwegians in the ninth and tenth centuries. They were, however, known and inhabited much earlier. It is possible that the Shetland Islands were the "ultima Thule" spoken of by Roman authors in the first centuries after Christ; but it is certain that the Romans at that time knew the Orkneys by the name of "Orcades:" whence it appears that the primitive root *Ork*, in the later Norwegian name of the islands, is very ancient, and probably of Celtic origin. Before the arrival of the Norwegians, both the Orkneys and the Shetland Islands seem to have been inhabited by the same Pictish or Celtic race that was settled in the rest of Scotland. Of these older inhabitants memorials still exist in different kinds of antiquities of stone and bronze that are dug out of the earth, as well as in numerous ruins of castles, or Pictish towers, originally built of flag-stones laid together, without any cement of loam or mortar. There are also cairns and stone circles; the most prominent amongst which are the "Stones of Stennis," on each side of Brogar Bridge, in Orkney. They are, like Stonehenge and Abury circle in England, surrounded with ditches and ramparts of earth; and, after Stonehenge, must be regarded as amongst the largest stone circles in the British Islands. The immense masses of erect stones are remarkable evidences both of the strength and of the religious enthusiasm of the old Celtic inhabitants; and it is no wonder that they made in ancient times such an impression on the Norwegians, on their arrival at these islands, as to induce them to call the promontory on which the largest circle stands "Steinsnes" (Stones-naze) and the adjoining firth, "Steinsnesfjördr" (Stones-naze Firth, now Loch of Stennis).

No sooner had the Scandinavian Vikings settled themselves, in the ninth century, securely in these islands, than they became a central point for the Northmen's expeditions not only to the British Islands, but also to Iceland and Greenland. Thus when Floke Vilgerdesön, or "Ravnefloke," went on a voyage of discovery from Norway to Iceland, he landed on Hjaltland, or Shetland, in a bay which obtained from him the name of "Flokavágr." This bay must probably be sought on the east coast of Mainland, about Cat Firth (Kattarfjörðr); for in its neighbourhood lay the Loch of Girlsta (originally "Geirhildarstaðir"), which is said to have obtained its name from the circumstance of Floke's daughter, Geirhilde, having been drowned in it during her father's short visit to the country. By degrees the islands became the rendezvous of a great number of discontented Norwegian emigrants, who, to avoid the new order of things, had withdrawn themselves from their old paternal home, and from this distant place of refuge continually harassed the coasts of Norway.

This induced King Harald Haarfager to undertake an expedition against the Orkneys and the Shetland Isles, as well as against the Hebrides, on the west coast of Scotland; all of which he succeeded in subjugating. He gave the Orkneys and the Shetland Isles, as an earldom under the crown of Norway, to Ragnvald Möre-Jarl's family. This family produced some great men, who extended their dominion over large tracts in the adjacent kingdom of Scotland. The islands continued, however, to be the resort of many malcontent and fugitive

Norwegians. The renowned Ganger-Rolf, the founder of the royal Norman house, is said to have dwelt a long time on them before he undertook his expedition against Normandy. When King Erik Blodöxe, Harald Haarfager's son, was driven with his queen, the atrocious Gunhilde, from Norway, he fled to Orkney, whence he carried devastation far and wide. Subsequently he obtained a kingdom in Northumberland; but, after his fall, his sons again sought the Orkneys; where they remained till they succeeded in obtaining the kingly power in Norway. Snorre Sturlesön states, that after the fall of this dominion, Gunhilde again fled to Orkney, where her daughter, Ragnhilde, had married a member of the Earl's family. Ragnhilde trod entirely in her mother's footsteps by occasioning dissension, and even murder, in the family of the Earl. Somewhat later the Orkneys were visited for a time by Kalf Arnesön, so well known in the more ancient history of Norway, who, at the battle of Stiklestad in 1030, was one of the chief leaders of the peasant army against King Olaf, the saint. He came to the Orkneys just in time to take part in a severely-contested naval battle, fought in the year 1046, near Rödebjerg (Rauðabjörg) in Pentland Firth, between the Jarls Thorfin and Ragnvald Brusesön. Kalf supported Thorfin with six long ships, and thus decided the victory in his favour.

The older history of the islands exhibits an almost uninterrupted series of bloody combats between members of the Norwegian Jarl's family. This, however, did not prevent them from making violent inroads on the coasts of Scotland and Ireland. Long after the Vikings' mode of life had ceased in the Scandinavian North, it continued to be preserved in these islands. This was not only owing to their remote situation, opposite hostile coasts, and to their characteristic independence, but also to the population having inherited the old Viking spirit, and carefully preserved the ancient Norwegian institutions. As long as Norwegian jarls ruled, Norwegian laws, customs, and habits, as well as the Norwegian language, were absolutely paramount in the islands. The connections which the jarls and other powerful leaders maintained with Scotch and Irish chiefs, and which often resulted in intermarriages between their families, do not seem to have had much effect on the Scandinavian national character of these island colonists. It was not till the beginning of the fourteenth century, when the male line of the old Norwegian jarls had become extinct, and when the Scotch Lord Saint Clair, who had married a daughter of Magnus, the last jarl, had obtained possession of the earldom, that the ancient liberties, customs, and manners of the inhabitants, began to be seriously threatened; nor did it suffice to protect the islands against the progress of Scottish influence, that they continued to be under the supreme authority of Norway. When, at length, the Danish-Norwegian king, Christian the First, on the occasion of the marriage between his daughter Margaret, and the Scotch king, James the Third, in the year 1469, pledged to Scotland the Orkneys and the Shetland Isles as part of Margaret's dowry, the last tie was severed that bound those countries to their Scandinavian friends. The Scottish kings and their successors, who also ascended the English throne, acknowledged indeed the right of the Danish-Norwegian kings to redeem the islands; but they continually found subterfuges to prevent its being exercised. The lawful claims of redemption, repeatedly urged by Denmark in the sixteenth and seventeenth centuries, were perfectly fruitless. The islands were too important, and far too conveniently situated with regard to Scotland, for Great Britain to give them up, without being compelled by the last necessity. The undoubted right of the Danish-Norwegian kings was forced to give way to the superior power and political influence of the British sovereigns.

The conduct observed towards the Norwegian population of these islands after their union with Scotland was quite as unjust as their separation from Norway and Denmark, and assuredly far more revolting to all proper feeling. A large part of the inhabitants had till then been in the free possession of their lands as freeholders, or "udallers" (Odelsmænd), and had likewise possessed their old Norwegian laws and privileges, which should of course have been respected when the islands were pledged to Scotland. But the Scotch nobles, who, partly as vassals, partly as royal lessees, obtained the government of the islands, took care to destroy all traces of the ancient liberties and Scandinavian characteristics of the people. The resistance of the islanders was fruitless. In the year 1530 they took up arms under the command of their governor, Sir James Sinclair, in order to oppose the appointment of a crown vassal over the islands.

The Earl of Caithness himself, who had been dispatched against them, fell, with five hundred of his men, in a sanguinary action near the "Stones of Stennis." But though the islanders thus asserted their rights for a short period, the Scotch regents soon afterwards succeeded in establishing crown-vassals in the islands.

Among these vassals none has left behind him a more despised or hated name than Earl Patrick Stuart, who from 1595 to 1608, or about thirteen years, oppressed the islands in the most shameful manner. He violently deprived the holders of allodial farms of their right of possession, and converted almost all the freeholders into leaseholders. He arbitrarily changed the weights and measures, so that the taxes and imposts became intolerable. Law and justice were not to be procured, for the Earl's creatures everywhere occupied the judgment-seats. To appeal to Scotland was no easy matter, as Lord Patrick's soldiers guarded all the ferries. In the Orkneys the Earl compelled the people to build him a strong fortress at Kirkwall, and in Shetland another at Scalloway; from which places armed men ranged over the country, to punish and overawe the malcontents. The ruins of these castles form a still-existing memorial of "the wicked Earl Patrick," who, for his tyranny, was at length recalled to Scotland, accused of high treason, and beheaded.

The Scottish kings, it is true, now promised the islanders that they should have relief in their need, and that no vassal of the crown should be placed over them. But this promise was not kept; and so far from the islanders again recovering their lost freedom, the feudal system of England and Scotland continued to take firmer root in the islands. Oppression stalked on with regular and steady step until it arrived at such a pitch that not only did the Norwegian laws and liberties disappear, but the islands themselves, with some few exceptions, became the private property of a few individuals. The successors of the mighty Vikings, descended from kings and jarls of Norway and the North, who in winter dwelt as chiefs, or at least as freemen, in roomy mansions, whilst in the summer they gained glory and booty in their long ships, are now in general obliged to content themselves with inhabiting as leaseholders, or rather as annual tenants, a poor cottage on a small piece of land, where, by hard labour, they are able to gain, at best, a very frugal subsistence. Their dwellings, particularly in Shetland, are of the most wretched description. The walls are formed of small unhewn stones, with turf and sea-weed thrust into the interstices, and, instead of a chimney, the smoke escapes by a hole in the roof. Within the house there are generally sleeping-places in the thick stone wall; but men and cattle live together in friendly harmony in the same apartment. The fire burns freely on the floor, and envelopes all in a dense smoke. If the people seek their living on the sea by fishing, it is usually in boats belonging to the proprietor of the estate, who consequently receives a large share of their profits. The condition of the common people in the Orkneys, and in the Shetland Isles, is certainly not at all enviable, even in comparison with that of their Scandinavian kinsmen on the poor and more remote Faroe Islands and Iceland; although commerce is still limited and oppressed there by a monopoly which was soon abolished in the Orkneys and Shetland Isles after their separation from the united Norwegian-Danish kingdoms. But in spite of all their calamities, the inhabitants of the Faroe Isles and Iceland have for the most part preserved to our times that freedom of landed property which they inherited from their forefathers.

SECTION VI

Shetland.—The People.—Songs.—Sword-Dance.—Language.—Names of Places.—Tingwall.—Burg of Mousa.—Tumuli.—Bauta Stones.[9]

9. Partly from S. Hibbert, P. A. Munch, and Chr. Plöyen.

If the present originally Norwegian population in the Orkneys and Shetland Islands possessed, on the whole, any strongly-marked Scandinavian characteristics, they would naturally occur most in the islands farthest towards the north. But the oppressions and political changes that have occurred there have done their work so thoroughly, that even the Shetlanders no longer bear in their character and natural disposition any strongly-marked feature of their Norwegian origin. The only ones remaining are, perhaps, their love of the sea, and their skill in contending with its dangers. Even their bodily frame has, through many years of want and debasement, lost much of its strength and nobleness. In the parish of Coningsburgh, in Mainland, precisely where the largest and strongest-built people are to be found, the Scandinavian population are said to have kept themselves most free from mixture. The inclination for disputes and fighting amongst the people of Coningsburgh is well known in Shetland. This trait is, at all events, more Scandinavian than moroseness and want of hospitality to strangers, which are almost unknown in the North, but which in the last century were alleged to be vices of these same men of Coningsburgh. It was said that they would not willingly give a traveller a night's lodging, and that directly at day-break they awoke him, saying:—"*Myrkin i livra; lurein i liunga; timin i guestin i geungna*;" that is, "It is dark in the smoke-hole, but it is light on the heath, and for the guest it is now time to depart." That this sentence, which was written down in the year 1774, consists of old Norwegian words, though in a corrupted form, is quite evident.

The Shetlanders still retained, in the last century, many of the customs of their Scandinavian forefathers. Thus surnames were given both to sons and daughters, according to the genuine Scandinavian custom, from the father's Christian name. The eldest son, for instance, of Magnus Anderson was called Anders Magnuson, and all the other sons had likewise the surname of Magnuson; whilst the daughters, in like manner, were all called Magnus-daughter, of course with different Christian names. Even the Norwegian language is said to have been spoken at that time by some few old persons in the most remote islands. The traditions and songs handed down by their forefathers still lived among the people, whose poets and poetical feeling have been celebrated from the earliest times. It was customary to revive the memory of former days by festal assemblies, in which the youth of both sexes danced to songs ("Visecks") and ballads, as they did in ancient times throughout the North, and as is still the custom in the Faroe Isles. At Yule time (Christmas), which was the chief festival, and the beginning of which was always announced at daybreak by playing an ancient Norwegian melody, called "the day-dawn" (*Dan.*, Daggry), all kinds of merriment took place. A favourite amusement was the so-called sword-dance, the origin of which may be traced with sufficient certainty to the times of the heathens. The Vikings were frequently very dexterous in playing with naked swords, throwing several at once into the air without allowing them to fall to the ground. This practice was easily converted into a dance, performed by several men with drawn swords; and consisting of many windings and figures calculated to develope a dexterous agility, which, in those warlike times, must naturally have excited a lively interest among the spectators. Later in the middle ages the sword-dance in the Shetland Isles lost by degrees the wildness of its character, the number of dancers being limited to seven, representing the Seven Champions of Christendom, viz., St. James of Spain, St. Denis of France, St. Anthony of Italy, St. David of Wales, St. Patrick of Ireland, St. Andrew of Scotland, all under the command of St. George of England, who both opened and closed the dance by reciting some English verses appropriate to the occasion.

All this, however, is now much changed. In the farthest island towards the west, that of Papa stour ("Papey stœrri," the great Pap Island, in contradistinction to the neighbouring Papa little, "Papey litla"), a last shadow of the old warlike sword-dance is occasionally to be seen. Instead, however, of being clothed in armour or shirts of mail, the dancing knights have shirts of sackcloth; and, in place of huge swords, they brandish straightened iron hoops, stripped from some herring-cask. The old Norwegian songs are no longer heard. Of the ancient Norwegian popular language the only remains are partly a few words, which, however, appear conspicuously in the English dialect now used; and partly a peculiarly sharp pronunciation, with a considerable rising and sinking of the voice, not unlike the vulgar pronunciation in the Faroe Isles. The old Norwegian words are particularly employed for certain objects and implements which have been in use from time immemorial.

Thus, for instance, the hole through which the smoke escapes (*Dan.*, Lyre) in the roof of houses covered with flat turf (flaas) is sometimes still called by the name of "livra" (in the Færoic language "ljowari"). The high seat for the mistress of the house is called, in remote districts, "hoy-saede" (*Dan.*, Höisæde); her "bysmer," which serves her for weighing, exactly agrees, both in name and nature, with the "Bismer" common in the North. The hand-mill, which is fast disappearing, is called as in the Danish part of north England, "qvern." The turf-spade, called in the Faroe Isles "torvskjæri" (*Dan.*, Törveskjærer), is here named "tuysker." The land-tax also, according to Scandinavian fashion, is paid in "merk" and "ure" (Mark and Öre). The outlying fields are called "hogan," "hagan" (*Old Norsk*, "hagi," an inclosed field). The deep-sea fishery (*Dan.*, Hav) is called "the haaf;" the fishing itself, "haaf-fishing" (*Dan.*, Havfiskerie); and the necessary lines, "tows" (*Dan.*, Touge). To the present day the Shetlanders use, in these fisheries, boats imported from Norway, which are peculiarly suited, by their construction, for the high seas and rapid currents on the coasts of Shetland. The dress worn by the fishermen when out at sea bears a striking resemblance to that of the Faroe men. The head is covered with a cap knit in the form of a night-cap, and ornamented with the most motley colours. They wear a coat of tanned sheep-skin, reaching down to the knees, where it generally meets a pair of huge and capacious skin boots, very carefully sewed. On land the Shetlanders use only a simple kind of shoe called "rivlins," consisting of a square piece of untanned cow-hide, covering little more than the sole of the foot, and fastened with a fishing-line or a strip of skin. The men of Faroe have similar shoes, called "skegvar," which, however, are far better made.

But what particularly reminds the Scandinavian traveller in Shetland of finding himself in a country formerly altogether Norwegian, is the names of places, all of which bear the impress of their Norwegian origin. This remark applies to the names of the islands themselves, as well as to the names of towns, farms, promontories, and bays existing in them. They, of course, resemble, in a great degree, the old Scandinavian names of places farther south, in Scotland and England. Thus, for instance, a fiord is generally called "firth" (fjorðr); a creek "wick" (*Dan.*, Vig); a holm, or small island, "holm;" a promontory, or naze, "ness;" a valley, "daill," or "dale." But it is peculiar to these districts, that the forms of names of places which occur most frequently in the old Danish part of the north of England, namely, those ending in *by*, *thwaite*, and *thorpe*, are extremely rare in Shetland, and in the rest of the old Norwegian possessions in Scotland. Of those in *by*, only a few instances are to be found; those in *thwaite* are still more rare; and those in *thorpe* are not to be met with at all. On the other hand, these districts possess several Scandinavian names of places which are also most frequently found in the old Norwegian colonies in the north and west of Scotland, but which are perfectly unknown in the old Danish part of the north of England. For instance, a small bay (*Dan.*, Vaag) is called "voe" (vágr); whence, on Mainland, we find "West-voe," "Aiths voe" (the bay by the tongue of land), "Lax-voe" (Lax, or Salmon-bay), "Selia-voe" (sildavágr, the "Silde Vaag," or herring-bay), "Hamna-voe" (hafnarvágr, the Havne Vaag, or harbour bay), together with others. A still smaller bay, navigable only by boats, is called "gjo," or "goe" (*Old Norsk*, gjá, an opening or cleft). For the rest, many farms have names with such endings as *seter* (Old Norsk, *setr*), *ster* and *sta* (Old Norsk, *staðr*, a place); and also *busta*, *buster*, and *bister* (contracted from "bolstaðr," a dwelling-place); whence, for instance, Kirkbuster (formerly Kirkjubólstaðr); all of which names agree just as well with those found in the Faroe Isles, Iceland, and the mother-country, Norway, as the names of places in the north of England ending in *by*, *thwaite*, and *thorpe*, agree with those in the corresponding mother-country, Denmark. Although the difference between the present traces of Danish colonization in England, and of Norwegian in Scotland, is not considerable, still it may be recognised in this manner. In consequence of the remote situation of the Shetland

Isles, the names of places, in spite of all revolutions, remain so much the same, that the old political and religious institutions of the islands are visible, as it were, through them. In the south part of Mainland lies the farm of Howff, where in ancient times there was certainly a "Hof," or house of God; and far northwards, near Hillswick (formerly Hildiswik), is the promontory of Torness (Þórsness), which probably once had a Hof for the god Thor. Nor far from thence is the Lake Helgawater (Helgavatn), or the holy water. Heathenism, however, lasted but a short time in the islands. The Irish Christian priests (*Old N.*, "Paper")—the memory of whom still lives in the names of the islands Papa (Papey), as Papa stour (great) and Papa little—seem to have worked indefatigably; insomuch that the Norwegian king Olaf Tryggvesön was able, at the close of the tenth century, to introduce Christianity throughout the islands. In place of the old god-houses there speedily arose a number of chapels or small churches, consecrated to different saints: viz., to the Norwegian saints, St. Sunifva (the daughter of an Irish king who suffered shipwreck in Norway), St. Olaf, as well as, at a somewhat later time, to St. Magnus, the patron saint of the Orkneys, after whom a great bay on the north-west coast of Mainland is to the present day called St. Magnus' Bay. St. Magnus seems also to have been the patron, or rather the chief saint, of Shetland; at least, the principal church in Shetland is consecrated to him. This church did not stand in Lerwick, the present chief town in Shetland, which has risen far later in the south-eastern part of Mainland, on the site of an old sea-side town near Bressasound (formerly "Breiðeyjarsund"). It lay about four miles to the north-west of Lerwick, in the parish of Tingwall; where, as the name (Þingavöllr) denotes, the chief *Thing* of the islands was held for centuries, and where, in heathen times, the chief place of sacrifice undoubtedly existed. The parish of Tingwall comprises one of the prettiest and best-cultivated valleys in Shetland. The old *Thing* place is still to be seen near the church, in a small holm, or island, in a lake, connected with the land by a row of large stepping stones. Secure against a sudden attack, here sat, when the island was free, the "foude" (*Dan.*, Foged), or magistrate, with his law-officers, whilst the multitude of the common people stood round about on the shores of the lake, and listened to what passed. Popular tradition says that the church was at that time a free place, or sanctuary, so that a person condemned to death was entitled to a pardon, if he could succeed in running from the holm over the stones, and reaching the church without being killed by the people. If this was really the case the commonalty must consequently have had power to pardon a convicted person by suffering him to escape into the church.

During the holding of the chief *Thing*, which in the olden times was generally accompanied with great sacrificial offerings, as well as with fairs and all sorts of merry-making, a multitude of persons always assembled, and a great many tents and booths were erected, both at the *Thing* place itself and in the immediate vicinity. Hence it undoubtedly arose that about three miles to the west of Tingwall, near a bay of the sea, there was a collection of *Skaaler*, or wooden booths; whence the present Scalloway (Skálavágr) which, next to Lerwick, is the most important trading place in the islands.

In Mainland alone there were at least seven lesser *Things*, under the jurisdiction of the chief *Thing* in Tingwall. The names of five of these are still preserved in Sandsthing (Sandsþing), Aithsthing (Eiðsþing), Delting (Dalaþing), Lunziesting (Lundeiðisþing), and Nesting (Nesþing); but the two other names, which are known from records, Rauðarþing—probably the most northern parish, Northmavine—and Þveitaþing (the most southern parish?), have disappeared. Special *Things* were, of course, also held on the larger islands, such as Yell ("Jali") and Unst ("Aumstr," "Örmst"); but it is certainly very incorrect to infer, as many persons do, from some stone circles near Baliasta, close by Unst, that the chief *Thing* of the islands was held there in the most ancient times of heathenism.

These stone circles belong simply to low graves encircled by stones, like those so frequently found in Norway, and whose date is of the latest period of heathenism, or what is called the iron age. Skeletons have been found in several similar graves in Shetland; and at different times urns containing burnt bones and ashes have also been discovered, together with other distinct traces of their having been burial-places. For the rest, barrows or tumuli, bauta stones, runic inscriptions, and similar monuments and antiquities of the heathen times, are by no means frequently to be met with; the reason of which must naturally be sought in the short duration of heathenism in these islands. The remains of only a single insignificant runic stone, and that of the Christian æra, have been discovered near Crosskirk, in the north of Mainland. The numerous round towers, or castles, of loose flag-stones laid together, which are often built on islands in lakes, and are called by many "Danish burghs," are, as before stated, of Pictish or Celtic origin. They have no resemblance whatever to the old fortresses in the Scandinavian North; whilst, on the other hand, buildings entirely corresponding with them are to be found in the Celtic Highlands of Scotland, and on the coasts of Ireland. The most that can be said is that the Norwegians availed themselves of these buildings after their conquests and settlements in these districts. Thus the remains of a tower are to be seen on a holm in Burra Firth (Borgarfjörðr, or Borgfjord, *i.e.* Castle fiord), in the west of Mainland, which may have been inhabited in the beginning of the twelfth century by the chief Thorbjörn, whom the Earls Magnus and Hakon attacked and killed in "Borgarfjörðr." The ground plan of the ruin (after Hibbert) shows how the chambers were disposed in the thick stone wall.

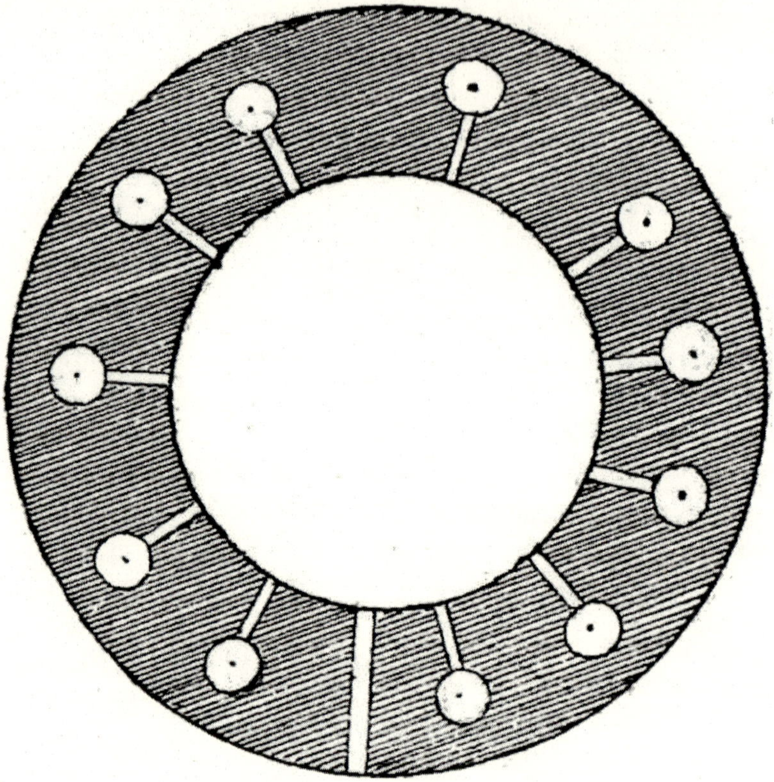

[+ +] Tower: Burra Firth

Another ancient Celtic tower, which tradition decidedly states to have been occupied by Norwegians, and which, on that account, has a particular interest for a Scandinavian, lies on the little island of Mousa (the ancient "Mösey"), close to the sound that separates the island from the south-eastern coast of Mainland. The tower is, fortunately, the best preserved one of the kind in the British Islands. It rises to the height of between forty and fifty feet, like an immense and perfectly round stone pillar, but bulging out towards the middle. Its appearance from without is quite plain, and no other opening can be perceived in the wall than the entrance-door, which even originally was so low that it was necessary to creep through it. To attack the tower, even when the door stood open, was not easy, and the bulging of the wall in the middle rendered the scaling of it almost impossible. The entire tower is about fifty feet in diameter, and consists of two concentric stone walls, the innermost of which encloses an open space of about twenty feet wide. The two concentric walls are each five feet thick, and stand at a distance of five feet from each other. The small space between them formed the habitable part of the tower. From the open yard we ascend a stone staircase, and, before we reach the top, seven divisions or stories are passed, separated by large flag-stones, which form a ceiling for one story and a floor for the next. In the different compartments, which quite encircle the tower, are small square openings, or air holes, one above the other, and looking out into the inner yard. The annexed drawings and sections (taken from Hibbert's description of Shetland), which represent the tower in its evidently original state, will serve to explain still more clearly the nature of this simple, yet remarkable, building.

[+ +] Tower: Mousa

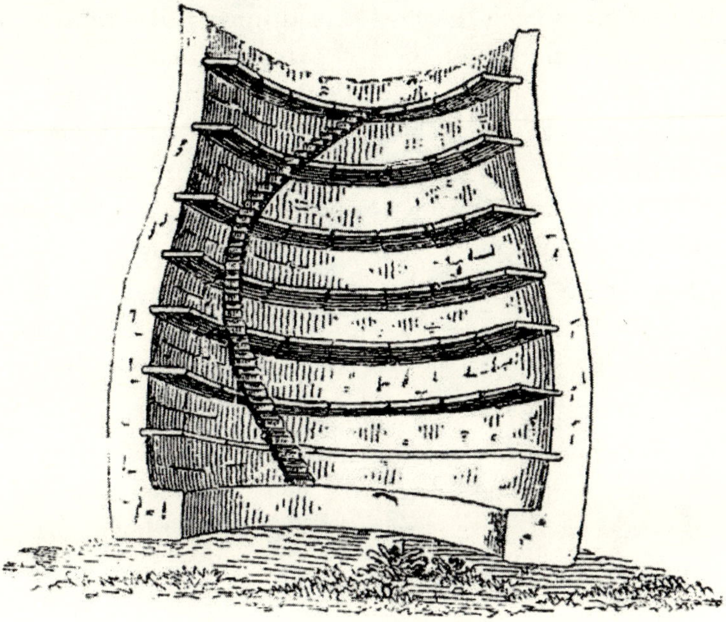

[+ +] Tower: Mousa - Interior

This tower appears to have stood deserted as early as the tenth century. Whilst Harald Haarfager reigned in Norway, a distinguished Norwegian Viking and merchant, Björn Brynjulfsön, carried off his beloved Thora Roaldsdatter (Roalds-daughter) from the fiords. He brought her first to his father's house; but, as his father would not permit him to celebrate his marriage there, he fled with her in the spring, on board his ship, and sailed westwards. After suffering much from storms and heavy seas, the couple landed at last on Mösey, and took up their temporary abode in the castle there, whither they brought the whole of the ship's cargo. In "Möseyjarborg," Björn celebrated his marriage with Thora, and dwelt there through the winter. But next spring he learned that King Harald, at the entreaty of Thora's friends, had exiled him from Norway; and that commands had even been sent by Harald to the jarls and chiefs in the Orkneys, the Hebrides, and in Ireland, to put him to death. He therefore again put to sea, and landed safely with his Thora in Iceland.

A few centuries later, the chief Erlend Junge fled from the Orkneys with Margaret, mother of the Jarl Harald Maddadsön, who was as much celebrated for her beauty as for her wantonness, and shut himself up with her in "Möseyjarborg." The Jarl Harald, who had opposed their marriage, set out in pursuit of them, and blockaded the castle for a long time, in order, if possible, to cut off their supply of provisions, and thus compel them to surrender; for, by force, says the Saga, the castle could scarcely be taken. But Harald at last became weary of the siege, and concluded an agreement with Erlend that he should have Margaret to wife on condition of swearing fealty to him as jarl.

This old and venerable tower has, therefore, not only been the scene of sanguinary battles and deeds of cruelty, but its strong walls have also afforded a secure asylum to sincere and all-sacrificing love.

SECTION VII

The Orkneys.—"Þingavöllr."—Monuments of the Olden Time.—Kirkwall.—St.

Magnus Church.

The Orkneys, on account of their greater fertility, and of their lying nearer to Scotland, were in ancient times, as indeed they are at present, of much more importance than the distant Shetland Isles. As the chief seat of the Norwegian jarls, they formed the central point of the Norwegian power in the north of Scotland. According to the Sagas, most of the many Danes and Norwegians who settled on the islands to the north of Scotland, resorted to the Orkneys; by which means, the jarls who governed them were enabled easily to assemble large fleets, and to man them with picked Scandinavian warriors. It was chiefly, therefore, Norwegians from the Orkneys, who, under the command of the jarls of Orkney, made such extensive conquests in the territories of the Scottish kings.

Jarl Sigurd the Stout (*Dan.*, Digre), who, as before mentioned, was married to a daughter of the Scotch king, Malcolm the Second, and Jarl Thorfin, his son by King Malcolm's daughter, pre-eminently distinguished themselves by bold Viking expeditions into the neighbouring countries, and particularly by their conquests on the Scotch coast. They extended these as far south as Moray; nay it is even said that at times they went as low as to the Firth of Forth. Thorfin was the last of the jarls of Orkney in whom the old Scandinavian Vikings' spirit lived and stirred. His power was greater than that of any of his predecessors or successors; for he ruled, say the Sagas, over no fewer than eleven earldoms (Jarledömmer) in Scotland, over all the Hebrides, and a large kingdom in Ireland. But after the many warlike expeditions, raids, and incendiarisms, in which he had played a part, he at length became penitent, and undertook a journey through Denmark and Saxony to Rome, where the pope gave him an indulgence for his sins. After his return, he governed his kingdom peacefully till his death, which took place about the year 1064.

Notwithstanding that a new and Christian æra had irresistibly established itself under this fierce Viking, the Orkneys continued for more than a century after his death to foster men who were Christians only in name, but in reality, both in their way of thinking and conduct, were heathen Vikings. Svend Asleifsön, who, in the middle of the twelfth century, lived on the little island of Gairsay (Gareksey), close to the north-east side of Mainland, occupies a prominent place among these Vikings. He was surrounded by a band of eighty men, with whom in the winter he remained at home in his mansion, living well on the booty that had been won. In the spring, after seed-time, he set out with them on expeditions to the Scotch, English, and Irish coasts. In the autumn he returned home for a short time, in order to gather the corn into his barns; and then again set out and harried the before-mentioned countries until the beginning of winter. On one of these autumnal Viking expeditions he even took Dublin; but whilst he fancied himself secure, the inhabitants suddenly fell upon and killed him, together with a great number of his men, who defended themselves with the utmost bravery.

In consequence of these important Viking expeditions, as well as of the greater life and bustle which prevailed in the Orkneys, not only are more historical accounts preserved of them than of the Shetland Isles, but they likewise exhibit more conspicuously how the warlike spirit of the Scandinavian population, when it began to be curbed by Christianity and the abandonment of piratical expeditions, preyed upon itself, and exhausted its strength in sanguinary internal conflicts. Memorials of this are found on almost all the islands. In going from Shetland, the first island made after passing Fairhill, and when approaching the proper group of the Orkneys, namely, North Ronaldshay ("Rinansey"), was the scene of a terrible revenge taken by Jarl Einar on King Harald Haarfager's son, Halfdan Haaleg (Long-legs), who had murdered Einar's father, Ragnvald Mörejarl, in Norway. Jarl Einar is said to have avenged his father in the same manner as, according to the Saga, the sons of Regner Lodbrog punished their father's murderer, King Ella of Northumberland; namely, by cutting a blood eagle on Halfdan's back. At Lopnes ("Laupandaness"), in the neighbouring island of Sanday ("Sandey"), Jarl Einar Sigurdsön was killed in the following century (the eleventh) by Thorkel Fostre, so called because he had brought up, or fostered, Einar's brother, subsequently the famed Thorfin Jarl. Not long afterwards, Thorfin's nephew, Jarl Ragnvald Brusesön, was killed by the same Thorkel on Little Papa Island ("Papey"), to the north-west of Sanday. Thorkel and Thorfin had previously surrounded and set fire to the house, wherein the jarl was with his men. The jarl's corpse was then conveyed to and buried on the neighbouring isle of Papa Westray ("Papey hin meiri," the Great Pap Island), adjacent to Westray ("Vestrey") and the most northern of all the Orkneys. Thorkel Fletter, surnamed the restless, was burnt in his house in Eday ("Eiðey"), in the twelfth century; and in the year 1137 the Jarl Paal was surprised by Svend Asleifsön on Rowsay ("Rolfsey"), and carried away prisoner to Athol, in Scotland. About twenty years previously (1110) the celebrated jarl, Magnus Erlendsön, was attacked and murdered by his kinsman, Jarl Hakon Paalsön, on the adjacent island of Egilshay, ("Egilsey"). In honour of Magnus, who was afterwards canonized, and became the patron saint of the Orkneys, a church was built on Egilshay, which still exists, though in a somewhat altered form.

Between the last-named islands and Mainland are the small isles Enhallow ("Eyin helga," the holy isle) and Wire ("Vigr"). On the latter Kolbein Ruga had, in the twelfth century, a castle, the site of whose ramparts can still be clearly distinguished. But Mainland itself is naturally the island with which the most numerous and remarkable memorials of the Norwegian dominion are associated. For centuries numberless Vikings' fleets constantly rode at anchor in its bays and in the adjacent straits; and almost every spot on the island is famous in the Orkneyinga Saga as having been the residence of some distinguished man, or the scene of some important historical event. The numerous Norwegian names of places ending in *wall* (vágr), *wick*, *firth*, *ness*, *buster*, *toft*, *holm*, and so forth, which are everywhere met with in the island, do not, however, merit particular consideration, since they resemble those in the rest of the Orkneys and Shetland Isles; yet they serve to establish that the Norwegians must have superseded here, no less than in the other islands, the older Celtic population. We soon discover that the vicinity of the Orkneys to Scotland, and their brisk intercourse with that kingdom, as well as with England, have contributed, both in Mainland and in the surrounding islands, to do away with many of those names of places which are still found in Shetland as witnesses of the old Norwegian judicial institutions. Thus we should look in vain in Mainland for that "Þingavöllr," or Tingvalla, which anciently was the chief *Thing* place of the island, as is expressly mentioned in old records. We should be just as unsuccessful in finding traces of the lesser *Things*, which, in Shetland, as we have seen, can almost all be still pointed out in the names of places; and this notwithstanding we know for a certainty that the Orkneys had a court of justice in common with Shetland, till the year 1196 at least; from which time Shetland was governed by its own laws. The same powerful Scottish influence has likewise effaced in the Orkneys most of the few Norwegian words, customs, and manners which still sustain a feeble existence in the remote islands of Shetland. The Norwegian language, some vestiges of which might be traced, in the last century, in the parish of Haray (Herað), has left behind it only a peculiar singing pronunciation, and some few characteristics in the English language now in use there; thus, for instance, in addressing a person, the nominative and accusative *thou* and *thee* are used, instead of *you*. The present language of the Orkneys is almost a purer English than that of the Scotch Lowlands; which is a natural consequence of English having begun at a later period to be the ruling language in the islands. The present population of Mainland, together with the other inhabitants of the Orkneys, has undeniably preserved a certain Scandinavian appearance; and English civilization has, among other things, both sharpened the people's innate inclination for a maritime life, and increased their coolness towards, not to say ill-will and contempt for, the Gaelic Highlanders. On the whole, however, Scandinavian characteristics are by no means conspicuous among the people. English civilization, and Scotch-English institutions, have been introduced to such a degree into Mainland, and thence into the other islands, that a traveller would not know he was in the chief country of the former mighty Norwegian jarls, unless he were able to decipher the frequently transformed names of places; or, above all, unless he had such a general knowledge of the island's history and antiquities that he could apprehend, and in some degree interpret, the hints given by silent monuments of the brilliant but long-departed age of heroes.

The memory of the warlike life of heathenism is conspicuously preserved in Mainland by the many large barrows, or tumuli, which meet the eye on all sides. It is, indeed, certain that several of these—viz., what are called the "Picts' houses," which form in their interior stone chambers, covered by small flag-stones laid over one another—must be ascribed to the older inhabitants of the island; yet enough remain which we may with good reason attribute to the Norwegians and Danes. They are not, like those tumuli, or "cairns," which are found most frequently in the north of Scotland, a mass of small stones heaped together without any filling-in of earth, but are formed, like our Scandinavian barrows, of earth thrown up to a very considerable height. As in Scandinavia, they are met with mostly on hills, and near the firths or seacoasts, whence there is an uninterrupted view of the sea. To the ancient Northman it was evidently an almost insufferable thought to be buried in a confined or remote corner, where nobody could see his grave or be reminded of his deeds. The greater chief a man was the more did he desire that his "barrow" should lie high and uninclosed, so that it might be visible to all who travelled by land and by sea. United with this desire to live in the memory of posterity, the Viking certainly also indulged the secret belief, that his spirit, or ghost, would at times arise from the barrow to look out upon that beloved sea, and to refresh itself, after the gloomy closeness of the grave, with the cool breezes which play upon its bosom.

Some of the largest and most prominent barrows in the Orkneys are found about the middle of Mainland. To the west of the deep fiord in the middle of the east coast, (formerly Örreðfjord "Aurriaðfjördr," *i. e.* Trout firth, but now called Firth), and cutting its way northwards far into the land, is the before-mentioned Loch of Stennis, with its famous old Celtic stone circles. But the largest of these, which lies on the ridge of a naze, or promontory (from Old N. "Steinsness"), is encompassed by twelve considerable, and partly perhaps Norwegian or Scandinavian, barrows; amongst which two in particular, to the north-east and north-west of the circle, are distinguished by their size and circumference. As the Saga informs us that it was on Steinsnæs that the chief, Einard Klining, at the instigation of Erik Blodöxe's daughter, Ragnhilde, killed her husband Jarl Haavard, it is not impossible that one of the last-named large barrows may be the jarl's grave. At all events it is natural enough that the Norwegians should have had a predilection for being buried on that lofty promontory, which was regarded even by the earlier inhabitants of the island as a holy place, and had been adorned by them with a truly imposing circle of immense blocks of stone. Future excavations will doubtless more clearly show which of the barrows are really Norwegian; but this much is certain—that the naze, with the circle of stones and the surrounding barrows, as well as the view of the three immense monumental stones, placed erect in a semicircle on the opposite side of Loch Stennis, afford a prospect not only interesting to the antiquarian, but which must strike every beholder.

Here and there, on Mainland, we meet with graves of the heathen times, which are not at all uncommon in the Orkneys and Shetland Isles. They are, however, of much lower elevation than those previously mentioned, and in general rise very little above the surface of the soil. In some of these, as in Shetland, besides urns, containing burnt bones and ashes, bodies have at times been found that have been buried without being burnt; together with swords of the Scandinavian kind before described, heads of lances, daggers, and knives; as well as bone combs, bowl-formed brooches of brass, and various other ornaments, evidently of Norwegian workmanship.

Just as the barrows, or grave hills, in Mainland, indicate by their peculiar size that in the heathen times the island was the chosen place of assembly for the mightiest men in the Orkneys and Shetland Isles, so also do the monuments of the early middle ages show that it continued to maintain its former pre-eminence after heathenism had ceased. Farthest towards the north-west, in the parish of Birsay, (Birgisherað), are to be seen considerable remains of the old castle, inhabited in the most ancient times by the jarls. Near the coast lies the Island of Brough (Burgh) of Birsay, on which also are seen traces of fortifications that have served to protect the jarls' castle on the side of the sea. In the neighbourhood of this castle, Jarl Thorfin built a church, called Christ Church, in which both he and Jarl Magnus were buried. The latter, however, being afterwards canonized, his body was taken to Kirkwall. In the twelfth century, Bishop Wilhelm, the first bishop of the Orkneys, had his throne in this church. In Orphir

("Orfjara"), on the south coast of the island, was another castle where the jarls usually dwelt, until, together with the bishops, they fixed their abode at Kirkwall.

This town, which lies close to an excellent harbour, and opposite the Island of Shapinsay, has for about seven hundred years been the capital of the Orkneys and the Shetland Isles. It seems, however, to have existed even earlier, as a village, or small trading place. Its name, "Kirkjuvágr" ("Kirkevaag," *Eng.* Church-bay), since corrupted into Kirkwall, was derived from a church which stood there. The elevation of the town to be the residence of jarls and bishops took place in the twelfth century, after Jarl Ragnhild had built a large cathedral there, to which he caused to be conveyed the body of St. Magnus, the patron saint of the island, to whom the cathedral was consecrated. Thus the body of the saint effected for the town what its excellent harbour had not been able to accomplish. In the parish of St. Ola', within the town, there was formerly also a church consecrated to St. Olaf, the patron saint of Norway, but it has long since been demolished.

The traveller cannot but dwell, when in Kirkwall, on the remembrance of the departed splendour of the island, as he views the proud ruins of the jarls' castle, which, however, in its last form was not built till the fifteenth century, and of the bishops' castle, in which King Hakon Hakonsön of Norway died on the 16th of December, 1263. But what is still more striking to him who has leisure to examine it thoroughly, is the magnificent Church of St. Magnus, incontestably the most glorious monument of the time of the Norwegian dominion to be found in Scotland. Only one other cathedral church in all Scotland, namely, St. Mungo's, in Glasgow, has in its most essential parts escaped perfectly uninjured from the violent religious commotions produced by the Reformation. The annexed sketch (partly after a drawing by Billings) will, at least, better serve to convey an idea of the remarkable appearance of this cathedral than any detailed description. Its length is 230 feet, its breadth 55 feet, or, if the transepts be included in the measurement, 101 feet, and its height about 50 feet. The arched vaults of the nave rest on 28 pillars, of which the four, in particular, that bear the tower are distinguished by their size and tasteful forms.

According to the Orkneyinga Saga, Jarl Ragnvald, by the advice of his father Kol, made a vow to St. Magnus that he would build a splendid church in his honour, if he (Ragnvald) succeeded in gaining the mastery over the islands. He obtained the dominion of them in the year 1137, and immediately afterwards began to lay the foundation of St. Magnus' Church. "At first," says the Saga, "the work went on so rapidly that subsequently there was not done near so much in four or five years. Kol was the person who, in fact, defrayed the expenses of the building, and determined how everything was to be. But by degrees, as the work proceeded, the expenses became burthensome to the jarl, whose pecuniary means were much exhausted. He therefore asked his father what he should do? Kol advised him to alter the law by which, upon the death of the owners, the jarls had hitherto succeeded to all the allodial land in the islands, so that the heirs had to redeem it, which they found very hard. The jarl, therefore, summoned the inhabitants to a *Thing*, and offered to sell them their right of Udal, so that they should no longer be obliged to redeem it. The matter was easily arranged on both sides. The jarl obtained a mark for every acre throughout the islands, so that there came in money enough for the building of the church, which is very handsome."

[+ +] St. Magnus' Church

History, however, as well as the building itself, teaches us that the whole church, as it now stands, was by no means the work of Kol and Ragnvald. For, first, it is known that the pillars farthest towards the east and west, marked in the annexed ground plan with the faintest shade, belong to additions made at a far later period (viz., as late as the sixteenth century); and secondly, it is not even decided whether Kol and Ragnvald built the whole of the remaining part of the church, the transepts included, or whether they built only that part of the present choir which, from the two eastern pillars of the tower, comprises the six nearest pillars to the east, marked on the ground plan with the darkest shade. Between this last-named portion of the choir, which is undoubtedly the oldest part of the church, and the portion lying to the west, whose pillars on the ground plan have a rather lighter shade, there is a perceptible difference of style.

That zealous and skilful archæologist, Sir Henry Dryden, Bart., of Canons Ashby, to whom I am indebted for the original of the following ground plan, likewise did me the favour to give me, among several large drawings, a very excellent, but here very reduced, section of that part of the choir which is certainly known to have been built by Kol and Ragnvald. The section is taken from the middle of the nave, and represents a part of the northern side walls nearest to one of the pillars of the tower. It enables us to form an idea of the very considerable size of the church, and of the importance of Kol's and Ragnvald's labours, as well as readily to perceive in what style the church was originally built. This style, which in England is called the Norman, was indeed already somewhat obsolete in more southern districts at the time when St. Magnus' Church was built; but it was quite natural that, so far northwards, it should be retained somewhat longer, especially as the architect was a native of the still more northern country of Norway.

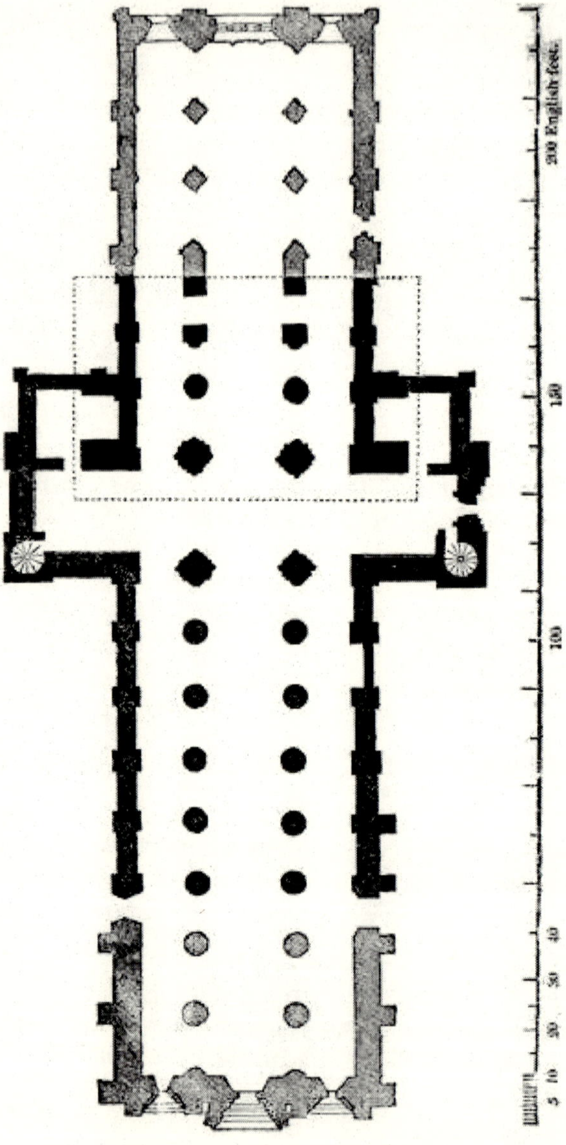

[+ +] St. Magnus' Church - Floorplan

[+ +] St. Magnus' Church - Interior

The next considerable portion of the cathedral which might possibly have been built by Kol and Ragnvald, or at least about their time, and which includes the transepts, the two western pillars of the tower, and the six pillars (three on each side) farther towards the west, has, indeed, like the very oldest part, round arches. But in these, as well as in the whole architecture, a much later style is clearly visible. It is, as we have said, doubtful whether this part of the church is also to be ascribed to Kol and Ragnvald. "Supposing that it is (says Sir Henry Dryden, in a letter accompanying the drawings), I explain the difference of scale and workmanship thus. Ronald began a church on a *much* smaller scale than the present St. Magnus. He became short of money, alienated seignorial rights in Orkney, got plenty of money, and went on with the church on a larger scale, and with better workmen than before. But (adds Sir Henry), though I spent eighteen weeks at the building, and have thought over the thing many times, I cannot make out the history of the building to my own satisfaction. There is no doubt that there is a great deal of copying in it; *i. e.*, of building at one time in the style of an earlier one. In Scotland the semicircular arch is used in all styles, down to the year 1600." In the additions made to St. Magnus' Church to the east and west, in the sixteenth century, round arches are also found between the chief pillars.

In the winter of 1263-1264 the body of the Norwegian king Hakon Hakonsön was deposited in the cathedral; and somewhat more than twenty years afterwards the Norwegian princess Margaret (the maid of Norway), daughter of King Erik, the priest-hater, and of Margaret, daughter of the Scotch King, Alexander the Third, was buried in it. Upon the death of Alexander, her mother's father, in 1289, Margaret, though only seven years of age, became queen of Scotland, but died in Orkney on her passage from Norway, in 1290. The cathedral naturally received the dust of most of the Norwegian jarls, bishops, and other mighty men, so long as the Norwegian dynasty lasted; but for their monuments we now seek in vain. By the alterations and rebuilding in the interior of the church they have all been long since destroyed.

For a Scandinavian, the church derives its greatest interest not only from the fact that it was founded, and partly built, by a Norwegian jarl, but more particularly from the circumstance that a Norwegian chief, the layman Kol, is expressly stated to have been the person "who was chiefly answerable for the building, and determined how everything should be." For we thus find on the British Islands, and far towards the North, a manifestation of the same desire to build splendid churches and convents, which farther southwards, as for instance in Normandy, so vividly animated the Christian descendants of the emigrant Vikings. The oldest part of St. Magnus' Church will, on a close inspection, show not a few resemblances to several of the nearly contemporary, but somewhat older, Norman churches in Normandy.

SECTION VIII

Pentland Firth.—The Highlands.—Caithness.—Sutherland.—Fear of the Danes.

The Orkneys are separated towards the south from the most northern part of the Scotch Highlands by a firth about eight miles in breadth, called Pentland Firth (*Old N.*, Petlandfjörðr, the fiord of the land of the Picts?). The maelstrom, or whirlpool, in this firth, where the currents from the North Sea and Atlantic Ocean meet, is at least as violent and dangerous as the "Röst," so famed in ancient times, between the Orkneys and Shetland. Even in calm weather the meeting currents raise the waves to an astonishing height, so that at times the whole firth is one sheet of white foam. If it happens that the current runs hard against the wind, or if a severe storm blows, it would not be advisable for any vessel to venture out into the firth. In the gales of winter, particularly from the north-west, the sea rises to such a height where the huge swell of the Atlantic is inclosed between the Orkneys and Scotland, and beats against the coast with such force, that the foam is driven far into the country, even over cliffs that stand more than four hundred feet above the sea! The Island of Stroma (*Old N.*, "Straumsey"), which has obtained its name from the current, lies about the middle of the firth; and by the eastern entrance of it are the Islands of Pentlandskerries (*Old N.*, "Petlandsker;" or *Danish*, "Pentlandskjære;" *Eng.*, sunken rocks off the Pentland Firth), near which the waves form whirlpools that are still called by the inhabitants "Swelchies" (or Svælg: *Old N.*, "Svelgr;" *Eng.*, gulf).

The old Sagas, indeed, expressly point out the dangers of the Pentland Firth. Thus, when Olaf Trygvesön came from the West to the Orkneys with the intention of Christianizing the islands, he was obliged to run into the harbour of Asmundarvág (now Osmondwall) in the south of Hoy, because Pentland Firth was not navigable; and on the return of King Hakon Hakonsön from the Hebrides in 1263, one of his ships was lost in the Röst, and another escaped only with the greatest difficulty. Nevertheless the ancient Norwegians and Danes navigated this dangerous firth regularly, and do not seem to have considered it as forming any real boundary between the Orkneys and Scotland. At an early period the Norwegians had settled themselves along the south coast of the Pentland Firth, and founded colonies there which soon became so preponderatingly Norwegian that they might almost be regarded as inseparable parts of the Orkney jarldom. On this account the two most northern counties of Scotland, both of which united originally bore the Gaelic name of Catuibh, are still called after the original Norwegian forms, "Caithness" (*Old N.*, "Katanes," the naze of Catuibh) and "Sutherland" (*Old N.*, Suðrland), or the land in the south; that is, as regards the Orkneys. It would be perfectly inexplicable, in any other way, why the north-western part of Scotland should be called the south land, or Sutherland. It is, moreover, a remarkable proof of the Norwegian origin of these names, that even the present Gaelic inhabitants do not adopt them, but always call Sutherland, after the old fashion, "Catuibh." For the sake of distinction, however, they call Caithness "Gallaibh," or the stranger's land, because so many Norwegians immigrated to, and settled in, that county in preference to Sutherland.

The district of Caithness, or, as it was often called in ancient times, "Næsset," forms a real naze, shooting out into the sea in a north-eastern direction. Its farthest point towards the north-east is called Duncansby Head (formerly "Dungalsnýpa"), from the neighbouring Duncansby (formerly "Dungalsbœr"). The broadest bay on the north coast trends in between the promontories of Dunnet Head and Holburn Head; the latter of which, by protecting Thurso Bay from western and north-western gales, renders it a tolerably good harbour, in a place where good harbours are scarce on this northern coast. Supposing, now, that we land in the Bay of Thurso, by the town of that name, we soon discover the outlet of the rivulet called Thurso Water (*Old N.*, "Þorsá," or Thorsaa, Thor's rivulet), which has given the easily-recognised Scandinavian name both to the town and bay. The town and its immediate environs afford a great number of Norwegian memorials. The Norwegian king Eistein imprisoned the Orkney jarl Harald Maddadsön in Thurso itself. Close to the eastern side of the town stands a more recent monument, "Harald's Tower," erected over the body of Jarl Harald, who fell there in a battle in 1190.

Not far from thence is the mansion called Murkle (formerly "Myrkhóll"), where, in the tenth century, Ragnhilde, the daughter of Erik Blodöxe and of Gunhilde, caused her husband, Jarl Arnfin, to be murdered. Immediately to the west of the town, near Scrabster ("Skarabólstaðr"), are to be seen the ruins of the palace formerly inhabited by the bishops of Caithness and Sutherland. In the twelfth century Bishop Ion was blinded and mutilated there, at the instigation of Jarl Harald. Five miles west of Scrabster, and close by a foaming waterfall, stands the mansion of "Forss," by the river Forss Water. The rivulet called Thorsaa runs through a valley in ancient times called Thorsdal ("Þórsdalr"), adjoining another valley "Kálfadalr," or Calf-dale (either the present Calder or Cuildal), in which Jarl Ragnvald was attacked and killed by Thorbjörn Klærk. In the "Dales of Caithness" (probably near Dale and Westdale, by Thurso Water) a battle was fought in the tenth century between Jarls Ljot and Skule, in which the latter fell.

Similar memorials present themselves everywhere on the promontory, with the exception, however, of the most western and more mountainous part, adjoining the frontiers of Sutherland. This district is still inhabited by a Gaelic population, the remnant of the ancient inhabitants, as is sufficiently testified both by the Gaelic names of places and the Gaelic language of the people. In Caithness, as well as everywhere else in the British Isles, it has been the fate of the Gaels or Celts to be driven to the poor and mountainous districts, whilst more fortunate strangers have taken possession of the fertile plains. The whole of the northern and eastern part of Caithness is a rather flat and open country, over which the sea wind sweeps freely without being intercepted by woods. Fertile and well-cultivated arable land is mingled with heaths, marshes, and small lakes. Wherever the soil is capable of cultivation, both on the coasts and in the interior, a great number of undoubted Norwegian names of places are still found scattered about, of the selfsame form as those in Orkney and the Shetland Isles: as, for instance, those ending in *toft* (as Aschantoft, Thurdystoft, formerly "Þorðarþupt") *seter* ("setr"), *busta*, *buster*, or *best* (originally "bolstaðr"); but particularly in ster (staðr). The bays, which are mostly small and narrow, are generally called *goe* (from "gjá," an opening). The larger ones are called *wick* (Viig); whence the town of

Wick, the most important hamlet in Caithness, derives its name; but they are never called, as in the islands lately mentioned, *wall* ("Vágr," or "Vaag"). Here and there a mighty barrow lifts its head, and sometimes—as, for instance, near Barrowston, parish of Reay—so extremely near the coast of Pentland Firth, that the spray washes over it. In general we shall not be mistaken in imagining that we have found in such barrows the last resting-places of the daring Vikings, who, not even in death, could endure to be far separated from the foaming maelstrom.

At times the common people dig up in these mounds pieces of swords and various kinds of ornaments, especially the peculiar bowl-formed brooches, of a sort of brass, which are very frequently discovered in the Scandinavian North, and particularly in the Norwegian and Swedish graves of the times of the Vikings. These are never found in England; and in Scotland they are discovered only in the Orkneys and Sutherland, as well as in some of the Western Islands, where the Norwegians also settled.

[++] Brooch

Tall bauta stones are to be seen in several places in Caithness, to which some legend about "the Danes" is generally attached; they now stand in a leaning position, as if mourning over the departed times of the heroic age. A monument of a somewhat later period, according to tradition that of a Danish princess, who suffered shipwreck on the coast, was also formerly to be found in a churchyard near Ulbster. Danish fortifications, consisting partly of square towers, once existed along the coast, principally near the navigable inlets; but these also have now, for the most part, disappeared.

With several intervals, Caithness was subject to Norwegian jarls until some time in the fourteenth century, or for about as long a period as Orkney and the Shetland Isles. After that time, however, it does not seem to have been oppressed to such a degree as those islands; which circumstance, in conjunction with the originally great number of Norwegian settlements in the country, is the cause that even in the present day we are not referred only to inanimate memorials of the ancient Norwegian population. The present living inhabitants bear a decided and unmistakable impress of their Norwegian descent. The language in the plains of Caithness, and in the open valleys, is the same dialect of the English as is spoken in Orkney and the Shetland Isles, because the transitions from Norwegian to English have been the same. The people have in some parts, as in the parish of Wick, pure Scandinavian names: Ronald (Ragnvald), Harold, Swanson (Svendsen), Manson (Magnuson), and others; and their tall and personable figures, as well as their light hair and broad faces, render them a striking contrast to the shorter and more swarthy Highlanders. As the descendants of an old Gaelic and of an old Norwegian population adjoin one another in Caithness, we have an excellent opportunity of observing, on a small scale, how the Norwegians and Danes have actually implanted in the British Isles a more seafaring spirit and greater nautical skill. Even to the present day the Gael, in Caithness, as well as throughout the Highlands, has a decided aversion to the sea, nay, a downright fear of its dangers. It is pretty well known that in general, and except on the most urgent necessity, one should not venture out into the Pentland Firth in boats steered and rowed by Gaels or Highlanders; for, in the event of a storm, all steady command is speedily lost, and gives place to anxious irresolution. The descendants of the old Norwegians, on the contrary, who are familiar with the sea from childhood, and amongst whom lies Wick, the most important fishing station in Scotland, show themselves precisely in the hour of danger the worthy sons of their forefathers, the ancient Vikings. It is only the man at the helm who speaks, and he gives his orders in a few decisive words. He is punctually obeyed, and the misfortune is said to be rare, if his coolness, joined to his knowledge of the sea and its currents, do not gain the victory over the violence of the storm and the turbulence of the billows. This seafaring population of Caithness do not, like the Highlanders, disdain to resort to fishing, in order to bring home the riches of the sea. As their soil, moreover, is by no means barren, and as they have naturally greater activity and more inclination to work than the Highlanders, as well as, through their English dialect, greater facility in their traffic with the more southern districts, it is not to be wondered at that the prosperity of Caithness manifests a great and constant progress. We may even justly assert that the descendants of the Norwegians in Caithness are in a far more fortunate situation than their kinsmen in the Orkneys and Shetland Isles.

In ancient times, a Norwegian population speaking its native language, was undoubtedly spread over the whole eastern coast of Caithness, as well as over several districts of Sutherland. But the English language, which in our times has superseded the Norwegian, ceases to be the common language of Caithness immediately to the south of the parish of Wick. A line drawn from Clyth Ness, in a north-western direction to the before-mentioned mansion of Forss to the west of Thurso, will indicate, as near as may be, the boundary between Gaelic and English. If, however, we travel southwards from the parish of Wick, through the parish of Latheron, where the common language is already Gaelic, we, nevertheless, pass a great many villages and farms bearing Norwegian names; as, for instance, Lybster and Forse (by a waterfall). The mountains here begin to be higher, and to stand closer and closer together towards the sea. At length, after passing the deep valley of Berrydale (*Old N.*, "Berudalr"), and the beautiful wood-crowned banks of its river, we ascend the steep mountain ridge called "the Ord of Caithness," which runs boldly out into the sea, and forms a natural boundary between the narrow projecting promontory of Caithness and the broader Sutherland.

The first large valley in Sutherland to the south of this mountain ridge is Helmsdale, which is watered by a river of no mean size. That Helmsdale is a Norwegian name (in the Sagas "Hjalmundsdalr") is at once evident from the present Gaelic inhabitants calling the valley in pure Gaelic, "Strath Ullie," or with a strange confusion of language, Strath Helmsdale; for as Strath signifies in Gaelic a valley or dale, the word *dale* is added both at the beginning and end. It is a similar repetition which we so often hear when the "Orkney Isles" are spoken of, in the original language "Orknö," but which, translated as now used, is Orknö Öerne (or the "Orkney-islands-islands"). Along Helmsdale River several places are met with whose original Norwegian names are still to be discerned; as, for instance, Eilderabol, Gilaboll, Dviaboll, and Leiraboll. All these have the ending *bol*, which is peculiar to a number of Norwegian names of places in Sutherland and in some of the Hebrides; but which, in Caithness, the Orkneys, and Shetland Isles, as well as in Lewis and several of the Hebrides, appears in the longer form of "bolstaðr." To the north-west of Helmsdale are the vales of Kildonan, which run up as far as the Vale of Strathmore in Caithness. Here, it is supposed, on the frontiers of Caithness and Sutherland, lay "Eisteinsdalr," so famed in history as the spot where the Scotch king William encamped in the year 1198. It is, however, very uncertain whether "Easterdale" in Strathmore be in any way connected with the name of Eisteinsdal.

On leaving Helmsdale the coast opens, and fertile and beautiful fields begin to expand themselves. Past Midgarty and Wester Gartie (the middle and western Gaard, or farm, from *Old N.* "garðr"?) the road runs along the shore of the Bay of Dornoch (an arm of the "Breidifjördr," or broad firth mentioned in the Sagas, in which the Moray Firth is also included) to the little village of Brora, which is built on a considerable river, and where for a long period the only large bridge in Sutherland was to be found. It was possibly from this circumstance that the Norwegians gave the village its name ("Brúrá," the bridge rivulet). A river in Iceland is also still called Brúrá, from a bridge which crosses it. The ancient seat of the Earls of Sutherland, Dunrobin (Robin's tower, from *dun*, a tower), lies on the seashore, in the neighbourhood of Brora, surrounded by fine corn-fields and considerable tracts of woodland. The latter, however, were planted at a recent period. In the background rise considerable mountains, covered with heath. In this place, so highly favoured by nature both as regards scenery and fertility, the Norwegian jarls who ruled over Sutherland undoubtedly had one of their chief residences; as, for instance, Sigurd Jarl, a brother of Ragnvald Möre-Jarl, Sigurd the Stout (+ 1014), and his son Thorfin (+ about 1064). Norwegian antiquities, like those discovered in Caithness, are found in graves near Dunrobin, particularly the well-known bowl-formed brooches or buckles. In the neighbourhood several places with Norwegian names can be pointed out; for instance, just south of Dunrobin, in the fertile valley by the river Fleet, Mickle Torboll and Little Torboll (from *Thor* and *bol*); and on the coast, Skelbo, Skibo, and Embo (from *bol*, or perhaps more correctly from *bœr*, *bö*). Sigurd, the first conqueror of Sutherland, is said to have extended his dominion as far as Ekkjalsbakke. As *bakki* in the ancient language signifies the bank of a river, there cannot be the least doubt that Ekkjal is the river Oykill, which still forms the southern boundary of Sutherland. Sigurd himself is said to have been interred at Ekkjalsbakke. He gained the victory in a foray over the Scotch jarl Melbrigd, and cut off his head, which, in the overweening pride of his triumph, he hung to his saddle; but a sharp tooth that projected from the head chafed his leg, and caused a wound which proved his death. On different parts of the banks of the Oykill numerous barrows are seen, indicating the many battles that have been fought in ancient times on the frontiers of Sutherland. But nobody is able to point out the barrow of Sigurd Jarl; the tradition relating to it has vanished with the Norwegian population.

For the rest, names of places prove that the Norwegians had also settled themselves along the coast to the south of the Oykill. On the narrow naze called Tarbet Ness, between Dornoch and Cromarty Firths, are the villages of Arboll and Wanby, as well as the town of Tain, whose Gaelic name, "Bailed Dhuich" (or St. Duthus' Town), shows at once that "Tain" must be of foreign origin. Tain is, moreover, a corruption of "Þing," a *Thing*; and in like manner the somewhat considerable town of Dingwall, at the extremity of Cromarty Firth, was originally called "Þingavöllr," or *Thingwalla*; whence the remarkable fact is evident, that the Norwegians were once sufficiently numerous in these districts to have both an inferior *Thing*

(Tain) and a superior one (Dingwall). Dingwall, like Tain, besides its original Norwegian name, has also the Gaelic one of Inverphaeron. As the Norwegians, therefore, must have permanently possessed considerable tracts in these districts, it is clear that their settlements on the east coast of Scotland must have extended quite down to Inverness-shire and Moray. The before-mentioned stronghold of Burghead in Moray, which the Northmen maintained to the last extremity, lies pretty close to the east of Cromarty Firth, the inlet to Dingwall.

As the Norwegian language and other Norwegian characteristics have given way to the Gaelic tongue, manners, and customs, in the former Norwegian districts on the north coast of Scotland, from Clyth Ness in Caithness to Dingwall on the Firth of Cromarty, we can scarcely be surprised that the north coast of Sutherland, whose rocks and heaths offered much fewer allurements to the Norwegians than the fertile valleys and plains of the east coast, and which were therefore far less colonized by them, should have preserved distinct traces of these foreign conquerors only in a few names of places. A remarkable instance of the Gaelic language having expelled the Norwegian is to be found immediately on the borders of Caithness, in the valley of Halladale. In a river there are two waterfalls, of which the uppermost is called Forsinard, and the lower one Forsindin. In both these names the Norwegian "Fors" is not to be mistaken; but Gaelic terminations have in later times been added by the Gaels, so that Forsinard now signifies the upper Fors, and Forsindin the under, or lower, Fors. Halladale is likewise frequently called by the additional Gaelic name of Strath—"Strath Halladale."

This much, however, is clear, that the whole of the north and west coast of Sutherland was once colonized by Norwegians. Besides various names of places west of Halladale, which likewise end in *dale*, such as Armadale, Swordale, and Torrisdale, it is surprising that we should still meet with pure Norwegian names on four of the largest firths of the north-west of Sutherland; viz., on the north coast the "Kyle of Tongue" (from "túnga," a tongue of land, a naze), together with the adjoining village, Kirkiboll (Kirkebolet); further, Loch Eriboll, with the large farm of Eriboll (the *bol* on the Eir, or tongue of land, from the *Old N.* "eyri"); the Kyle of Durness, or Dyrnæs, with the *bol*, or dwelling, of Crossboll; and lastly, on the west coast, not far from Cape Wrath, Loch Laxford (Laxfjorden, or the Salmon Firth; *Old N.*, "Laxafjörðr"). "Loch" is the Gaelic name for a lake or firth, and consequently, in Loch Laxford, expresses tautologically the existence of a fiord or firth; just as the name "valley" is twice expressed in Strath Helmsdale and Strath Halladale. The last three of the above-mentioned firths seem to have been of much importance to the Norwegians. There is an excellent harbour in Loch Eriboll, which is still frequented by numerous ships. The neighbourhood round Loch Durnes afforded excellent opportunities for hunting the deer, particularly on Durnæs itself, which extends between Loch Durnes and the Atlantic up to Cape Wrath (*Old N.*, "Hvarf"), and which, still later in the middle ages, was celebrated for its excellent deer. Loch Laxford, which obtained its name from the salmon (Lax) in the river and at its mouth, is commonly known to the present day as one of the rivers in Scotland most abounding with that fish. Several isolated rocks in the sea by the coast of Sutherland are called, as in the Shetland Isles, "stacks;" and in several names of islands we meet with the Scandinavian *sker* or *skjær*; such as Skerroar (Skjæröerne, the rock islands); and in Loch Eriboll, Dhusker, Skerron, and others. A little island near the middle of the west coast is called Calva (*Old N.*, "Kálfey," or the Calf Island), a name frequently given by the Northmen to small islands that lay in the neighbourhood of a larger one (for instance, the Calf of Man). For the rest, Calva is one of the last decidedly recognisable Scandinavian names of places on the west coast of Sutherland. The real Norwegian population evidently ceased at Laxfjord. Norwegian names of places are scarcely to be found on the coasts of the Highlands to the south of Sutherland. The country there was so wild, rocky, and remote, that foreign conquerors could only with the greatest difficulty have maintained a position against the Highlanders, who were always prepared to make sudden and dangerous attacks from the mountains in the interior. Aware of this, the Norwegians seem to have limited themselves, on the western shores of the Highlands, chiefly to the levying of provisions along the coast, and to the plundering of cattle and other property. Round about the mouths of the Highland firths are still to be seen the remains of old castles, which the Scotch kings, and particularly Alexander the

Second, are said to have built, in order to prevent "the Danes" from making these devastating descents.

The memory of the conquests and predatory incursions of the Norwegians, or "Danes," is still preserved in a remarkable degree among the poorer classes in Sutherland, as well as in the rest of the Scottish Highlands. Numberless traditions are in circulation respecting the levying of provisions by "the Danes;" and barrows, or cairns, are not unfrequently pointed out, in which a Scandinavian prince, or king's son, killed by the natives whilst on some Viking expedition, is said to be buried. Besides the usual cruelties ascribed to the Danes in the traditions of the Lowlands, and of England, they are here accused, into the bargain, of having burnt the forests, and thus caused that want of wood which acts so injuriously on the climate of the Highlands. In proof of this it is adduced that roots and trunks of trees, sometimes perceptibly scorched, are discovered in the turf-bogs of the Highlands. It is not considered that similar discoveries are very common in other countries, as, for instance, in Denmark itself; where trunks of trees, especially firs, have been dug up, precisely as in the Scotch Highlands. They are the produce of vegetative processes in the pre-historical times; and the apparent scorching has been produced either by accidental fires, or more, probably, by the simple mode of felling trees in use among the aboriginal inhabitants of Europe; who, like certain savage tribes at the present day, for want of metal tools, were obliged to burn the trunks of trees which they wished to fell.

By way of amends, the Danes have now and then the honour of being regarded in the Highlands as having been the teachers of the natives. One of the first jarls of the Orkneys was, according to the legends, called by the name of Torf Einar, because he was the first who caused turf to be dug on a point of land (Torfnæs) in Scotland. This promontory, probably the present Tarbet Ness, was at all events either in Caithness or Sutherland; and it is certainly a remarkable coincidence, that the common people of that district still relate that "the Danes" taught them to burn turf. We likewise hear at times that "the Danes" taught the use of hand querns, or hand-mills; nay, even that the favourite national instrument of the Highlanders, the bagpipes, was originally introduced by the Danes. In short, if anything, whether good or bad, be of doubtful origin, it is frequently attributed to "the Danes."

But it is peculiar to the north-western and most remote districts of the Highlands, that the common people still harbour no small degree of dread lest "the Danes" should return, and repeat their cruel devastations. About thirty years ago (according to J. Loch, "An Account of the Improvements on the Estate of the Marquis of Stafford," London, 1820, 8vo.), English engineers were employed in measuring all the heights in Sutherland. This caused much sensation among the natives, who thought that these engineers were sent by the Danes to make maps and plans of the country, previously to the arrival of the Danish army. They imagined that the king of Denmark had an old feud with the Mackays, and that he was now coming to take a sanguinary revenge on the whole clan.

During my stay in Sutherland I had repeated occasion to convince myself not only that the fear of the Danes has not yet died away there, but also that tradition has connected with them things with which they had nothing whatever to do.

Close outside the town of Dornoch, on the east coast of Sutherland, there stands a stone pillar in an open field, which is simply the remains of one of those crosses so frequently erected, in Roman Catholic times, in market-places. As a matter of course, the arms of the jarls of Sutherland are carved on one side of the stone, and on the other are the arms of the town—a horse-shoe. Tradition, however, will have it that the pillar was erected in remembrance of a battle fought on this spot, in which the Jarl of Sutherland commanded against "the Danes." In the heat of the battle, while the Jarl was engaged in personal combat with the Danish chief, his sword broke; but in this desperate situation he was lucky enough to lay hold of a horse-shoe that accidentally lay near him, with which he succeeded in killing his adversary. The horse-shoe is said to have been adopted in the arms of the town in remembrance of this feat. In the cathedral church of Dornoch is a carved stone monument of the middle ages, representing one of the ancient bishops who once resided in Dornoch. He also is said to have fallen in the same battle, but my authority, the person who showed me over the church, added:—"I am proud to tell that the Danes were defeated."

Having employed myself in examining, among other things, the many so-called "Danish" or Pictish towers on the west and north-west coast of Sutherland, the common people were led to believe that the Danes wished to regain possession of the country, and with that view intended to rebuild the ruined castles on the coasts. The report spread very rapidly, and was soon magnified into the news that the Danish fleet was lying outside the sunken rocks near the shore, and that I was merely sent beforehand to survey the country round about; nay, that I was actually the Danish King's son himself, and had secretly landed. This report, which preceded me very rapidly, had, among other effects, that of making the poorer classes avoid, with the greatest care, mentioning any traditions connected with defeats of the Danes, and especially with the killing of any Dane in the district, lest they should occasion a sanguinary vengeance when the Danish army landed. Their fears were carried so far that my guide was often stopped by the natives, who earnestly requested him in Gaelic not to lend a helping hand to the enemies of the country by showing them the way; nor would they let him go till he distinctly assured them that I was in possession of maps correctly indicating old castles in the district which he himself had not previously known. This, of course, did not contribute to allay their fears; and it is literally true, that in several of the Gaelic villages, particularly near the firths of Loch Inver and Kyle-Sku, we saw on our departure old folks wring their hands in despair at the thought of the terrible misfortunes which the Danes would now bring on their hitherto peaceful country.

SECTION IX

The Hebrides.—The Northern Isles: Lewis and Harris; (Næs);

Skye.—Ossian's Songs.—Iona.

The rocky western coast of the Highlands south of Sutherland was not, as I before mentioned, permanently inhabited by the Norwegians. They had, indeed, regular settlements on the west coast, but these were on the islands. They were here secure from the sudden attacks of the Gaels, or Highlanders, who, generally speaking, would scarcely have ventured out on a sea which then swarmed with Vikings. The farther, therefore, the islands were from the mainland, so much the more secure would the Norwegian settlers be, and so much the greater, in effect, did their colonies become. By degrees they settled themselves on all the islands along the west coast, from Lewis to Man, which they called under one name, "Suðreyjar," or the southern islands, from their situation with regard to the Orkneys and Shetland Isles. Sometimes, however, they did not reckon Man among them, and then divided the rest of the islands into two groups, in such a manner, that only the islands to the south of Mull were called "Suðreyar," whilst Mull itself, and the islands to the north, obtained the name of "Norðreyar." The Irish, and the rest of the Gaels, on the contrary, after the conquest of the islands by the Norwegians, called them "Inis Gâl" (the foreigners' isles).

The most northern and largest of the northern isles was the extensive one which forms the present Lewis and Harris (the "Ljóðhus" of the Sagas). It is separated from Scotland by the broad, stormy, and troubled channel called the Minch. The southern part of it only, or Harris, where the mountains reach the height of between two and three thousand feet, can be called mountainous, for the rest of the island is rather flat, devoid of wood, and covered with heaths and moors. Some good arable land is, however, to be met with here and there along the coasts. Even in very early times this island was very densely inhabited by the Gaels, of which, among other things, some immense rows of stones, near Callernish, bear witness. In like manner, the Norwegians must, at a later date, have had considerable colonies in it. On this head we must not, of course, implicitly rely on the numerous traditions related by the common people about the landing of "the Danes," their rising power, and subsequent overthrow. But, what is more certain, the names of not fewer than about ten large lakes in the island still retain the Norwegian termination *vat* ("vatn," Vand, water); and three of the largest are called Loch Langavat (the long water). Several coves (Vige) in Harris are called *vagh* ("vagr"); as Groesavagh, Flodavagh; and in Lewis *wick*, as Sandwich (Sandvig; *Eng.*, Sand-bay), and Norwich (Nordvig; *Eng.* North-bay). To these may be added a great number of Norwegian names of places ending in *stra* or *sta* (staðr, stead); as Little Scarristra, Meickle Scarristra (Harris); Erista, Mangersta (Lewis); in *bost* (bolstaðr), as, in Harris, Nisibost, Hagabost, Chillibost; and in Lewis, Callbost, Habost, Luirbost, Crossbost, Melbost, Garrabost, and others (in all about thirteen). Further, we find such names as Laxay (Laxá, Laxaa; *Eng.*, Salmon river), Laxdale, Nether Holm and Upper Holm, Tong (túnga), &c. These Norwegian names of places are met with as well towards the south and west as on the east coast, where they are most numerous about Loch Seaforth (Sæfjörðr), and in the vicinity of the little town of Stornoway. But they are chiefly concentrated at one point, the most northern in the island, in a district which still retains the pure Norwegian name of "Ness."

On this Naze, or promontory, are the lakes Langavat and Steapavat; the valleys Dibidale, Eorodale, North Dell, and South Dell; the manors and towns Skegersta, Swainbost, Habost, Cross, and at the farthest extremity Oreby or Eoropie ("Eyribœr," the town on the Eir or Naze?); with the adjacent headland of Raven, which may possibly have been called after Odin's sacred bird. At all events, there is good ground for assuming, from these names of places, that the promontory had a pre-eminently Norwegian population, which, indeed, is unmistakably apparent even at the present day.

Throughout Harris and Lewis, for instance, the Gaelic inhabitants are small, dark-haired, and in general very ugly. But no sooner do we arrive at Ness, than we meet with people of an entirely different appearance. Both the men and women have in general lighter hair, taller figures, and far handsomer features. I visited several of their cabins, and found myself surrounded by physiognomies so Norwegian, that I could have fancied myself in Scandinavia itself, if the Gaelic language now spoken by the people, and their wretched dwellings, had not reminded me that I was in one of those poor districts in the north-west of Europe where the Gaels or Celts are still allowed a scanty existence. The houses, as in Shetland, and partly in Orkney, are built of turf and unhewn stones, with a wretched straw or heather roof, held together by ropes laid across the ridge of the house, and fastened with stones at the ends. The houses are so low, that one may often see the children lie playing on the side of the roof. The family and the cattle dwell in the same apartment, and the fire, burning freely on the floor, fills the house with a thick smoke, which slowly finds its way out of the hole in the roof. The sleeping-places are, as usual, holes in the side walls.

It is but a little while ago that the inhabitants of the Naze, who are said to have preserved faint traditions of their origin from Lochlin (called also in Ireland, Lochlan), or the North, regarded themselves as being of better descent than their neighbours the Gaels. The descendants of the Norwegians seldom or never contracted marriage with natives of a more southern part of the island, but formed among themselves a separate community, distinguished even by a peculiar costume, entirely different from the Highland Scotch dress. Although the inhabitants of Ness are now, for the most part, clothed like the rest of the people of Lewis, I was fortunate enough to see the dress of an old man of that district, which had been preserved as a curiosity. It was of thick coarse woollen stuff, of a brown colour, and consisted of a close-fitting jacket, sewn in one piece, with a pair of short trousers, reaching only a little below the knees. It was formerly customary with them not to cover the head at all. In a carefully compiled Scotch and English guide book (Anderson's Guide, 1842) it is stated, that "The islanders of the northern part of Lewis, with their long, matted, and uncombed hair, which has never been restrained by hat or bonnet from flowing as freely in the wind as their ponies' manes, and their true Norwegian cast of countenance, form living portraits of the ancient Norsemen. The other inhabitants are chiefly of Celtic origin." The difference between the descendants of the Gaels and of the

Norwegians is consequently so apparent that it is as striking to a Scotchman or an Englishman as to a Scandinavian.

It is said on the island that the inhabitants of Ness are more skilful fishermen and better sailors than the rest of the men of Lewis. However that may be, as a pretty numerous Norwegian population on it has long kept itself unmixed and distinct from the Gaels, it is not improbable that those men of Lewis who are related to have formerly harried Shetland, until they were entirely defeated in a great battle in Mainland, may have been inhabitants of Ness, who, after the custom of the ancient Norwegians, went on expeditions beyond sea, either to gain booty, or, more probably, to decide some old dispute by the sword. That men of Lewis, of Gaelic descent, who have never liked the sea, but, on the contrary, always feared it, should have ventured repeatedly, and in great numbers, so far as Shetland, altogether exceeds belief.

On the coasts of Lewis and Harris are several small islands, with still recognisable Norwegian names, such as Calvay ("Kálfey"), Pabbay ("Papey"), Skarpa (Skarpey), Scalpay (Skalpey), together with the places called Meathallybost, Bernera (Bjarnarey), and others. In the south-west there are three large islands in a row; North Uist, Benbecula, and South Uist (in the Sagas "Ivist"), where there are also evident traces of a Norwegian population. A small island to the west of North Uist is called Kirkibost (Kirkjubolstaðr); on Benbecula there are the lakes Loch Ollevate and Langavat, as well as the *Vaage*, or inlets, Uskevagh, Kenlerevagh, and Riavagh; and on South Uist there are likewise lakes and inlets called *vat* and *vagh*; to which may be added such names of places as Frobast, Kirkidale, Hillisdale, and lastly, a mountain called Heckla, probably from the well-known volcanic mountain in Iceland. In a bay in the middle of South Uist are the islands Calvay and Pabbay. There is still a great number of small isles on the coasts of these islands, whose names in a greater or less degree all betray their Norwegian origin; for instance, Grimsa ("Grimsey"), Barra ("Barey"), Lingay ("Lyngey"), Hellesay ("Hellisey"), Eriskay ("Eiriksey"), and others. The Norwegians must even have visited the little island of St. Kilda, which lies about eighty miles west of Lewis; at least, two of the often-mentioned and peculiarly Scandinavian bowl-formed brooches have been discovered on the island; one of them I have seen in the Andersonian Museum, in Glasgow. Similar brooches were also found, with a skeleton, in the island of Sangay, between Harris and North Uist.

To the east of North and South Uist is the large island of Skye ("Skið"), separated from the Highland mainland by a narrow sound ("Skiðsund"). Between its more northern part and the mainland, where the sea is broader, are the islands of Rona, Raasay ("Hrauneyjar"), Scalpa ("Skálpey"), Pabba ("Papey"), and Longa ("Langey"). Skye, towards the south, is remarkable for its numerous and lofty mountains, whose beautiful forms are visible at a great distance. Towards the north the island becomes gradually flatter and broader. In the west and north-west parts it is indented by deep firths, round which are to be found the most fertile districts in the island. The east coast, on the contrary, is not so capable of cultivation, as it has large tracts of moorland heath and sand. The Norwegians, therefore, advisedly chose to settle on the western and north-western firths, which, besides being more fertile, were not so exposed to the attacks of the Gaels as the eastern and south-eastern coast, which very nearly approach the mainland. Not a few Scandinavian names of places may be still clearly recognised near Loch Snizort, such as Scuddeburgh, Skabost, Braebost, and, near a waterfall, Forscachregin (the Norwegian *Fors* with a Gaelic termination). By Dungevan Loch are the inlets Kilmaluag and Altivaig, and the villages Husabost, Collbost, and Nisabost. By Loch Bracadale (the "Vestrifjorðr" of the Sagas) are Fors, Orbost, Collbost, and Eabost. By Loch Harporth, Carabost; and by Loch Eynort, Husedalebeg and Husedalemore; which latter, in a mixture of Norwegian and Gaelic, signify little and great Huusdal (Housedale); and, with a similar mixture, Ghionaforsenary. A little more inland is the valley of Tungadelebeg, where the Gaelic beg (little) is added to the Norwegian Tungadal.

From the frequent Gaelic terminations and corruptions of the Norwegian names, it is sufficiently evident that the Norwegian language has lost its former dominion in the island, and that the Gaelic has resumed its ancient pre-eminence. The western districts of Skye, as well as the previously-mentioned Norderöer, or northern islands, from Lewis to Barrahead (which last are often called under one name, "the Long Island"), are precisely those places in the Highlands where the Gaelic tongue is most unmixed, and where the greatest quantity of old Gaelic traditions and songs still survives among the people. It was here also, that a great number of the world-renowned songs of Ossian were first composed. It is true we no longer hear the people sing them, but there can nevertheless be scarcely any doubt, particularly if we regard the perceptible traces of the ancient metre in the Gaelic texts, that the so frequently and warmly disputed edition by Macpherson is really founded on ancient songs, although these may have been somewhat altered by lapse of time, and by a not very happy translation. They have quite a peculiar interest for the Scandinavian North, from the striking agreement in tone and spirit which they present to several of the songs of the Sagas and Edda. These last, again, afford a strong proof of the genuineness of those attributed to Ossian, since the songs of the Sagas and Edda, at the time when Macpherson published his Ossian, were either not at all, or but very imperfectly known, even in Scandinavia itself, not to speak of other countries. The real age of Ossian's songs is very uncertain, and very difficult to discover; but this much is clear, that they indicate a lively intercourse between Alba (Scotland) and Lochlin (Scandinavia), long before the times of the Vikings, and previously to all historical accounts of connections between those countries. We cannot, however, venture to conclude from this that the Orkneys, or any other part of Scotland, were at so early a period inhabited by a Scandinavian people. That such a colonization should really have taken place before the time of the Vikings, which began at the close of the eighth century, there are not only wanting historical and archæological proofs, but likewise all internal probability.

Mull ("Myl") is the largest of the most southern Norderöer, or northern islands, but it is not richest in memorials of the Northmen. In the narrow strait or sound ("Mylarsund") which separates the island from the mainland, there lies straight before Tobermory, the most important place in the island, the little island of Calve ("Mylarkálfr"); and somewhat farther south of Tobermory, on a rivulet by the coast, are the ruins of the palace of Aros (from "árós;" *Dan.*, Aarhus, the mouth of the rivulet or Aa), once frequently inhabited by the rulers of these islands, called "Lords of the Isles." Another river in Mull, well stocked with fish, was formerly called Glenforsay (Monro, "Description of the Western Isles," 1594), from the Norwegian "forsá" (Fosaa; *Eng.*, Waterfall-river), to which the Gaelic *glen* has since been added. With the exception, perhaps, of Assapoll (from *-bol*), in the south-west, the island has no Norwegian names of places. Of such names, however, several are to be met with on the islands west of Mull, particularly on Coll ("Kóln"), where we find Crossapull, Gisapoll (from *bol*), Arnabost (-bolstaðr), and Balehough; and on Tiree, Tyrvist, together with Kirkapoll, Heylipoll, Vassipoll, and Crossipoll. In the bay formed by Mull, towards the west, are found many small islands with originally Norwegian names, such as Ulva ("Ulfey"), together with Soriby, Gometra ("Guðmundarey"), and Staffa ("Stafey"), so famed for its stalactic caverns.

But of all the Hebrides, none is more renowned than Iona (Ithona, "the Waves' Island"), or Icolmkill, "the island with Columba's cells," which lies in the open Atlantic, near the south-west point of Mull. It is not distinguished either by size and fertility or by numerous and splendid ruins; it is now but an inconsiderable island, with some few remains of churches, conventual buildings, and ancient Christian sepulchral monuments. But about thirteen centuries ago it was the light of the western world; for, after St. Columba settled there, it became the central point whence Christianity diffused itself towards the east and north, over Scotland and the surrounding islands. Iona thus obtained such repute for sanctity, that it was said that a deluge which was to overwhelm Ireland, and the islands round about, would have no power to inundate it. Tradition adds, that, for this reason, the ancient Irish, Scotch, and Norwegian kings, besides many other chiefs and mighty men, both at home and abroad, chose Iona as their place of burial; and that at the commencement of the sixteenth century, no fewer than three hundred and sixty splendid stone crosses, or tombstones, were still to be found on the island, which, however, with some few exceptions, have now entirely disappeared.

According to an old description of the island, by Dean Monro (1594), there was to the north of the Scotch graves an inscription, which ran thus:—"Tumulus regum Norwegie," or, "the tombe of the

Kings of Norroway, in the quhilk tombe, as we find in our ancient Eriske cronickells, there layes eight Kings of Norroway, and also we find in our Eriske cronickells, that Coelus, King of Norroway, commandit his nobils to take his bodey and burey it in Colmkill, if it chancit him to die in the isles; bot he was so discomfitit, that ther remained not so many of his army as wold burey him there." By the kings of Norway here mentioned we must of course understand only the kings of the Sudreyjar, or southern islands, and the Irish kings of Norwegian descent. It is in itself very probable that these kings often desired to be buried in Iona, where the first bishops of the proper Sudreyjar, "the bishops of the isles," dwelt, and whose church of St. Mary was consequently the chief church in the islands. The tombs of the kings, however, can at present scarcely be pointed out with certainty; we only know that they must have been in the large and still visible burial-place consecrated to St. Oran. On this place there is likewise a little chapel consecrated to the same saint, which, according to the opinion of some, is of Norwegian workmanship—a point, however, which must be very doubtful.

In the chapel are to be seen the remains of a carved monument erected in the year 1489 to Lachlan Mackinnon (Mac Fingon), and on it, underneath the inscription, is a ship, which is still to be found in the family arms of the Mackinnons, but which is said to have been originally the heraldic bearing of the Norwegian kings in the Isle of Man.

[+ +] Monument - Boat Decoration

The Island of Iona was of special importance in ancient times, not only to Scotland, but to the Scandinavian North. From it Christianity was assuredly disseminated among the Norwegians in the Sudreyjar, or southern isles, the Orkneys, and the Shetland Isles; whence, again, it was often carried by Vikings and merchants to Norway and Iceland. In the latter place, where not a few men from the southern isles were among the first colonists, there was even a church dedicated to St. Columba. Whilst, therefore, heathen Norwegians plundered and destroyed the churches and convents of Iona, the Christian Norwegians seem to have respected its sanctity. The Sagas, which call it "Eyin helga" (the holy island), state, that the Norwegian king, Magnus Barfod (Barefoot), when in his first expedition to the Sudreyjar and Ireland, in the year 1097, he came to "the holy island," gave all the inhabitants a guaranty of peace and security, and allowed them to retain their possessions. It is also stated that "King Magnus opened the little Kolumkille Church, and went therein; but that he directly locked the door again, and said that no one should dare to enter; and since that time the church has never been opened."

SECTION X

The Sudreyjar, or Southern Isles.—Cantire.—Islay.—Man.—Names of Places.—Runic Stones.—Kings.—Battle of Largs.—"Lords of the Isles."—Tynwald in Man.

Iona was not always accounted one of the northern isles. Farther towards the north, on the north-west coast of Mull, are the islands of Treshinish, and among them a steep rocky island, called Cairnburg, which is said to have formed, at all events at times, the boundary between the northern and southern isles, or Sudreyjar. Cairnburg is accessible only at one spot, and by its height above the sea it forms an important stronghold, which in former times was often numerously garrisoned. The Sagas, which call the island "Bjana," or "Bjarnarborg," state that it was one of those strong fortresses in the southern isles, the surrender of which was in vain demanded by King Alexander the Second of Scotland, from the Norwegian tributary king, Ion Dungadson; and tradition still tells that "the Danes" often fought for the possession of this important place.

"The Sudreyjar" (in which, among the larger islands, were included Colonsay, Oransay, Jura, Islay, Arran, Bute, the Cumbr Islands, and likewise the Peninsula of Cantire) are, strictly speaking, far from being so numerous as the northern islands; but in general they are distinguished from these by a richer and more fertile soil, which is the result of their more southern and more protected situation. This remark applies particularly to the charming islands of Arran ("Hersey"), Bute ("Bót"), and the Cumbr Isles (Kumreyar), which lie eastwards of the Peninsula of Cantire ("Satiri"), at the mouth of the Firth of Clyde; and which, together with the rocks, heaths, and moors of the Highlands, possess the woods and corn-fields of the Lowlands. They also enjoy a fine climate.

But although these last-mentioned islands were often under the dominion of Norwegian kings and jarls, they do not appear to have been inhabited by a settled Norwegian population; at all events, Norwegian names of places have disappeared from them. It is probable that they lay somewhat too near the hostile coasts of Scotland, and somewhat too far from the larger Norwegian colonies, for Norwegian settlers steadily to maintain upon them a position against the Gaels; nay, the Norwegian name, "Kumreyar," the Cumbr Islands, seems to indicate that Cimri or Gaels dwelt upon them.

Names of places on the Peninsula of Cantire, on the contrary, where we find Smerbys (from *by*), Killipol (from *bol*), Torrisdale, and the pure Norwegian Skipness, but more particularly on the islands outside the Peninsula, near the west coast of Argyle, indicate a very considerable Norwegian colonization. Not only have several of the small islands Norwegian names, as Scarba ("Skarpey") and Lunga ("Langey"), but the largest and most fertile of them, Islay (the "Il" of the Sagas), which Dean Monro as early as 1594 found to be fruitful, full of good pastures, abounding with large deer, having many forests, excellent hunting, and a river called Laxay (the pure *Old N.* "Laxá") in which many salmon were caught ("*with ane water callit Laxay, whereupon maney salmon are slaine*"), still exhibits various traces of decidedly Norwegian settlements. On its east coast, as is usually the case with the Hebrides lying nearest to Scotland, few or no Norwegian names of places are found; but in the middle of the island is Nerby; by Loch Indal, Lyrabolls, Scarabolls, Conisby, Nerabolls, and Elister; and by a rivulet, Skeba ("Skipá;" *Dan.*, "Skibeaaen," or the ship rivulet); whilst on the west side of the island we find Olista, Culaboll, &c. This agrees very well with the accounts that the kings and jarls of the Sudreyjar of Norwegian descent had one of their chief residences in Islay; for it was quite natural that they should surround themselves with countrymen on whose courage and fidelity they could rely. The island abounds, moreover, in traditions and pretended memorials of "the Danes." Near the bay of Knoch are two large upright stones, called "the two stones of Islay," under which it is said that the Danish princess, Yula, after whom the island is named, lies buried. In various parts of the island are shown what are called "Danish" castles, encampments, and fortifications. It is also stated (see Anderson's Guide), that there is a circular mound of earth on the island, with terrace-formed steps, which may possibly have once been used by the Norwegians as a *Thing* place, like a similar one in the Isle of Man.

The chief seat of the Norwegian power on the islands was, however, still more southward than Islay, namely, the Isle of Man (the "Mön" of the Sagas), which lies in the Irish Channel, to the south-west of Solway Firth, about midway between the coasts of Cumberland and Ireland. A peculiar dialect of the Gaelic tongue, called Manx, is spoken throughout this island, and the inhabitants have in general the same appearance as their Gaelic neighbours in Wales, Scotland, and Ireland. But no other of the western islands affords so many and such incontestable proofs of its having once had a very wide-spread Norwegian or Scandinavian population, who spoke their own language, and who, through a long series of years, must have been the predominant race.

The highest mountain in the island, which is about 2000 feet high, is called "Sneafell" (*Norw.*, Sneefjeld; *Eng.*, Snow-mountain). On the east side is the rivulet and town of "Laxey" (Laxaa); in the south-east is the long naze, "Langness." To these may be added the bay called Derby Haven, which the Norwegians called "Rognvaldsvágr," whence the neighbouring Ronaldsway derived its name. There are also the inlets of Perwick and Fleswick; the islands Calf of Man, Eye (Oë), and Holm, near the town of Peel; and, lastly, the villages Colby, Greenaby, Dalby, Kirby (Kirkeby), Sulby, and Iurby (formerly "Ivorby"—Ivarsby?), &c. The proportionately large number of names of places ending in "by," which suddenly appear in Man, in contrast to the more northern islands, with their pure Norwegian names of places ending in "bol" and "bolstaðr,"—which, it must be observed, are not to be found on Man,—is a sort of proof that it received some colonists from the neighbouring old Danish Cumberland, by which means a mixed Norwegian-Danish population arose in the island.

The antiquary is much surprised to find on Man not merely one, but several of those runic stones, with genuine Scandinavian inscriptions, which he may have sought for in vain in England and Scotland. The different districts of the island contain altogether about thirty ancient sculptured monuments or sepulchral crosses; and of these at least thirteen have once had runic inscriptions, which in great part are still preserved. It is remarkable enough that these runic inscriptions are found exclusively in the more northern half of the island (at Kirk

Andreas, two; at Kirk Michael, four; at Kirk Braddan, one; and at Kirk Onchan, five); whence we may, with some degree of probability, conclude that, at the time when these runic stones were erected, the Scandinavian language was the most prevalent one in the northern part of the island. The chronicles, indeed, state that the Norwegian, Godred Crovan, who conquered Man in the year 1077, retained the southern part of the island for himself and his followers; but the before-mentioned runic stones are certainly older than Godred's conquest. The inscriptions on the stones have hitherto been copied and explained only in a very imperfect manner; but since casts in plaster have been taken of them, their interpretation has become incomparably easier and more simple. I have myself closely examined and compared them in two places (at Edinburgh, in the Museum of the Society of Scottish Antiquaries, and at Canons Ashby, in England, the seat of Sir Henry Dryden, Bart.); and I have since had an opportunity to renew my examination of all of them, in conjunction with the learned Norwegian professor, P. A. Munch, to whom I am indebted for several very important hints relative to their correct interpretation (amongst others that the rune ᚭ, which in most inscriptions signifies *o*, must in these always be read as *b*).

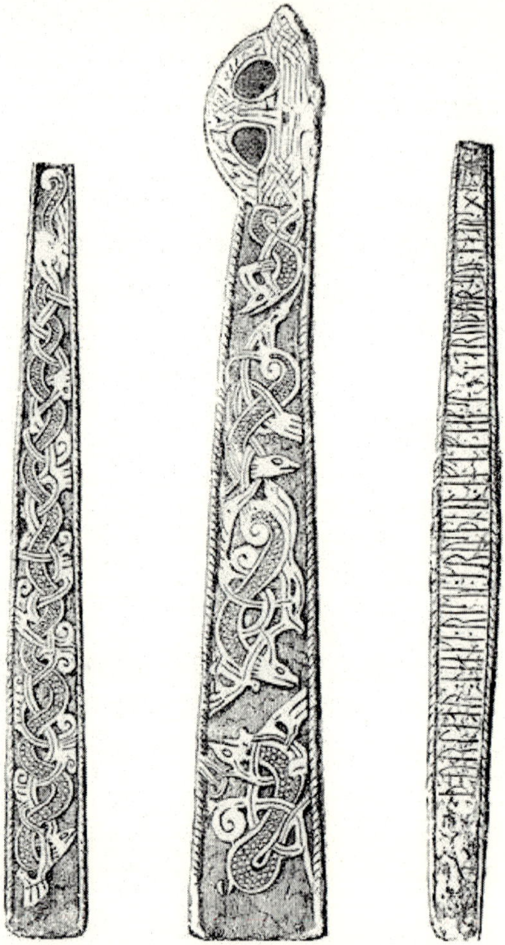

[+ +] Stones - Runic Inscriptions

The annexed cut, after a plaster cast, represents one of the finest and best preserved runic stones in Man, namely, at Kirk Braddan, about the middle of the island.

The stone is fifty-seven inches high, eight inches broad at the base, and when the cross was whole, had a breadth of twelve inches at the top. Both its broad and one of its narrow sides are ornamented with serpents ingeniously interwoven, whilst the fourth side has the following runic inscription:

"Thurlabr Neaki risti krus thana aft Fiaks ... bruthur sun Jabrs."

("Thorlaf Neaki erected this cross to Fiak ... brother, a son of Jabr.")

Another extremely well-preserved monumental cross, on which are carved various scrolls, animals, birds, and other things, such as horses, a stag, cows (?), swine, &c., stands in Andreas churchyard, and has the following inscription:—

"Sandulf ein suarti raisti krus thana aftir Arin Biaurg kuinn sina."

(*i. e.*, "Sandulf the Swarthy erected this cross to his wife Arnbjörg.")

[+ +] Monument: Andreas Churchyard

(The drawing of this monument, as well as those of the following inscribed stones, is borrowed from W. Kinnebrook's "Etchings of the Runic Monuments in the Isle of Man," London, 1841, 8vo. But the faulty inscriptions in that book are here corrected.)

[+ +] Monument: Kirk Michael

In the middle of the village of Kirk Michael, close to the northern corner of the churchyard, is a stone not less richly sculptured than the preceding one, with all sorts of figures of stags, dogs, serpents, horses, horsemen, &c., which are placed round a large cross covered with interlacings, or scrolls. The inscription on it runs thus:—

"Jualfir sunr Thurulfs eins Rautha risti krus thana aft Frithu muthur sina." (Or, "Joalf, son of Thorolf the Red, erected this cross to his mother Frida.")

At the end of the inscription is carved the figure of a man (probably Joalf), with a shield on his arm and a lance in his hand. (See the annexed cut.)

The language of the inscriptions, as well as the Scandinavian names which appear in them,—as Thorlaf, Arnbjörg, Frida, and particularly the names compounded after the genuine Scandinavian fashion, as Sandulf the Swarthy, and Thorolf the Red,—sufficiently prove that these monuments were erected by Northmen, or Norwegians, to their relatives who had died in the Isle of Man. A piece of runic stone in the wall of Michael's Church bears the name of Grim the Swarthy ("Grims ins Suarta"); and in some similar fragments of inscriptions near Kirk Onchan we find the names of Thurid ("Thurith raist runir," *i. e.*, Thurith engraved runes) and Leif ("tra es Laifa fustra guthan son Ilan"). The well-known Scandinavian name, Asketil, is also found on the remains of a runic inscription in the museum in Douglas ("p. Askitil vilti i trigu——aithsaara siin;" *i. e.*, whom Asketil deceived in security, contrary to his pledge of peace). At the same time, however, we may infer from names like Neaki, Fjak, and Jabr, that the Northmen must, when these inscriptions were written, have already mingled with the original Gaelic inhabitants of Man. A stone at Kirk Michael, which is ornamented with a finely sculptured cross, on the sides of which are seen a stag, a dog, a harper, and two figures apparently in an attitude of prayer, has a Norwegian inscription with purely Gaelic names, such as Mal Lumkun and Mal Muru:—

"Mal Lumkun raisti krus thana eftir Malmuru fustra sin...;" (*i. e.*, "Mal Lumkun erected this cross to his foster father Malmor.")

[+ +] Monument

Some hitherto inexplicable fragments of inscriptions at Kirk Onchan may also possibly contain Gaelic words. The Manx runic stones bear, both in form and workmanship, a striking resemblance to the previously-mentioned sculptured monuments in the Lowlands, and on the north-east coasts of the Highlands. Yet several of the Manx stones exhibit certain peculiarities; as, for instance, the singular scale-covered serpents surrounded with interlacings, which do not appear in a similar form on the Scotch monuments. But as these serpents and interlacings very much agree with ornaments on different antiquities of the heathen times found in Scandinavia, and, as the language of the runic stones is pure Scandinavian, there is every reason to conclude that the splendid specimens on Man were carved by Norwegians, who, though they imitated the monuments in vogue in Scotland, frequently allowed their own characteristically fantastic ideas to display themselves in peculiar devices. This view is confirmed in a remarkable manner by a few Manx runic inscriptions, the real interpretation of which was first given by Professor Munch. On the stone at Kirk Michael, represented below, is the following inscription:—

"Mail Brigdi sunr Athakans smith raisti krus thana fur salu sini sin brukuin Gaut girthi thana auk ala i Mann." *i. e.*, "Malbrigd, son of Athakan (the) Smith, erected this cross for his soul.... Gaut made this (cross) and all on Man."

According to this, Gaut, who, to judge from the name, was a Norwegian, erected all the crosses which, it must be observed, were at that time on Man. Another inscription perfectly agreeing with this, though taken from a very much defaced and broken monument near Kirk Andreas, on which has been carved a cross with many scrolls (delineated in Kinnebrook's work, No. 8), runs as follows:—

"... thana af Ufaig fauthur sin in Gautr girthi sunr Biarnar ..." "(N. N. erected) this (cross) to his father Ufeig, but Gaut Björnsön made it."

Gaut's surname, here given, further proves his Norwegian, or Scandinavian, descent. From the language and manner of writing in the Manx inscriptions still extant, we may assume that, with the exception perhaps of some few pieces at Kirk Michael (Mal Lumkun's inscription) and Kirk Onchan (Leif inscription), which, according to Professor Munch's opinion, are of a somewhat later period, all these inscriptions were from the artist-hand of Gaut Björnsön. It is even probable that several of the other sculptured stones in Man, which are not known to have had inscriptions (particularly at Kirk Onchan,

Kirk Braddan, and Kirk Lonan; see Kinnebrook, Nos. 16, 17, 20, 22, 23), were carved by Gaut, or at least by a Northman. At all events, they are somewhat different from the corresponding stones in Scotland; and some of them (Onchan, 20, and Braddan, 23) prove themselves to be genuine Norwegian runic stones, by the same peculiar figures of dragons and serpents as on those before described.

The circumstance that those sculptured monumental stones in Man, which are Norwegian, have both runic writings and peculiar representations of figures, certainly affords a strong corroboration of the opinion before expressed, that the sculptured monuments, generally so finely executed, which are found on the east coast of Scotland, are in fact, though called "Danish," not Scandinavian, but Scotch. As, on the other hand, the runic stones in Man have expressly preserved the name of the person who made them—the Norwegian skilled in runes, Gaut Björnsön, who imitated and altered the Scotch models with great expertness and taste—it is clear that the Norwegians in the remote Western Isles must not be regarded, any more than their kinsmen in the Orkneys and in England, as merely rude barbarians, living only for plunder, war, and bloodshed, and having no feeling for anything higher and nobler. The discovery of Gaut Björnsön's name may be regarded as an instructive addition to the proofs before adduced, that the cathedral in Kirkwall was originally founded, and partly erected, by a Norwegian layman, the chieftain Kol; as well as that there existed at the same time in England a considerable number of Danish, or Scandinavian, coiners. Of the latter, as we shall see, there were likewise several employed by the Norwegian-Danish kings in Ireland. For the rest, these characteristic Scandinavian runic writings suffice to show that, with regard to the civilization then prevailing, the Norwegians or Danes settled in these districts were by no means deficient in education. The Northmen on the Isle of Man were, besides, at a very early period, Christians. Almost all the Manx runic stones are ornamented with the Christian cross; and on a defaced piece of such a monumental stone at Kirk Onchan we even find the words Jesus Christ ("Jsu Krist"). From the language of the inscriptions there is reason to suppose that they were for the most part engraved in the eleventh century. We cannot, therefore, doubt that Christianity must at that time have been already disseminated among the Scandinavian population in the Isle of Man. There was a bishopric in the island in very ancient times; and we learn from history, as well as from the names of the bishops, that in the eleventh, twelfth, and thirteenth centuries several of them were of Norwegian descent; for instance, in 1050-1065, Roolwer (Rolf?); 1077-1100, Aumond M'Olave; 1181-1190, Reginald, or Ragnvald; 1203-1226, Reginald (son of a sister of King Olaf, of Man); and his successor, John Ivarsön. Unfortunately there is a gap in the chronicles of the bishops of Man from about the year 700 to the year 1025. Had they been perfect, we should possibly have been able to find Scandinavian bishops in the island even earlier than 1050.

The Norwegian monuments in the Isle of Man already mentioned are in themselves numerous and considerable enough to convey an idea of the power which the Norwegians must have possessed there. At all events, the names of places and the runic stones contribute in a high degree to strengthen and illustrate the assertion of the Chronicles, that Norwegian kings and jarls held confirmed dominion in the Sudreyjar, or southern isles. When, in the ninth century, the Norwegian king Harald Haarfager succeeded in subjecting the Orkneys and the Sudreyjar, he is said to have appointed a viceroy or jarl in Man. During the tenth and eleventh centuries a long series of Norwegian kings ruled there, whose descent is clearly shown by their names; *viz.*, Godred (Gudröd), Reginald (Ragnvald), Olave, Hacon, Harold, &c. In the eleventh century the connection between these kings and the Norwegian or Scandinavian kings of Dublin was so particularly close, that either the same, or at all events nearly-related kings, reigned over both Man and Dublin.

The kings of Man were tributaries of Norway, and acknowledged the supremacy of that country, although in reality they ruled independently. At that time their dominion extended over the rest of the Sudreyjar. But in the year 1077 the Norwegian, Godred Crovan, succeeded in conquering Man, after a battle near "Scacafell," or Skyhill, in which King Fingal, a grandson of Sygtrig (Sigtryg), "King of the Danes in Dublin," fell, as well as Sygtrig Mac Olave, the actual Danish king of Dublin. Godred Crovan now assumed the government of the islands, and appears to have declared himself perfectly independent of Norway. Subsequently he conquered Dublin also, as well as the province of Leinster in Ireland. In order to maintain Norway's right of supremacy, the Norwegian king, Magnus Barfod, shortly afterwards undertook an expedition to the west. He committed great havoc along the firths of Scotland ("Skotlandsfirðir," or the coasts by the Caledonian Sea), and in the Sudreyjar, Man, Anglesey, and Ireland, and regained the kingdom which his forefathers had possessed. According to a treaty with the Scottish king Malcolm, all the islands lying to the west of Scotland, which Magnus could approach with sailing vessels, were to belong to Norway. King Magnus accordingly caused his ship to be hauled over the narrow isthmus (Satíriseið) which connects the peninsula of Cantire with the mainland, and which to the present day is called by the Gaels "Tarbet" (a place over which vessels can be dragged). The King himself sat at the helm, and thus acquired the peninsula, besides all the Western Islands. Having appointed his son Sigurd king of the Sudreyjar, he returned home to Norway, where, with several of his followers, he adopted the dress generally worn in the Western Isles. "They went about the streets with bare legs, and wore short coats and cloaks; whence Magnus was called by his men Barfod, or Barbeen" (Barefoot, or Barelegs), says the Icelandic historian, Snorre Sturlesön, who, as is well known, lived in the first half of the thirteenth century. It is remarkable enough that this is the oldest account extant of the well-known Scotch Highland dress, whose high antiquity is thus proved.

The Jarl Ottar, who after Magnus Barfod's expedition was made governor of Man, was expelled by the inhabitants of that island ("Manverjar"), who chose in his place another jarl named Macmanus (or Magnusön). But a civil war now broke out in the island, and as King Magnus Barfod fell in Ireland in 1103, when on a fresh expedition to the Western Islands, Godred Crovan's family regained the Manx throne. It appears, however, that they acknowledged the supremacy of Norway; at all events, the previously distinct bishoprics of the Sudreyjar (founded in 838) and of Man were united after Magnus Barfod's expedition, and connected more closely than ever with Norway, by being subjected to the archbishopric of Trondhjem. From 1181 until 1334 the bishops of the Sudreyjar ("Episcopi Sodorenses") were consecrated by the Archbishop of Trondhjem. In the year 1380 the bishopric of Man was again separated from that of the other Sudreyjar; but the subsequent bishops of Man have retained to the present day the old title of bishop of Sodor (and Man), taken originally from Suðreyar.

About the same time that the proper Suðreyar were, with, regard to ecclesiastical matters, united with Man, many of them were, as to secular government, separated from that island; although, since the time of Harald Haarfager, all had been governed by the same kings. Jarl Somerled, who was related in various ways to the Norwegian chiefs on the islands, had assumed the dominion of Cantire, Argyle, and Lorn (the "Dalir i Skotlandsfirdi" of the Sagas). After a naval battle, in the year 1156, with the Manx king, Godred Olavesön, Jarl Somerled compelled Godred to resign to him all the Sudreyjar from Mull to Man, which possessions afterwards remained in his family ("Dalverja-Ætt"). His youngest son, Dugal, the founder of the family of the Mac Dougals of Lorn, obtained Argyle and Lorn, whilst Cantire and the islands were assigned to his eldest son Ragnvald, or Reginald. Meanwhile Godred Crovan's successors reigned over Man, and frequently, as it seems, over the islands to the north of Mull likewise, and particularly Lewis. They constantly sought to strengthen their diminished power by forming alliances with royal families, and other powerful races in Ireland, Scotland, and Norway. Thus King Harald Olafsön, whose father King Olaf Godredsön had, in the year 1230, repaired to King Hakon Hakonsen in Norway, and taken the oath of allegiance to him, married King Hakon's daughter Cecilie; but on the voyage home from Norway in 1248, the royal couple perished in the dangerous Somburg Röst, to the south of Shetland, together with the Manx bishop, Lawrence, and a numerous retinue of Manx chiefs. Harald's brother, King Ragnvald, was shortly afterwards murdered by the knight Ivar, and was succeeded on the throne by his youngest brother Magnus, who was the last of Godred Crovan's descendants, and above all the last Norwegian who filled the throne of Man.

The Scotch kings had long been aiming at the expulsion of the Norwegians from the north and west of Scotland. Alexander the Second (1214-1249) repeatedly sent ambassadors to King Hakon, in Norway, offering to purchase the right of that kingdom to the Norwegian possessions in Scotland; but as they did not succeed,

Alexander declared that he would not rest till he had planted his banner on the farthest point of the Norwegian dominions in Scotland. But whilst he lay with part of his army at the island of Kerrera ("Kjarbarey"), not far from Mull, he fell sick and died, after which the army was disbanded. However, his successor, Alexander the Third (1249-1289), zealously prosecuted the plan for the expulsion of the Norwegians. The Scots having at length begun to ravage the Sudreyjar, and particularly the Isle of Skye, with fire and sword, King Hakon, when the tidings reached Norway, equipped a large fleet, and issued orders for an expedition to avenge the attack that had been made on his dominions.

Accordingly, in 1263, he sailed with a large and well-appointed force to Elwick ("Ellidarvik") on Shapinsay, in the Orkneys, and thence to Ragnvaldsvaag ("Rögnvaldsvágr") under South Ronaldshay, near Pentland Firth. He had despatched several ships before him to the Sudreyjar, whose crews devastated the coasts of Sutherland, particularly the district around the firth of Durness ("Dyrnes"), where they destroyed a castle and burnt more than twenty mansions. The King then sailed to the before-mentioned isle of Kerrera, where he assembled his fleet, consisting of about 200 ships. King Magnus from Man, and King Dugal from the Sudreyjar, joined him there; but Ion, the other king of the Sudreyjar, or, as he was called in Scotland, Ewen, was exempted by King Hakon from fighting against the Scots. King Hakon permitted his men to devastate the islands and coasts of the Firth of Clyde. Some of his chiefs sailed up Loch Long ("Skipafjörðr"), and hauled their ships over the narrow strip of land, called Tarbet, into Loch Lomond ("Lokulofni"), whence they harried the surrounding district of Lennox ("Lofnach"). Meanwhile verbal messages passed between the Norwegian and Scottish kings, but without leading to any reconciliation. The time was thus whiled away till late in the autumn, when King Hakon anchored with his fleet under Cumbrey in the Clyde, opposite the hamlet of Largs. Here he was assailed by such a furious storm, that his Norwegians, unacquainted with the equinoctial gales on the west coast of Scotland, imagined that the tempest had been evoked by witchcraft. Some of the King's ships were driven ashore near Largs, when the Scots immediately began to attack them. As the Scotch king had in the meantime arrived on the spot with a large army, a fierce battle took place on the plain near Largs (3rd of October, 1263), in which the Norwegians, who were exhausted by their endeavours to save their ships, and who on account of the storm could not avail themselves of their whole force, were overpowered. King Hakon then sailed with the remainder of his fleet round Cape Wrath to "Goafjörðr" (undoubtedly the excellent harbour in Loch Eribol in Sutherland), and after suffering much from violent storms and tempests, at length again reached Ragnvaldsvaag in the Orkneys. He now prepared to pass the winter in Kirkwall, where, however, he shortly afterwards died (16th December, 1263).

The battle of Largs, the last combat in these western regions between the kings of Scotland and Norway, was of a decisive character. The kings in Sudreyjar and Man, who could now no longer venture to reckon upon adequate protection from Norway, submitted to the dominion of the Scotch king. King Magnus Hakonsön, of Norway, found it most advisable (1266) to cede Norway's supremacy over the Sudreyjar and Man to the Scottish crown for the sum of 4000 marks sterling and a yearly tribute of 100 marks. But the Scots did not obtain immediate possession of Man. King Magnus died there in 1265, and was buried in the convent of Russin, near Derby Haven ("Rögnvaldsvágr"), which one of his forefathers had founded, or at all events enlarged, in 1134, and which already contained the bones of several Norwegian kings, chiefs, and ecclesiastics (as, for instance, of Bishop Reginald, + 1225; King Olave Godredsön, + 1237; and the chief Gospatrick, + 1240). With Magnus the family of Godred Crovan became extinct; but the powerful knight Ivar assumed the dominion of Man; and it was not till the year 1270 that the Scots, who had landed in Ragnvaldsvaag, succeeded, in a hard-fought battle, in killing Ivar, together with a great number of the leading men of the island, who had fought desperately for their independence.

Thus was terminated the actual Norwegian dominion over the Sudreyjar. As the battle of Largs considerably contributed to this event, it is no wonder that this battle, and above all King Hakon's expedition, still figure in Scottish traditions. On the battle-field near Largs—where human bones, as well as "Danish axes" and swords, are often found—are still to be seen two almost unique barrows or tumuli, the most remarkable in Scotland, being about 25 feet high, and nearly 20 feet broad at the top, in which the Norwegians and Scots who had been slain are said to have been buried. One of the mounds, which stands just at the back of the town, and close to the shore, is probably the grave of the Norwegians; for the Sagas, whose accounts agree on the whole so exactly with the localities that they must have been derived from eye-witnesses, relate that King Hakon, the day after the battle, buried his dead on the coast, in the neighbourhood of a church. The other mound stands on the plain, a few thousand paces farther off. According to the statements of the common people, on the day of the battle, blood flowed instead of water in a little rivulet or beck that runs past "Killing Craig." A number of smaller barrows and scattered stones, formerly to be seen on the plain, were likewise ascribed by tradition, though certainly without reason, to the same battle. They undoubtedly belonged to a far more ancient time; as is also the case with an excellent silver-gilt brooch found near Hunterston, about three miles from Largs, which was at once said to have been lost by some Norwegian who fled from the field of battle. There is a short Scandinavian runic inscription scratched on the back of it; but, from what has hitherto been deciphered, it would rather seem to denote the name of a Scotchman than of a Norwegian. Professor Munch reads, and certainly with good reason, as follows:—

"Malbritha a dalk thana" ... or, "Melbrigd owns this brooch."

In workmanship, moreover, it resembles the contemporary Irish and Scotch more than Scandinavian ornaments.

The remembrance of this last expedition of the Norwegians is scarcely less vivid in several of the harbours which King Hakon visited with his fleet; as, for instance, Lamlash ("Melasey"), in Arran; Sanda ("Sandey") near the south point of Cantire, where are shown the remains of a chapel and a churchyard, in which are said to repose the bones of many Danish and Norwegian chiefs; also in Gigha ("Gúdey"); Kerrera ("Kjarbarey"), with its "Danish" fort "Gylen;" and lastly, in Kyle Rhee (the King's Strait), and Kyle Akin (Hakon's Strait?), in the straits between the Isle of Skye and Lochalsh, on the coast of Ross-shire. According to a tradition, which is, however, entirely without foundation, King Hakon, in his flight from Largs, was attacked in this strait and killed, together with a great number of his followers. With similar exaggeration the Scots relate that all the Norwegians round about in the Sudreyjar were killed after the battle of Largs. On one of the islands near Barra was shown, not long since, and perhaps is even still, a heap of human bones, as the remains of the last Danes murdered there. On Lewis there is the following tradition—that when the Danes were quartered round about in the island, and were very troublesome on account of their oppressions, the Gaels laid a plan to murder them. The "fiery cross" was circulated through the island, with this brief announcement: "marbhadh ghach then a Bhuana;" that is, "every one shall kill his guest." The strangers, who had not time to assemble together, were thus murdered one by one.

It cannot admit of a doubt that the Norwegians on the Sudreyjar, who for centuries had taken fast root in the islands, and become mixed with the families of the Scotch chiefs, could not thus disappear all at once without leaving a trace behind them. In Lewis, as I have before proved, vestiges of a Norwegian population still exist. The best refutation of the tradition is, however, the circumstance that with the exception of Man, the Sudreyjar continued to be governed by the same chiefs who had ruled the islands under the Norwegian dominion; and who, being descended from Somerled himself, were in a great degree of Norwegian extraction. Somerled's successors also continued, after the old fashion, to defy the Scotch kings, who often sought in vain to subdue the bold "Lords of the Isles," so famed in song and legend. Sometimes they declared themselves independent, and sometimes they were compelled to yield to the superior force of the kings, and acknowledge them as their feudal lords; until at length, but not before the sixteenth century, the power of these island chieftains was entirely subdued. Even to the present day many Highland clans assert that they are descended from the Danes, or Norwegians. This much is at all events certain, that several clans have Scandinavian blood in their veins, as appears clearly enough from the names of Clan-Ranald (from Reginald or Ragnvald) and Clan-Dugal (from Dubhgall, "the dark strangers," the usual name for the Danes); both which clans, it is expressly stated, are descended from Somerled. To these may be added the clan of Macleod in Skye, whose chiefs still commonly bear the pure Norwegian names of "Torquil" and "Tormod."

But the enduring influence of the Norwegian dominion in the Sudreyjar is best established by the fact that since the battle of Largs, the Isle of Man, through all the vicissitudes of fate, and after passing by sale into the possession of the English crown, has uninterruptedly retained its peculiar position as a kingdom, having its own originally Norwegian or Scandinavian constitution, and its annual assemblies on the identical Thing-hill, Tynwald (or, as it was formerly called Tingualla, "Þingavöllr"), from which, about a thousand years ago, the Norwegians governed the Sudreyjar. Although the British Parliament makes laws for England, Ireland, and Scotland, they are of no validity in the Isle of Man, unless they are in accordance with the ancient laws and liberties of the island, and, after being confirmed by its own Parliament, are proclaimed from Tynwald Hill.

The Manx Parliament, whose origin is lost in the mists of remote antiquity, but whose establishment is usually ascribed to the Danish king Orry (Erik?), who settled in the island in the beginning of the tenth century, consists of the three "estates" of the island: 1st, the king, or superior lord; 2nd, the governor and council; 3rd, the twenty-four representatives of the island ("Keys, or Taxiaxi"). The upper house, or council, consists of the bishop, two superior judges ("deemsters"), and six other of the highest officers in the island. The representatives in "the house of Keys" fill up vacancies themselves, and hold their seats for life, without being in any way responsible to the people for their votes.

This aristocratic mode of election reminds one of the time of the Norwegian conquest, when the Norwegians made themselves lords over the natives. The *Thing*, or Tynwald Court, which can be assembled by the governor at any time whatever, possesses, according to old Scandinavian custom, both the judicial and the legislative power. The house of Keys is the first, and the Council the second court of appeal for certain causes, after they have been tried by the inferior courts in the island. The Council can reject proposals for laws brought in by the house of Keys, and the king again can reject the united proposals of both houses. On the other hand, what all the three estates have agreed on becomes a law ("a Tynwald act"); but it is not in force until it has been proclaimed from Tynwald Hill.

This hill, which stands in the midst of a valley on the west coast of the island, close to the northern side of the town of Peel, is said to have been originally raised with earth taken from all the seventeen parishes in the island. It forms four terraces, or steps, the lowest of which is eight feet broad, the next six feet, the third four feet, and the topmost six feet. There are three feet between every step, or terrace, and the circumference of the hill is about 240 feet. It is covered with green sward. (See Cumming. "The Isle of Man." London, 1848.)

Once a year, on St. John the Baptist's Day, the governor of Man, attended by a military escort, sets out from Castle Town, and, together with the Tynwald Court, attends divine service in St. John's Chapel, situated a few hundred paces from the hill. After the service, the whole court repairs in solemn procession to the hill, whence all the laws that have been passed in the course of the year are proclaimed in English and Manx. The procession then returns to the chapel, where the laws are signed and sealed.

Amongst all the Scandinavian Thing-hills, or Thing-walls ("Þingavellir") that can be traced in the old Danish part of England, in the Norwegian part of Scotland, as well as in the Orkneys and Shetland Islands, and which also formerly existed in Iceland, Norway, and throughout the North, Tynwald in Man is the only one still in use.

It is, indeed, highly remarkable that the last remains of the old Scandinavian *Thing*, which, for the protection of public liberty, was held in the open air, in the presence of the assembled people, and conducted by the people's chiefs and representatives, are to be met with not in the North itself, but in a little island far towards the west, and in the midst of the British kingdom. The history of the Manx *Thing* court remarkably illustrates that spirit of freedom and that political ability which animated the men who in ancient times emigrated from Norway and the rest of the Scandinavian North.

THE
NORWEGIANS IN IRELAND

SECTION I

Nature and Population of Ireland.—The "Danish" Conquests.—Traditions about the "Danes."—Political Movements.

Ireland may still be justly called the chief land of the ancient Celtic tribes. Long after the Britons and Caledonians had been driven out by the Romans and Anglo-Saxons, and obliged to fly to the remotest mountain districts of the west, their Irish kinsmen retained firm possession of the whole large and fertile country of Ireland. Subsequently, it is true, the Irish also were compelled to give way before the conquests of the Norwegians and English; yet they continued to inhabit the greater part of the country in vastly superior numbers; and even in the districts conquered by foreigners, which were mostly confined to the sea coasts, they dwelt intermingled with the new immigrants. In spite of the attempts of the English to subdue and annihilate the nationality of the Irish, they continued to preserve throughout the middle ages their ancient language and their characteristic manners and customs. With all their power the English have not even been able to root out the Roman Catholic religion, which to the present day forms the predominant church of the Irish. It is only in later times that they have succeeded in gaining a firmer footing in Ireland than they previously possessed. The English language and customs are continually making greater progress towards the west; and the Irish, who can no longer withstand England's power, seek in great numbers, like their kinsmen in Scotland, a new asylum in America. The struggle is the more severe in proportion as the Irish are more numerous than the Celtic population in Scotland and England. The last violent throes of the once powerful and mighty Irish nationality now fearfully agitate Ireland, which has been so long and so severely tried by oppression, pestilence, and famine.

One of the most active causes of the misfortunes of Ireland and the Irish is, however, the same that occasioned the ruin of the Celts in England and Scotland; namely, that they could never sincerely unite together. They have always abandoned themselves too much to eastern indolence and quiet, regardless of the march of civilization, and neglecting to avail themselves sufficiently of the rich resources afforded by their native land. For, although it is true that there are considerable tracts of boggy land in Ireland, and that many districts are but little capable of cultivation, yet in the main Ireland is exceedingly well adapted for agriculture. The neighbourhood of the Atlantic produces mild breezes, which permit neither frost nor snow to be of long duration, and consequently promote a rare and luxuriant vegetation. In few countries do we behold so many creeping plants, and such beautiful and verdant fields and pastures, as in Ireland, which, from its green meadows, has obtained the appropriate name of "the Emerald Isle." The land is intersected by rivers partially navigable, abounding in fish, and its coasts are washed by a sea—which not only from its rich fisheries, but from the facilities which it affords to navigation, particularly towards America—might, if properly used, become an inexhaustible source of wealth.

From time immemorial Ireland was celebrated in the Scandinavian North for its charming situation, its mild climate, and its fertility and beauty. The "Kongespeil" (or "Mirror of Kings"), which was compiled in Norway about the year 1200, says that "Ireland is almost the best of the lands we are acquainted with, although no vines grow there." The Scandinavian Vikings and emigrants, who often contented themselves with such poor countries as Greenland and the islands in the North Atlantic Ocean, must therefore have especially turned their attention to "the Emerald Isle," particularly as it bordered very closely upon their colonies in England and Scotland.

But to make conquests in Ireland, and to acquire by the sword alone permanent settlements there, were no easy tasks. The remote situation of Ireland, so far towards the west in the Atlantic Ocean, was of itself no slight defence. With the exception of certain tracts, principally on the east coast, the land is full of mountains, which everywhere afford secure retreats from an invading enemy. In our days Ireland, the second of the British Isles in point of magnitude, has a population of between six and seven millions, chiefly of ancient Irish, or Celtic origin; and in ancient times, when the Celts were entirely independent, and absolute masters of the country, the population does not appear to have been much less numerous. Ireland, moreover, distinguished itself by adopting Christianity, together with its accompanying civilization, at a very early period, which, however, was not able to put an end to the cruel and sanguinary disputes that raged between the different tribes composing its population. Thus the proportionately few and scattered Norwegians, who could reach Ireland only by sea, and who could derive assistance only from their countrymen settled upon the coasts of England and Scotland, had to contend with a numerous, and by no means unwarlike people, inhabiting an extensive and mountainous country. To obtain assistance in the hour of need from their own Scandinavian home was, on account of the great distance, a physical impossibility. When, therefore, we consider that neither the Romans nor the Anglo-Saxons ever obtained a footing in Ireland, although they had conquered the adjacent country, England; and when we further reflect upon the immense power exerted by the English in later times in order to subdue the Celtic population of Ireland, and the many centuries which elapsed before they even partially succeeded, we cannot help being surprised at the very considerable Scandinavian settlements which, as early as the ninth century, were formed in Ireland, and at the great influence which the Norwegians, according to the concurring evidence of the Irish and Scandinavian chronicles, must for more than three centuries have exercised in all the most important places in the country.

On his first entrance into Ireland, a Scandinavian traveller will be immediately reminded of the ancient dominion of his countrymen. It cannot possibly escape his observation what a striking part the Norwegians—or, as they are there exclusively called, the "Danes"—play in the popular legends and traditions of Ireland. That, like the north-western districts of Scotland, it should have best preserved the popular life of ancient times with its songs and legends, must, it is true, be ascribed to its remote situation. Everywhere, even far in the interior of the country, we are shown Danish raths (mounds and entrenchments), and among others the so-called "Danes-cast," a long ditch and rampart in Ulster. "Danish cooking-places" are also pointed out, consisting of small circular spaces set round with stones, and bearing traces of embers and burnings, some of which are met with scattered about on heaths and moors. In the ancient copper mines in the south of Ireland roundish stones with a dent round the middle are now and then dug up, which it is evident were used in former times in working the mines. These stones are called by the common people "Danes' hammers." In like manner they generally call most of the antiquities that are dug up, whether weapons or ornaments, "Danish." Tales calculated to awaken horror of outrages of the Danes are connected with all these pretended Danish memorials; and the farther we travel into the remote western districts, the more terrible are the tales we hear of the distress and cruel oppressions which the inhabitants endured under their Danish conquerors. Nevertheless the Irishman has preserved, like the Englishman, the remembrance of the Danes' contempt of death, and irresistible bravery. "That might even frighten a Dane," says the Irishman at times, when speaking of some desperate undertaking. A kind of superstitious fear of the redoubted Danes seems in some places to have seized the common people; at least it is an acknowledged fact, that in several parts of the country they continue to frighten children with "the Danes."

Similar ideas about the Danes are to be met with even among the more enlightened portion of the people. Not long ago, it was a firm belief among many educated men in Ireland, that there were still families in Denmark, who could not forget the dominion they had formerly exercised in Ireland, and who bore a title derived from the large estates which their forefathers had once conquered and possessed there. It was likewise commonly supposed that the Danes had carried with them from Ireland a great number of manuscripts, which were said to be preserved in one of the large collections of books in Copenhagen; as if, forsooth, it had been one of the chief aims of the bold and dangerous expeditions of the ancient Norwegians at that remote period, to carry off scientific treasures, and above all, manuscripts written in Irish, and, consequently, in a language that was for the most part entirely incomprehensible to them. In the last century in particular, and at the beginning of the present one, the Irish literati attributed to the Danes, or rather to the Norwegians, much to which, strictly speaking, they could have no valid claim.

The remarkable round towers, whose stone walls are built in so workmanlike a manner, and which are so evidently of Christian origin—being erected both as belfries, and as places of security for the clergy, and certainly *against* the Scandinavian Vikings and conquerors—were nevertheless proclaimed to be "Danish towers." The large stone rooms, or sepulchral chambers in cairns, that are found in several places in Ireland, as at New Grange, and which have so striking an agreement with the sepulchral chambers in the Scandinavian North and other countries, dating from the pre-historical, and so-called stone age, were also called "Danish;" although we know from the Sagas, as well as from the Irish chronicles, that the "Danes," or rather the Norwegian Vikings, sometimes opened these sepulchres, in order to take out any treasure that might have been buried in them by the natives. In several other instances, the Irish were not disinclined at times to regard the Norwegian conquests in a somewhat too favourable light.

Recently, however, they have gone to the opposite extreme. Everything possible that is bad, but nothing whatever that is good, is ascribed to the Scandinavian conquerors. In Ireland, as in Scotland and England, it is at present the commonly received opinion that the Norwegian conquerors did nothing but plunder and burn, and thus annihilated a very considerable civilization, which had prevailed in Ireland for centuries before the Norwegian expeditions. The "Danes" are, besides, accused of subverting the independence of Ireland, and of being the sole cause of her subsequently coming under the dominion of England. It is remarkable enough, however, that the Irish appear entirely to forget that the fault must be ascribed to themselves. They were so divided, and at such variance with one another, that, in spite of their vast numbers and boasted civilization, they could not unite to resist a mere handful of Scandinavians, who came from a great distance across the sea, and still less could they resist their powerful English conquerors.

This change in the opinions commonly received in Ireland concerning the Danish conquests has been effected more particularly by the late political movements. It is but little known in the Scandinavian North that, since the Repeal agitation in Ireland, the Danish conquests and the Danish name have been used as a constant and effective means of agitating against England; yet such is actually the case.

When O'Connell stepped forward as the mouth-piece of Irish nationality, to revive the ancient independence of Ireland, and if possible to restore its Parliament, by means of a repeal of the Union, it was of course important for him to awaken in his countrymen a feeling of freedom. With this design, he looked to Ireland's earlier history, and particularly to the period when she formed an independent kingdom. But that time was extremely remote. As early as the close of the twelfth century the English had firmly established themselves in Ireland; and the Danes before them had, for several centuries, held dominion over the most important places in that country. Had O'Connell, therefore, wished to dwell on the time of Ireland's real independence, he must have reverted to a period more than a thousand years ago. But he shrewdly foresaw that the vast and uneducated mass of the people, whom he chiefly wished to agitate, would not be able to follow him. He therefore chose historical events that lay nearer, and of which the remembrance still lived among the people; and, as his chief aim was to irritate the Irish against the "Saxons" (or English) he laid great stress upon the glory which the Irish had acquired in former combats against that people, as well as against the "Danes," who had preceded them in conquering Ireland.

Nothing could have been better adapted to O'Connell's object than the traditions and exaggerated notions about the Danes, still so widely diffused among the poorer classes in Ireland. O'Connell knew how to flatter with dexterity the vanity and self-love of the Irish, by representing how great a triumph they had achieved in former times, by driving out the Danes, and annihilating a dominion founded with so much bravery and wisdom. If he drew no direct conclusion from this, he let it, however, be sufficiently seen, that as the Irish were formerly able to expel their Danish conquerors, there was nothing to prevent them from chasing the hated "Saxons" from their coasts. At one of his great meetings, held on Tara Hill, where the ancient Irish kings were crowned in the time of Ireland's independence, he reminded his countrymen that their forefathers had, in the year 978, gained on that spot a considerable victory over the "Danes."

On the coast about three miles to the north-east of Dublin, is the plain of Clontarf, where, in the year 1014, a great battle was fought between the "Danes" and the Irish. This battle, one of the most sanguinary in all the wars of the Norwegians and Irish, was gained by the latter, whose king Brian Boroimha (or Brian Boru) fell just as victory declared for his army. A victory over the Danes like this must naturally always occupy a prominent place in the historical reminiscences of the Irish; and their historians throughout the middle ages, and down to our own times, have accordingly dwelt with extreme complacency on the description of the bravery of the Irish and of their king. But it did not suffice O'Connell and his followers to adhere to historical realities. They followed the chroniclers of a later period, by whom the victory of Clontarf has been delineated in far too brilliant colours. In songs, pamphlets, and speeches, the battle of Clontarf was now represented as having completely annihilated the Danish power in Ireland, and saved her independence and freedom. According to these accounts, not a single Dane or Norwegian would seem to have remained in Ireland after the battle. Brian Boroimha (Boru) was extolled to the skies, as a martyr for the deliverance of his country from the yoke of the oppressors. And in the intoxication of enthusiasm thus produced, his portrait, together with a picture of the battle of Clontarf, was distributed among the people in immense quantities, and at the very lowest price. On the tickets of members of the Repeal Association, which were ornamented with the names of the most important national triumphs of the Irish, as well as with portraits of the victors, the battle of Clontarf, and Brian Boru's portrait, stood at the top.

When at length this representation of the battle of Clontarf, as one of the most important fought by Ireland for liberty, had been so impressed upon the common people that it seemed an event which had only recently taken place—and which, at least in the lively imaginations of the Irish, might possibly enough be repeated—O'Connell gave out that he would hold a great repeal meeting on the plain of Clontarf. Everybody knew beforehand that the real meaning of O'Connell's speech was, that just as the Irish, with Brian Boroimha at their head, had formerly defeated the Danes on that very place, and thus saved Ireland's freedom, so should they now in like manner follow O'Connell (who, besides, gave himself out for a descendant of Brian Boru[?]), and make every sacrifice to wrest back their lost independence from English, or "Saxon," ascendancy. The English government, however, forbade the meeting, and indicted O'Connell. But the same extravagant notions respecting the national importance of the battle of Clontarf naturally continued to be generally received; and that not only amongst the adherents of O'Connell, or "Old Irelanders," as they are called, but also among the members of a political party, the "Young Irelanders," which has arisen since, and whose aim it is to sever the connection with England by open force. In the seditious songs of both these parties the Danes and the English generally share the same fate, as the war-cry, "The Saxon and the Dane," constantly forms the burthen of the songs. It is but very rarely that an Irish repealer (for instance, Mr. Holmes) dares venture to express an opinion that it would probably have been no detriment to Ireland if the "Danes" had remained settled there. This, when explained, means that the Danes would never have been so dangerous to the independence of Ireland as the English have since become; and that the Irish, united with and assisted by the Danes, would certainly have had a fleet capable of resisting any attacks of their powerful English neighbours(?).

SECTION II

Irish and Scandinavian Records.—Finn Lochlannoch.—Dubh-Lochlannoch.—The

Names of the Provinces.

One of the many complaints made by the Irish against the Danes, and particularly of late, is, that by destroying Irish civilization they likewise choked the vigorous germs of a national literature, which, in consequence of the early introduction of Christianity, had begun at a very early period to take root among the Irish people. The existence of a literature, particularly like the ancient Irish, in the vernacular language of the country, must of course always afford a strong proof of a certain degree of education among the people. During the late political agitation in Ireland, the old Irish literature, of which various remains are still preserved, was therefore extravagantly extolled, with the view of proving how glorious and enlightened was the age of Ireland's long-vanished independence.

Whatever opinion may be formed of the remaining relics of this ancient literature, which are mostly limited to chronicles in the form of annals, and a few old songs, it is at all events agreed that they are of very peculiar importance as regards a knowledge of the Norwegian and Danish expeditions. It is true that the Scandinavian Sagas and chronicles contain many accounts of the achievements of the Norwegians in Ireland, both in war and peace; but the Irish records of them are still more copious. The oldest Irish chronicles relate almost as much to the battles of the Norwegians and Danes with the Irish, as to the internal state of Ireland. A singular chronicle in Irish, of the close of the eleventh century, about "the Wars of the Irish and the Northmen," was discovered a few years ago. It contains not only a complete account of every battle between the Irish and Northmen, down to that of Clontarf, but also various information respecting the settlements of the Norwegians in Ireland, their mode of warfare, weapons, &c. That this chronicle must have been composed not long after the battle of Clontarf, is proved by the fact that it is referred to as an old record in another Irish work, called "The Book of Leinster," written in the first half of the twelfth century. The above-named ancient chronicle—the publication of which, by that distinguished Irish scholar, Dr. Todd, cannot be far distant—will, in conjunction with the rest of the Irish accounts relative to the Norwegian expeditions into Ireland, afford an excellent opportunity for comparison with the narratives of our Scandinavian Sagas. Meanwhile we have already sufficient information at hand to compare the accounts of the conquerors and the conquered—a method by which the historical truth will evidently come forth more clearly than if we were obliged to adopt exclusively the one-sided statements of either party.

The Irish accounts are, however, far from being always perfectly trustworthy. They not only reflect the customary hatred and prejudices of the Christians against the heathen Northmen, but frequently bear the stamp of being derived from early poetical legends. They relate how several Irish saints, as St. Columkill, St. Berchan, St. Kieran, and St. Comgall, had long before predicted the coming of the Scandinavian heathens and their barbarous proceedings. They likewise depict how terribly the heathens devastated and plundered unhappy Ireland. People were everywhere killed or maltreated; churches and convents were plundered, burnt, and desecrated. Thus the heathen chief Turges' (Thorgils') wife, Odo, sat on the altar of the conventual church in Clonmacnois, and on it, as on a throne received the homage of the assembled people. At the same time the Danes everywhere endeavoured to settle themselves in the country. They launched ships even on the lakes, with which they coerced the people dwelling around their shores. In the tenth century (continues the Irish scholar Duald Mac Firbis, in his unpublished treatise respecting "The Fomorians and Lochlanns," written about A.D. 1650) "Erinn was filled with ships (or adventurers), viz., the ships of Birn, the ships of Odvin, the ships of Grifin (or Grisin), the ships of Suatgar, the ships of Lagmann, the ships of Earbalbh, the ships of Sitric (?), the ships of Buidin, the ships of Bernin, the ships of the Crioslachs, the ships of Torberd Roe, the ships of Snimin, the ships of Suainin, the ships of Barun, the ships of Mileadh Lua, the ships of the Inghean Roe (Red Maiden). All the evils which befel Erinn until then were as nothing;

for the Galls spread themselves over all Erinn, and they built Cahirs (Caers) and Cashels (or Castles), and they showed respect to no one; and they used to kill her (Erinn's) kings, and carry her queens and noble ladies over the sea into bondage.

"A fleet the like of which was never seen, came with Jomar More, grandson of Jomar, and his three sons, viz., Dubhgall, Cualladh, and Aralt; and they took Inis Sibtonn in the harbour of Limerick, and forced submission from the Galls who had come before.

"The Galls then ordered a king on every territory, a chief on every chieftaincy, an abbot in every church, a bailiff in every town, a soldier in every house, so that not one of the men of Erinn had power over anything of his own from even the hen's clutch to the hundred milch cows. And they dared not show their kindness nor generosity to father, mother, bishop, ollave, spiritual director, those in sickness nor disease, nor to the infant one night old. If there was but one cow in the possession of any one of the men of Erinn, her broth should be given to the soldier the night that no milk could be procured from her. And an image of gold, or silver, of Fionndruine (a carved ornament of white metal) for the king's rent every year, and the person who would not be able to pay that should go himself into bondage, or have his nose cut off."

As the Irish chronicles give in this manner embellished and exaggerated pictures of the victories and power of the Norwegians in Ireland, so also they frequently depict the defeats of the "Danes" in colours that are too vivid. The ancient chronicle before mentioned concerning "The Wars of the Irish and the Northmen" states, for instance, that some time before the battle of Clontarf a desperate conflict took place at Glennmama, in the neighbourhood of Dublin, between the Irish king, Brian Boroimha, and the Danes in Dublin; with which latter were united the inhabitants of Leinster, who had shortly before entered "the Danish precinct of Dublin." King Brian was victorious in the battle; "*and then there was not a threshing-spot from Howth to Brandon in Kerry without an enslaved Dane threshing on it, nor a quern without a Danish woman grinding on it.*"

Very different are the accounts given by the Scandinavian Sagas relative to the Norwegians in Ireland. It was to be expected that the

Irishman, endowed with a southern vivacity, and at the same time thrown into deep anxiety by the Norwegian expeditions, should have regarded them in quite a different light from the tranquil Norwegian himself, who in the conquests in Ireland beheld only a repetition of what was occurring at the same time in so many other countries. The Scandinavian accounts are in general shorter than the Irish, and confine themselves merely to the relation of single events. Ireland is usually treated of incidentally, nay almost accidentally. According to the Sagas, we should almost be inclined to think that the dominion of the Norwegians in Ireland was much less in extent and duration than was actually the case, so little have the writers of them thought of magnifying their countrymen's renown at the expense of historical truth. What, therefore, the Sagas, and the rest of the Scandinavian chronicles relate about Ireland is, for the most part, very trustworthy, and at all events agrees with the representations at that time current amongst the Irish themselves. It is quite evident that the writers of the Sagas had either been in Ireland, or at all events derived their knowledge from men who knew the country well, either through Viking expeditions or trading voyages. The accuracy with which different places in Ireland are described affords a very remarkable proof of this. Thus the ancient seat of royalty "Teamor," or Tara, which is also celebrated for its delightful situation, is mentioned in the "Kongespeil" under the name of "Themar;" and it is added that "the people knew no finer city on the earth." In the same place it is further stated that the town and castle sunk suddenly into the earth, because a king pronounced an unjust judgment—a tradition common in Ireland to the present day.

Places in Ireland mentioned in the Sagas, but which formerly could not be traced, have recently been pointed out by the aid of the Irish records. The "Kongespeil" states, for instance, that Saint Diermitius had a church on a small island, "Misdredan" or "Inisdredan," in the lake "Logherne." This island is evidently "Inisdreckan" in Lough Erne, where formerly St. Diermitius actually had a church. Subsequent transcribers of the book have clearly enough transformed Inisdreckan into Inisdredan, Misdredan, &c. The same has been the case with the celebrated King Brian Boroimha's castle, which, by a mistake in copying, is called in the Sagas "Kanntaraborg" or "Kunjáttaborg," instead of "Kanncaraborg." Brian Boroimha's castle, so celebrated in the Irish songs and legends, was called in Irish "Ceann-Caraidh" (pronounced Cancara), and was situated on the river Shannon, not far from Limerick. To the Irish Cancara the Norwegians, therefore, only added the Scandinavian termination "borg." Again, it is stated in the Sagas that one could sail from Reykjanæs in Iceland to "Jöllduhlaup" in Ireland, in about eight days, or, according to some readings, even in a much shorter time. Formerly this place was sought on Lough Swilley, near Cape Malin, in the north of Ireland. But Jöllduhlaup, which signifies "the course or breaking of the waves," is merely a translation into Icelandic of the Gaelic name "Corrybracan" (Coire Breacain), whereby the Gaels denote a whirlpool between the little island of Rathlin (or Raghrin) and the north-easternmost part of Ireland (the county of Antrim). That the ancient Icelanders designated this precise spot in Ireland is owing in all probability to the circumstance that the island of Rathlin was in the olden times the chief station in the passage from Ireland to Scotland, and as such the rendezvous for a number of merchants and other travellers. Lastly, Snorre Sturlesön relates that in the beginning of the eleventh century a desperate naval battle was fought between the Orkney jarl Einar and the Irish king "Konofögr," in Ulfrek's, or Ulfkel's, Fiord, on the coast of Ireland. The situation of this fiord, or firth, was entirely unknown until it was lately discovered that in a document issued by the English-Irish king John in the year 1210, the Firth Lough Larne, on the east coast of Ireland, about fourteen miles north of Belfast, was at that time still called "Wulvricheforð," which agrees most accurately with the Icelandic name "Ulfreksfjörðr." By a remarkable coincidence, a skeleton was dug up a little while previously just on the shores of Lough Larne, together with a pretty large iron sword, having a short guard and a large triangular pommel at the end of the hilt; the form of which sword (as I shall prove) was not Irish, but pure Scandinavian, like that of the swords used towards the close of heathenism in the North. There is every probability that the skeleton and sword belong to one of the Scandinavian warriors who fell in the above-mentioned battle, and who was afterwards buried on the shore. Thus both the exhumed antiquities, and the lost but re-discovered name of the place, contribute to corroborate the credibility of Snorre Sturlesön's account.

Both the Irish and the Scandinavian records agree that Norwegians and Danes were settled in Ireland at a very early period. The Vikings are said to have ravaged its coasts for the first time in the year 795; and in the ninth century many of them were already settled in the country. Amongst the men who, at the close of the ninth and beginning of the tenth century, first colonized Iceland, several Irishmen, or rather descendants of Norwegians settled in Ireland, are mentioned; as, for instance, Thormod and Ketil Bufa, Haskel Hnokkan, who was descended from the Irish king Kjarval, besides others. Intermarriages between the Norwegians in Ireland and the native Irish seem to have taken place from the very first; which explains the circumstance that many men in Iceland bore at an early period Irish names, such as Kjaran and Niel or Njäll.

The Norwegians in Ireland, like their Danish kinsmen in England, were obliged to begin by settling on the coasts; whence, both by warlike and peaceful means, they gradually extended their dominion over the country. Besides this continually-increasing and more peaceful colonization, roving Scandinavian Vikings continued their attacks in different parts of Ireland, whereby the power of the Irish was considerably weakened. A pause took place, however, in the tenth century, both in the expeditions of the Vikings, and in the progress of the Scandinavian settlements in Ireland. It is even stated that for about forty years "the strangers" (the Galls) were entirely driven out of the country; but this is probably an exaggeration. This diminution of the power of the Norwegians in Ireland occurred about the same time with the decrease of the Danish power in England, and appears to have been produced by the same causes; namely, internal commotions in the mother-country of Scandinavia, which prevented the sending of such ample assistance as previously to the colonists in the British Islands.

Subsequently, however, the Norwegian dominion in Ireland became doubly powerful; and the Irish were so far from being able to expel the strangers, that, notwithstanding the numerical inferiority of the latter, they were often masters in the country. It was evidently

Norwegians rather than Danes who settled in Ireland, although not a few of the latter were mixed with them. In later times all the Northmen in Ireland are included under the common name of "Danes." But the best and oldest Irish chronicles distinguish, as it has been previously remarked, between the light-haired "Finn-Lochlannoch," or "Fionn-Lochlannaigh" (the Norwegians), and the dark-haired "Dubh-Lochlannoch," or "Dubh-Lochlannaigh" (the Danes); or, what is the same, between Dubgall ("Dubh-Ghoill") and Finngall ("Fionn-Ghoill"). The above-mentioned chronicle of "the Wars of the Irish and the Northmen," which draws a clear distinction between the Norwegians and Danes, expressly says that the Danes were only one of those tribes that made expeditions of conquest to Ireland. We even learn from the Irish chronicles that the Norwegians and Danes often fought between themselves for the dominion in Ireland. For instance, it is stated in the Irish annals in the year 845: "the Dubhgalls (the Danes) came this year to Dublin, sabred the Finngalls (the Norwegians), destroyed their fortresses, and carried away many prisoners and much booty with them." Similar intestine disputes are mentioned in other places of the annals; yet, as might be expected, the Danes appear still more frequently as fighting in alliance with the Norwegians. On the flat shores in the middle of the eastern coast of Ireland, between Dublin and Drogheda, which are called Finngall, or "the strangers' land" (from "finne," a land, and "gall," a stranger), and which in ancient times were colonized chiefly by Norwegians, is a small town called Baldoyle. In old documents this town is named "Balidubgail," the Dubhgalls' or Danes' town ("bal," a town). We have thus an existing proof that the Danes also were once actually settled in Ireland. The Dubhgalls are likewise said to have settled in the districts nearest to the south and west of Dublin.

For the rest, among the names of places in Ireland which remind us of the Norwegian dominion, we must in particular specify the names of three of Ireland's four provinces, viz., Ulster (in Irish "Uladh"), Leinster (Irish, "Laighin"), and Munster (Irish, "Mumha," or "Mumhain"); in all of which is added to the original Irish forms the Scandinavian or Norwegian ending *staðr*, *ster*. It might even be a question whether the name "Ireland" did not originally derive this form from the Northmen. On this head we have, at all events, a choice only between the Northmen and the Anglo-Saxons, for to the present day the Irish themselves still call the country Eirinn or Eiri. The termination *land* is entirely unknown in their language.

That the Northmen, and especially the Norwegians, should have been able to give to the three most important provinces of Ireland the names which they still bear, sufficiently indicates that they must have been settled there in no inconsiderable numbers, or that they must at all events long have ruled these districts, which is also confirmed by the statements of the Irish chronicles. But in general we shall seek in vain among the names of places in Ireland for traces of such an extensive Scandinavian colonization as existed in the North of England. All circumstances clearly show that the Northmen in Ireland were proportionately less settled in the rural districts than in the towns. In consequence of the remote situation of Ireland, its extent, and the magnitude of its population, they were exposed in the rural districts, when at some distance from the coast, to much more danger than in the towns, where they could better assemble their forces behind ramparts and ditches. It is a very striking circumstance that the chief strength of the Norwegians lay precisely in those towns which have since continued to be the greatest and most important in Ireland. The Norwegian dominion in Ireland had quite a peculiar character, having been divided into several small and scattered kingdoms, each comprised in a town, or even only part of a town, with at most an inconsiderable adjacent tract of land. That such kingdoms should subsist for several centuries, and even long after the Danish dominion had ceased in England, is certainly one of the most remarkable, and, with regard to the civilization of the Northmen, most pregnant facts in the history of the Scandinavian emigrants.

SECTION III

Norwegian Kings.—Limerick.—Cork.—Waterford.—Reginald's Tower.—Dublin.—Thengmotha.—Oxmantown.

According to trustworthy historical evidence, the Norwegians and the Danes, or the Ostmen, as they were called in Ireland (from having come originally from the east), principally fixed their abodes in Dublin, Waterford, and Limerick, where, as early as the ninth century, they had founded peculiar Scandinavian kingdoms. They were also settled in considerable numbers in Wexford, Cork, and several Irish cities, so that they had possessed themselves, by degrees, of the best-situated places in the east, south, and west of Ireland, both for navigation and for intercourse with the rich countries of the interior.

The central point, however, of the real Norwegian power was the present capital, Dublin. This considerable city, which is said to contain at present more than three hundred thousand inhabitants, lies on both sides of the river Liffey, near the spot where it discharges itself into the Irish Channel. It is surrounded by a charming and fertile country. Anciently, however, and especially before the arrival of the Norwegians in Ireland, it seems to have been comparatively insignificant, both as regards extent and population. Yet even at that time it was, probably by means of its fortunate situation, and its connections with the neighbouring countries, the most important place in Ireland, which, at that early period, did not possess any very large towns. But as Dublin and its vicinity was at all events one of the most attractive points on the east coast of Ireland, some of the first Scandinavian kingdoms were founded there. About the middle of the ninth century a celebrated Norwegian Viking, Olaf the White, is said to have taken Dublin, and made himself king of the city and district. After the death of Olaf in a battle, two sons of King Harald Haarfager (Fair-hair), of Norway, arrived there, namely, Thorgils, called by the Irish Turges, and Frode; who, by means of the sword, likewise won for themselves thrones in Dublin. Subsequently to them, again, as the

Irish chronicles relate, there landed three brothers, Olaf, Sigtryg, and Ivar, who became kings in Dublin, Waterford, and Limerick. From that time Norwegian kings reigned in those places, with but few interruptions, for full three centuries.

There would certainly be some cause to doubt of so extensive a Norwegian dominion in a country so remote as Ireland, as well as of the actual existence of so striking a number of Scandinavian, and especially Norwegian, kings of cities, if the names of a great number of them were not preserved; and that, too, not so much in the chronicles of the Norwegians themselves, as in those of the conquered Irish, who had no reason to exaggerate in this respect. Several of the Norwegian kings mentioned in the Irish chronicles are, besides, mentioned in contemporary Anglo-Saxon and Scandinavian records; whence it becomes doubly probable that the remainder of the Norwegian kings mentioned by the Irish actually reigned in the places indicated.

As the Irish chronicles thus not only give the most detailed accounts respecting the Norwegian kings in Ireland, but also the least partial ones in favour of the Norwegians, I have annexed a list of kings compiled by an Irish author from Irish records. We may see from this, although it is scarcely complete, that the Scandinavian names of the kings (such as Olaf, Ivar, Eistein, Sigtryg, Godfred or Gudröd (?), Ragnvald, Torfin, Ottar, Broder, Eskil, Rörik, Harald, and Magnus) appear in general clear and distinct through the somewhat altered Irish forms, whilst a few names, such as Gluniarand (which in Irish signifies Iron-Knee), Eachmargach, Maelnambo, and Gilalve, seem to be mere Irish translations, or at all events purely Irish transformations, of Scandinavian forms.

NORWEGIAN, OR SCANDINAVIAN, KINGS IN IRELAND.

(*From Lindsay, "The Coinage of Ireland," Cork, 1839.*)

A.—Kings of Dublin.

Anlaf (Olaf), 853.

Ifar (Ivar), 870.

Ostinus (Eistein), 872.

Godfred (Gudröd), 875.

Sihtric (Sigtryg), 893.

Sihtric, 896.

Regnald (Ragnyald), 919.

Godfred, 920.

Anlaf, 934.

Blacar (Blake), 941.

Godfred, 948.

Anlaf, 954.

Godfred, 960.

Anlaf, 962.

Regnald.

Gluniarand, 981.

Sihtric (deposed), 989.

Ifar, 993.

Sihtric (again), 994.

Anlaf, 1029.

Sihtric, 1034.

Anlaf, 1031.

Ifar, 1050.

Eachmargach, 1054.

Maelnambo, 1064.

Godred Crovan, 1066(?).

Godfred Merenach, 1076.

Gilalve, 1094.

Torfin, 1109.

Regnald, 1125.

Godfred, 1147.

Oicterus (Ottar), 1147.

Broder, 1149.

Askel, 1159.

Roderick, 1171 till about 1200.

B.—Kings of Waterford.

Sihtric, 853.

Ifar, 983.

Regnald, 1000.

Sihtric, 1020.

Regnald, 1023.

Commuanus, 1036.

C.—Kings of Limerick.

Ifar, 853, King of Dublin in the year 870.

Ifar, 940.

Olfin, 942.

Harold O'Ifar.

Magnus, 968.

More detailed accounts are wanting relative to the kings of Limerick and Waterford during the eleventh and twelfth centuries; though it is certain enough that they continued to reign there just as long as in Dublin. Nor can we at present discover many apparent or recognisable traces of the dominion of the Ostmen and their kings in the two places just mentioned. Still Waterford appears to have derived its present name from the Norwegians. The Irish called the town "Port Lairge;" to which name, however, modern Irish scholars would ascribe a "Danish" origin, as it is supposed to be derived from a Danish chief called Lairge, mentioned in the Irish annals in the year 951. The Norwegians, on the other hand called it "Veðrafjörðr," the resemblance of which to Waterford is not to be mistaken. Near the coast of this "fiord," which may have given name to the town, is still to be seen a monument, very rare in Ireland, of the ancient Norwegians' art of fortification, namely, a round tower, said to have been erected in the year 1003 by the reigning Norwegian king in Waterford, Regnald, or Reginald (Ragnvald), and which to the present day is commonly called "Reginald's Tower."

This tower, which in Irish was also called "Dundory," or the king's fortress, was afterwards used both as a fortress and a mint. After the English conquest of Waterford, Earl Strongbow used it in the year 1171 as a secure dwelling-place; and, among other prisoners, for a long time kept Reginald, the last king of the "Danes" in Waterford, imprisoned in it. The tower afterwards underwent several changes, till, in the year 1819, it (or at least the exterior) was restored to its original form, just as the following delineation of it (after Petrie) shows.

[+ +] Reginald's Tower

With regard to Dublin, however, the case is quite different. The series of kings there from the year 853 until about 1200, and consequently for almost three centuries and a half, is pretty complete. It was a natural consequence of the considerable power and influence possessed by the kings of Dublin, that their names were often mentioned in the chronicles in connection with important events both in Ireland and in the neighbouring countries. The Norwegian kings in Dublin knew how gradually to strengthen and extend their power, not only by arms, but also by a shrewd and able policy. They soon learnt how to avail themselves of the intestine disputes by which the Irish tribes and chiefs were divided. They joined one of the ruling parties, contracted marriage with the daughters of Irish kings and chieftains, and on their side gave Scandinavian women in wedlock to leading Irishmen. According to the old Irish book called "the Book of Lecan," the Irish king Congolaich (934-954) had a son, Mortogh, by Radnalt, daughter of the Dublin king Anlaf, or Olaf. At a somewhat later period a Norwegian king in Dublin, named Anlaf, was married to an Irish woman, Dunlath, who was mother of the Dublin king "Gluin-Jarainn" (Iron-Knee). Similar marriages between Norwegian and Irish royal families are often mentioned; even King Brian Boru, so adored by the Irish, was nearly related to the Norwegian kings. He was father of Teige and Donogh, by Gormlaith, or Kormlöd, a daughter of Morogh Mac Finn, king of Leinster. But Gormlaith was also married for a long time to the Dublin king, Anlaf, by whom she had a son, afterwards the celebrated king of Dublin, Sigtryg Silkeskjæg (Silk-beard); and thus Brian Boru's two sons Teige and Donogh—of whom Teige afterwards married Mor, a daughter of the "Danish" king Eachmargach of Dublin—were half-brothers of their father's enemy, King Sigtryg. "The Book of Leinster" says that Gormlaith was likewise mother of the Norwegian-Irish king Amlaff Cuaran (Olaf Kvaran); whilst the Irish chronicler, Duald Mac Firbis, mentions this same Olaf Kvaran as married to Sadhbh (Save), a daughter of Brian Boru, and that even "at the time when the battle of Clontarf took place." After this we are better able to understand how it happened that whole Irish tribes, with their kings at their head, so often fought in union with the Norwegians and Danes; since we learn that their mutual political interests were hound closer together by the ties of relationship.

On the other hand, the Norwegian or Scandinavian kings of Dublin and other parts of Ireland also constantly maintained connections, both of friendship and relationship, with their countrymen in England and Scotland, as well as in the mother countries of Scandinavia. It might, indeed, sometimes happen that Scandinavian kings or Vikings, from Man or the Orkneys, attacked, nay even conquered for a time, the Norwegian kingdom of Dublin, particularly when the Norwegians in Ireland were at variance with one another. But in general the Scandinavian colonists in the British Isles appear to have stood or fallen with one another. Numerous Scandinavian warriors from England, Scotland, and the surrounding islands, fought now and then in conjunction with the Norwegians settled in Ireland, against the native Irish. But the Norwegian kings in Ireland frequently supported their friends in England and Scotland against the Anglo-Saxons and the Highland Scots, and at times won kingdoms there by force of arms. Mutual marriages, also, were frequently made, whilst Scandinavian merchants and Vikings, for instance, dwelt in Dublin at the court of the Norwegian kings. Thus the Norwegian prince Olaf Tryggvesön, after having been christened at Dublin, stayed there for some time with the Norwegian king Olaf Kvaran, and married his sister Gyda.

Many accounts testify that the Norwegians in Ireland, at least in the cities, and especially Dublin, were powerful enough to maintain their language, and the rest of their Scandinavian characteristics, in spite of the Irish. The Icelandic bards, Thorgils Orraskjald and Gunnlaug Ormstunga, are expressly stated to have visited the court of the Norwegian kings in Dublin in the tenth and eleventh centuries, where they diverted the Scandinavian warriors with their national songs. Ancient Irish manuscripts contain proofs not only of the peculiar language, but also of the peculiar writing, of the Norwegians, or runes, which in Irish were called "Ogham na Loochlannach" or "Gallogham" (the Northmen's, or strangers', Ogham). Ogham was the name of a mode of writing then used by the Irish. There are also some traces of characteristic Scandinavian institutions among the Norwegians and

Danes in Ireland. In an Irish poem of the early middle ages, about the Norwegian chief "Magnus the Great," the Norwegians are called "the people with the twelve counsellors." This leads us to think that the Norwegians, like the Danes in England, must have employed in their judicial proceedings a sort of jury, consisting of twelve men of repute, an institution so foreign and striking to the Irish, that they were led to characterize the Norwegians by it. It is at least quite certain that the Norwegians in Ireland, as the Irish chronicles admit, kept themselves entirely separate from the Irish with regard to their ecclesiastical institutions, and that they likewise had their own assize place in Dublin, which bore the Scandinavian name *Thing*. A document of the year 1258 conveys a gift of some ground in the suburbs of Dublin, in "Thengmotha" (from "mote," a meeting), which the Irish publisher of it (the Rev. R. Butler) correctly explains by "the place of legal assembly in the Danish times of Dublin." The *Thing* place, which seems to have been not far from the present site of Dublin Castle, where the Norwegians had erected a strong fortress, gave to the surrounding parish of St. Andrew the surname of "de Thengmote."

One of the chief causes that the Norwegians in the Irish cities maintained uninterruptedly their Scandinavian characteristics, and consequently their independent power likewise, was that they not only lived in the midst of the Irish, but that, as Giraldus Cambrensis expressly intimates, they erected in every city a town of their own, surrounded with deep ditches and strong walls, which secured them against the attacks of the natives. They built a rather extensive town for themselves on the river Liffey, near the old city of Dublin, which was strongly fortified with ditches and walls, and which, after the Norwegians and Danes (or Ostmen) settled there, obtained the name of Ostmantown (in Latin, "vicus," or "villa Ostmannorum"), *i. e.* the Eastmen's town. Even the Irish chronicles, which attest that, as early as the beginning of the tenth century, the Norwegians in Dublin had well intrenched themselves with walls and ramparts, also state that in the art of fortifying towns they were far superior to the Irish. Ostmantown continued through the whole of the middle ages to form an entirely separate part of Dublin, and the gates of the strong fortifications with which it was surrounded were carefully closed every evening. The walls were at length razed, and Ostmantown, or, as it was now corruptly pronounced, "Oxmantown" (whence an Irish peer has obtained in modern times the title of Lord Oxmantown), was completely incorporated with Dublin. But to the present day the name of Oxmantown remains an incontrovertible monument of an independent Norwegian town formerly existing within the greatest and most considerable city of Ireland.

SECTION IV

Norwegian Names of Places, near Dublin.—Norwegian Burial—Places.—Norwegian Weapons and Ornaments.

The few Scandinavian names of places in Ireland are, with the exception of the previously-mentioned provinces, confined to the coasts, and there particularly to the names of islands and fiords. On the west coast there are only two rather doubtful ones; namely, Enniskerry, an island (the first part of which is the Irish *Inis*, an island, whilst the latter part seems to include the Scandinavian name "*Sker*," or *Skjær*, a reef); and the harbour, Smerwick. Several places on rivers are still called *Laxweir*, as for example on the Shannon near Limerick and Killaloe, where salmon are caught in a net stretched across the river. The word "Lax" (salmon) is unknown in the Irish language, but appears, as we have seen, in several Scandinavian names of places in Scotland. On the south coast, besides Waterford, we can mention at most only the Isle of Dursey (Þorsey?) with the small adjoining island of Calf. The greatest number of Scandinavian names appears on the east coast. In some names of places situated on the finest fiords we may trace the Scandinavian ending "fjörðr;" as, for instance, to the south of Dublin in Wexford (in Irish, Loch Garman), and to the north of Dublin, in Strangford and Carlingford (in Irish, Cuan Cairlinne). But in general, all the names of places of Scandinavian origin, or with Scandinavian terminations, are collected round Dublin as the central point.

At the southern entrance of the bay of Dublin is the Island of "Dalkey" (in Irish, "Delg Inis"), and at the northern entrance the high and rounded cape Howth (in Irish, "Ceann Fuaid," or "Beann Edair"), which in ancient letters is also called Hofda, Houete, and Houeth. This is clearly the Scandinavian "höfud," or "Hoved" (head), a name particularly suited to the place. In the immediate neighbourhood is also the old Danish town Baldoyle, and the district of Finngall, colonized by the Norwegians. Directly north of Howth rises "Ireland's-eye" (in Irish, "Inis Eirinn" and "Inis Meic Ness-áin"); and still farther to the north the islands of "Lambay" (in Irish, "Rachrainn") and "Skerries," or the Skjære (reefs). Close to the west side of Dublin is the little town of Leixlip, where there is a famed salmon-leap in the river Liffey. In old Latin epistles the name of Laxleip is translated by "saltus salmonis," which is plainly neither more nor less than the old Norsk "lax-hlaup" (*Dan.*, Laxlöb; *Eng.*, salmon-leap): which name reminds us again of the salmon fishery, so highly cherished by the ancient Norwegians. It is doubtful whether the county of Wicklow, which adjoins that of Dublin, derived its name from the Norwegians; though it is not improbable that it did, as in Irish it is called Inbhear Dea, but in old documents Wykynglo, Wygyngelo, and Wykinlo, which remind us of the Scandinavian Vig (Eng., *bay*) or Viking.

At all events the decidedly Scandinavian names of places around Dublin sufficiently indicate the predominance of the ancient Norwegians and Danes in that city. Discoveries made by excavations in and around Dublin have also, in recent times, very remarkably contributed towards placing this matter in a still clearer light.

In constructing a railway close by Kilmainham, now the most western part of Dublin, the workmen some years ago laid bare a number of ancient tombs. In these lay whole rows of skeletons, each in its own grave, and by the side of them many kinds of iron weapons and ornaments. Fortunately several of the specimens thus discovered were preserved, principally for the museum of the Royal Irish Academy in Dublin; by which means Irish archæologists had an opportunity of convincing themselves that these antiquities must be a good deal older than the English conquest of Ireland; yet that they are by no means of the kind usually found in Ireland, and belonging to the period of the Irish iron age. It is thus placed beyond all doubt, that they are not Irish remains, but derived from the Norwegians and Danes at that time settled in Ireland. The few illustrations here annexed will present to every Scandinavian archæologist mere well-known objects, corresponding so exactly with the antiquities of the iron age preserved in our Scandinavian museums, that we might even believe them to have been made by the same hands.

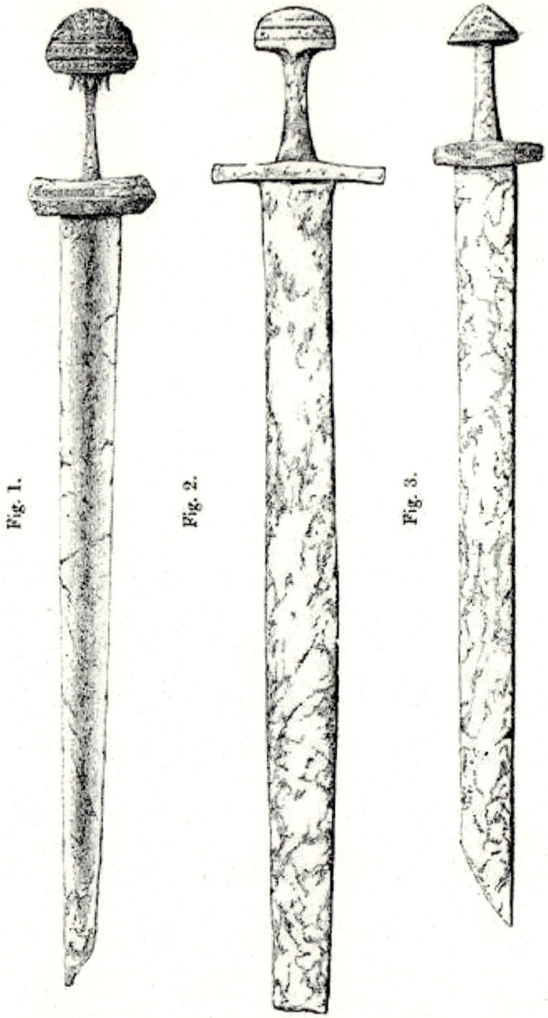

[+ +] Swords: Fig. 1 to Fig. 3.

The swords (Figs. 1-3), which very much resemble the Scandinavian swords found in England (described at p. 45,) are from twenty-four to thirty-two inches long. Some have two edges, others only one. The pommel and guard of the hilt are in several of them ornamented with very neatly inlaid pieces of gold, silver, and other metals. On one of them some engraved Latin letters have been found, which may also be seen on a sword of the iron age in the Museum at Copenhagen. Even the old Irish chronicles relate that the Norwegians placed inscriptions on their swords. Thus an ancient Irish poem says: "Hither was brought, in the sword sheath of Lochlan's king, the Ogham across the sea. It was his own hand that cut it." It is most probable that by the Ogham writing is here meant "the Norwegian's Ogham," or runes, with which, as our Sagas state, the old Northmen's swords were frequently ornamented.

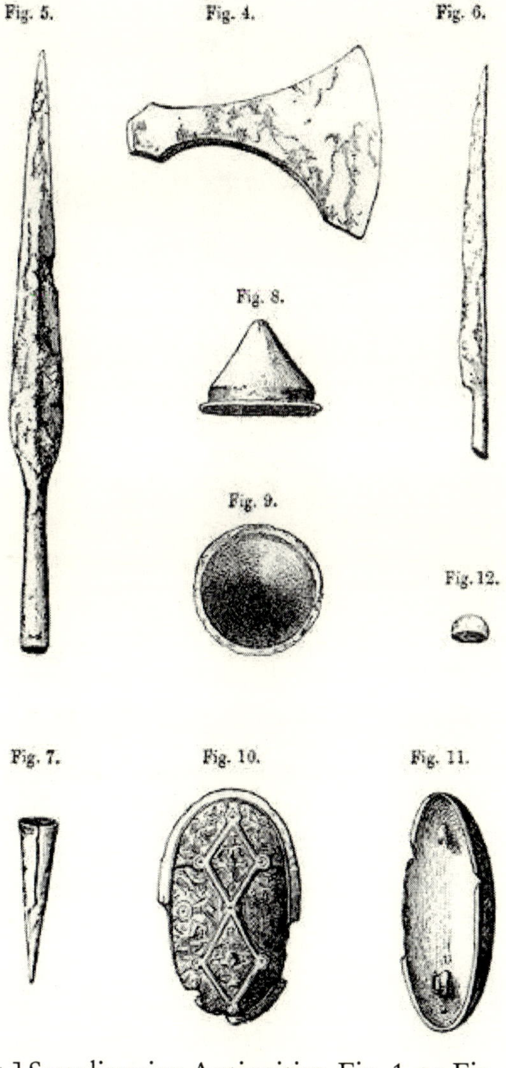

[+ +] Scandinavian Antiquities, Fig. 1. to Fig. 12.

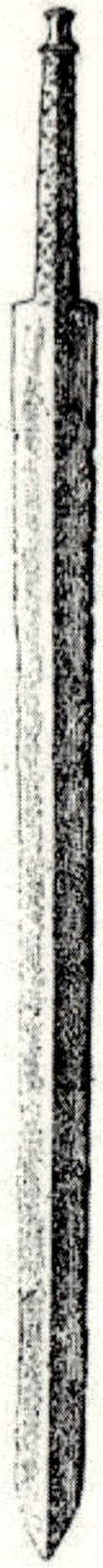

[+ +] Sword: Irish Iron

Several genuine Irish iron swords of that ancient period have been discovered in Ireland at various times, both in the river Shannon and in old Irish castle-yards, or on the sites of castles. They are much smaller than the Norwegian swords, and in general want both the guard and the large pommel at the end of the hilt, as the annexed figure of those most frequently found shows. On the whole the Irish iron swords are of an older and more imperfect kind, and very strikingly resemble the bronze sword used in Ireland in the age of bronze. On placing the short and ill-formed Irish sword by the side of the much larger, better, and handsomer Norwegian one, we may almost say that we obtain, as it were, a living image of the degenerate and miserably-equipped Irish people in comparison with the strong and well-armed Norwegians.

The Norwegian warriors who found their last resting-place at Kilmainham, were evidently buried with all their arms, from the renowned "Danish battle-axe" (Fig. 4), and the lance (Figs. 5, 6)—which must have been deposited with the entire shaft, since the ferrule (Fig. 7) has been found—down to the shield. But as the last was mostly of wood, nothing more remains of the whole shield than the large iron boss (Figs. 8, 9), which was placed in the middle, and which served to protect the hand which bore the shield.

Among all the things discovered at Kilmainham, scarcely any more decidedly indicate their Norwegian, or Scandinavian, origin, than the bowl-formed brooches (Figs. 10, 11), already mentioned when speaking of the coasts of Scotland, and which are not found in any other part of Ireland. There are also some very peculiar small bone buttons (Fig. 12), having a small hole in the flat side, penetrating the button for some way without entirely piercing through it. Buttons of this form have not been before found in Ireland, though they are very well known in the Scandinavian North. They are discovered in Sweden and Norway, in graves of the period of the iron age, or times of the Vikings. It is highly probable that in those times they served as men, or counters in some game, as they are generally found, especially in Norway, collected together in great numbers, and in conjunction with dice. To judge from the holes in the bottom, they have certainly been used in a sort of game of draughts; for, till late in the middle ages, nay, almost down to our own times, the Icelanders were accustomed to furnish their boards with small pivots, on which they placed the men, that they might not by any accidental shaking of the table be mixed with one another, and the whole game thus suddenly disturbed. The Irish also seem to have had a somewhat similar mode of proceeding at that time, as among a great number of things undoubtedly Irish, discovered at Dunshauglin, there was found a bone button or knob, certainly a draughtsman, which, instead of a hole, is furnished with a metal point at the bottom, by which it was evidently intended to be fixed in the board. But for the Scandinavian Vikings, who were so much at sea, and who, it seems, liked to while away the time by playing draughts, such a precaution was doubly necessary, as the rolling of the vessel would otherwise have thrown the draughtsmen together every moment. It is remarkable that at Kilmainham, as well as in Scandinavia itself, the draughtsmen are found deposited in the graves, by the side of the arms and ornaments of the warriors. This affords an instructive proof that the old Northmen must have been very fond of gaming; and consequently that the picture drawn by Tacitus of the passion of the ancient Germans for play, which at times even led them to gamble away their personal freedom, might apply to their neighbours, the Scandinavians.

We can scarcely err in referring the antiquities found at Kilmainham to the ninth, or at latest to the tenth century. The mode of burial is heathenish rather than Christian; and, as is known, the Norwegians settled in Ireland were converted to Christianity in the tenth century at latest, and probably still earlier. It is not at all probable that the graves are to be attributed to an isolated band of heathen Vikings, who came over at a later period, and who, after a battle, buried their dead on the field. The great number of graves, and the careful manner in which each is said to have been set or enclosed with stones, rather show that they were made in all tranquillity by the Norwegians and Danes, who at that time dwelt in Dublin, or its immediate neighbourhood, and who probably had a common burial ground there. Scandinavians appear also to have been buried in an adjoining churchyard, which at that time belonged to a convent dedicated to St. Magnen, but which afterwards became the burial-place for a hospital of the knights of the order of St. John, founded at Kilmainham. It has at length become one of the largest churchyards in Dublin. In corroboration of the conjecture that Scandinavians were buried in it, it may be mentioned that a tall upright stone with carved spiral ornaments stands there—a sort of monumental, or bauta-stone, under which, several years ago, various coins were discovered, minted by Norwegian kings in Ireland; and near them a handsome two-edged iron sword, with a guard and a longish flat pommel. Some have, indeed, thought that this sword must have belonged to Murrough, a son of Brian Boru, or to Murrough's son Turlough, as both these warriors, having fallen in the battle of Clontarf, are said to have been buried in this churchyard. This, however, is only a vague conjecture; whilst it is quite certain that the above-mentioned sword agrees most accurately in form with the many swords of the Vikings' times found in the North. There is, therefore, reason to suppose, that the sword was formerly deposited there with the body of a Norwegian warrior; and this supposition is strengthened by the discovery of the Norwegian-Irish coins.

Other old Norwegian, or Scandinavian burial-places, have been discovered in the Phœnix Park, near Dublin, where a pair of bowl-formed brooches were found near a skeleton. In making, a few years ago, some excavations in Dublin itself, in "College Green," which formerly lay outside the city, the workmen met with several iron swords, axes, lances, arrows, and shields, of the well-known Scandinavian forms. It is probable that this also was a burial-place similar to that at Kilmainham. With the exception of the burial-place on the coast of Lough Larne, the ancient Ulfreksfjord, no other decidedly Norwegian graves are hitherto known to have been discovered in Ireland.

Just as the proportionally numerous Norwegian graves near Dublin prove that a considerable number of Norwegians must have been settled there, so also do the peculiar form and workmanship of the antiquities that have been discovered in them afford a fresh evidence of the superior civilization which the Norwegians in and near Dublin must, for a good while at least, have possessed in comparison with the Irish. The antiquities hitherto spoken of only prove, indeed, that the Norwegians and other Northmen were superior to the Irish with regard to arms and martial prowess. But there are other Norwegian antiquities, originating in Ireland, and found both in and out of that country, which also prove that the Danes and Norwegians formerly settled there contributed, like their kinsmen in England, by peaceful pursuits, to influence very considerably the progress of civilization in Ireland.

SECTION V

Ancient Irish Christianity and Civilization.—Trade.—No Irish, but Norwegian Coins.—Sigtryg Silkeskjæg.—Norwegian Coiners.

Centuries before the introduction of Christianity into the Scandinavian North (in the tenth and eleventh centuries)—nay, centuries before the actual commencement of the Viking expeditions—the Irish people had been Christianized. At a very early period numbers of churches and convents were erected in Ireland, which was also celebrated for its many holy men. It was a common saying that the Irish soil was so holy that neither vipers, nor any other poisonous reptiles, could exist upon it. Numerous priests set out from Ireland as missionaries to the islands lying to the west of Scotland; nay, they even went as far as the Faroe Islands and Iceland, long before those islands had been colonized. Thus, when the Northmen first discovered Iceland (about the year 860), they found no population there; but on "Papey," in "Papyli," and several places in the east and south of the country, they found traces of "Papar," or Christian priests, who had left behind them croziers, bells, and Irish books; whence they perceived that these priests were "Westmen," or Irishmen; for just as the Irish called the Scandinavians "Ostmen," because their home lay to the east of Ireland, so also did the Scandinavians call the Irish "Westmen." The most southern group of islands near Iceland is called to the present day "Vestmannaeyjar" (the Westman Isles), because, at the time of their colonization, a number of Irish serfs, or Westmen, were put to death there for deceiving their masters.

Not even the Norwegian expeditions into Ireland, and the destruction of churches and convents by which they were accompanied, were able to annihilate the influence of the Irish clergy on the diffusion of Christianity in the north-western part of Europe. Not only were the Norwegians and Danes settled in Ireland and the rest of the Western Isles soon converted from heathenism by Irish monks and priests, but Christianity was communicated through these converts to many of their Scandinavian countrymen, who visited

Ireland partly as Vikings and partly as merchants. Thus the Norwegian king Olaf Tryggvesön was baptized by an abbot on the Sylling Isles near Ireland, or, as other Sagas state, "to the west over in Ireland;" whence we may probably conclude that the Sylling Isles are not, as was before supposed, the Scilly Isles near England, but the Skellig Isles on the south-west coast of Ireland, on one of which there was at that time a celebrated abbey. At all events, it is certain that Olaf Tryggvesön, during his long abode with his brother-in-law, King Olaf Kvaran, in Dublin, must, by his constant intercourse with the Irish Christians, have been strengthened in his determination to christianize Norway. Another proof of the influence of Christianity in Ireland on the North is, that an Irish princess, Sunneva, was at a later period worshipped as a saint in Norway. Her body is alleged to have been deposited in a large and handsome shrine over the high altar in Christ Church, in Bergen, and on the 8th of July the Norwegians celebrated an annual mass in her honour. Even in Iceland there is a fiord, or firth, on the north-west coast, called "Patreksfjorðr," after St. Patrick, the patron saint of Ireland.

As we have before stated, the commencements of a national Irish literature were also developed among the clergy at a very early period; which, together with the numerous ecclesiastical buildings in Ireland, prove that the Irish clergy of those times must have attained no mean degree of civilization, and that with regard to education they must, in certain respects, have been a great deal in advance of the heathen Scandinavians. But not to speak of the Icelandic literature—which developed itself in the remotest North immediately after the heathen times, and contemporaneously with the Norwegian dominion in Ireland, and which both in form and substance was undoubtedly far superior to the Irish—there is reason enough to doubt whether the Irish people of that time, although christianized, were really more educated or more advanced in true civilization than the certainly too much decried heathen Norwegians and their Scandinavian kinsmen. It is true, indeed, that the Norwegian Vikings made their way with fire and sword, that they destroyed a number of churches and convents in Ireland, and that in this manner they often occasioned the most violent intestine commotions, which for a time, at least, could not but tend to hinder the progressive development of Christian civilization.

But the Irish chronicles themselves teach us that the Christian Irish acted precisely in the same manner at the same period. In their mutual contentions they often burnt ecclesiastical buildings, plundered the shrines of saints, and maltreated the clergy, besides, as is well known, constantly perpetrating amongst themselves the most horrible butchery. Lastly, in Ireland, as in England, we must certainly distinguish between the Vikings, who came to the country for the sake of war and plunder, and the colonists, whose aim it was to obtain a new home in Ireland. The latter brought with them not only great skill in the forging and management of arms, as well as in building and navigating ships for expeditions, both of war and trade, but likewise had their own runic writing; and by the readiness with which they imbibed the newer Christian civilization, soon acquired the ascendancy in the most important Irish cities, so as to become perceptibly enough, not only the equals, but the superiors of the Irish.

What particularly warrants us in doubting the alleged early and extensive civilization of the Irish, is the very striking circumstance that, previously to the arrival of the Norwegians, they do not appear to have carried on any very great trade, or on the whole to have had any very extensive intercourse with the rest of Europe. This appears particularly from the fact that the Irish at that time (about the year 800) had not yet minted any coins of their own; although their Celtic neighbours in Britain and Gaul had for centuries—that is, from about the birth of Christ—minted a great number, mostly in imitation of Greek and Roman coins. And though the Romans, Franks, and Anglo-Saxons, after their conquests of France and England, had made very considerable coinages in those countries, we do not even find in Ireland any trace of the coins of these neighbouring people being brought over the sea in any considerable quantity before the period mentioned. Yet in other countries, where the minting of coins also came late into use—as, for instance, in the Scandinavian North—so great a quantity of older foreign coins, together with all sorts of foreign valuables, is continually dug up as to show that even at a very early period active connections of trade must have existed between the Northmen and more southern nations. Neither Phenician nor Celtic coins are known to have been found in Ireland, and discoveries even of Roman and the more ancient Anglo-Saxon coins are very rare.

That Ireland should have remained for so long a period and to so great an extent unconnected with the neighbouring nations, was undoubtedly caused partly by its remote situation, partly by the indolence of the Irish and the disinclination so general among the Celts to traverse the sea, to which an old author (Giraldus Cambrensis) expressly alludes. It must partly also be ascribed to the peculiar hostile position which the Irish were obliged to assume towards the Romans, Anglo-Saxons, and Franks; since these people having gradually conquered the Celtic countries, France and England, naturally only awaited a favourable opportunity to make themselves masters of Celtic Ireland also. According to this we might even, perhaps, regard the isolation of Ireland as a necessary system of self-defence adopted by the Irish.

But no sooner were the Norwegians and Danes settled in the chief cities of Ireland, than Irish trade and navigation obtained an extent and importance before unknown. An active commerce was opened with England and Normandy through the numerous and influential Scandinavian merchants settled in those countries, as well as, of course, with the mother countries of Scandinavia. In Ireland, therefore, as well as in England, Arabian coins, minted in countries near the Caspian Sea, are here and there found buried, which have evidently been imported by Scandinavian merchants. The Sagas mention regular trading voyages to Ireland from Norway, and even from Iceland; where there was, for instance, a man named Rafn, who was commonly called Rafn Hlimreksfarer (Eng., *Limerick trader*), on account of his regular voyages to Limerick (Limerick being called by the old Northmen, Hlimrek). The Sagas further mention, under the head of Ireland, "Kaupmannaeyjar" (Eng., *the merchant islands*), probably what are now called "Copeland Islands," on the north-eastern coast, where there may have been a sort of rendezvous for the ships of Scandinavian merchants. The Icelandic and Norwegian ships brought fish, hides, and valuable furs to the English and Irish coasts; whence, again, they carried home costly stuffs and clothes, corn, honey, wine, and other products of the south.

These accounts of the old Northmen, respecting their commerce in Ireland, are far from being unsupported. The Welsh author, Giraldus

Cambrensis, who visited Ireland during the English conquest, whilst the Ostmen were still living there in considerable numbers, says in plain words that they had settled near the best harbours in Ireland, where they built themselves towns, and that they had by no means come to the country as enemies, but with the design of carrying on a peaceful trade. He adds that for this reason the Irish chiefs, who clearly saw the importance and advantages of commercial connections with other countries, had not at first in any way opposed the establishment of these foreign towns in their country; but that, after the Ostmen had very much increased, and after their towns had become well fortified, the old dissensions between them and the Irish revived.

In perfect accordance with this are the statements of the Irish themselves respecting the many Scandinavian merchants in the towns of the Ostmen. An old Irish manuscript relating to the battle of Clontarf ("Cath Chluana Tarbh") states that, after the battle, "no Danes were left in the kingdom, except such a number of artisans and merchants in Dublin, Waterford, Wexford, Cork, and Limerick, as could be easily mastered at any time, should they dare to rebel; and these King Brian very wisely permitted to remain in these seaport towns, for the purpose of encouraging trade and traffic, as they possessed many ships, and were experienced sailors." Duald Mac Firbis also says in his chronicles that in his time (1650) "most of the merchants in Dublin were the descendants of the Norwegian-Irish king, Olaf Kvaran."

That the Norwegians and Danes must really have possessed themselves of the Irish trade, and given it a new impulse, clearly appears from the circumstance that the Norwegian kings in Ireland were the first who caused coins to be minted there. One of these coins, which formerly belonged to the Timm's collection in Copenhagen, but which is now in the collection of M. von Römer, in Dresden, seems (according to the opinion of that distinguished numismatologist C. J. Thomsen, of Copenhagen) to have been minted by a Scandinavian king of Dublin, as early as the eighth or ninth century. It is an imitation of the ancient Merovingian coins, and has a remarkable inscription on the obverse, half in runes and half in Latin letters, but which can scarcely be read otherwise than "Cunut u Dieflio," or, Canute in Dublin.

[+ +] Coin: Canute in Dublin

The Old Northmen call Dublin "Dýflin," whence the surrounding district also obtained the name of "Dýflinarskiri," as appears in the Sagas. This legible inscription encircles the bust of a royal warrior, clad in scale armour. On the reverse are seen the letters ENAE, and under them two figures, both turning their faces upwards in the same direction, and each extending a very large hand, whilst in their other hands, joined together, they hold a ring, as if they were taking an oath on the holy ring. They are, besides, represented as standing before, or sitting on, an elevated platform (perhaps an altar?), under which is a mark (∾) like the letter S placed on its side. These figures probably contain an allusion to some treaty concluded between an Irish king and the Scandinavian king Canute.

By the kindness of Mr. C. F. Herbst, of Copenhagen, I have been enabled to give a woodcut of this silver coin, the only one of its kind, and never before copied. The drawing was made from a cast taken in Dresden. If the preceding explanation, which is certainly by no means far-fetched, be the right one, we shall consequently have a proof that other Scandinavian kings, besides Olaf the White, the first-mentioned in the Sagas, reigned at a very early period in Dublin, if only for a short time. But all the rest of the Norwegian coins minted in Ireland are of the tenth, eleventh, and twelfth centuries. They are of silver, and undoubtedly coined in various towns of Ireland besides Dublin, as in Limerick, Cork, Waterford, and several other towns where the Ostmen had settled.

The most remarkable of all are the Dublin coins, especially those with the legend "Sihtric rex Dyfl," or, Sigtryg king of Dublin. It is true that there were several kings of Dublin of this name in the ninth, tenth, and eleventh centuries; but the coins alluded to, to judge from the impressions, all of which are imitations of contemporary Anglo-Saxon dies, and especially of those of King Ethelred the Second, must for the most part have belonged to Sigtryg, surnamed "Silkbeard," who reigned in Dublin at the close of the tenth and beginning of the eleventh century, and who was one of those who fought the battle of Clontarf against Brian Boru. It is very remarkable that on Sigtryg's coins, as well as on several of the Danish coins minted in the north of England, we find not only the Latin title "Rex," but also the Scandinavian "Cununc" (king), as, for instance, on the annexed coin (in Mr. C. F. Herbst's collection), which has never before been copied:—

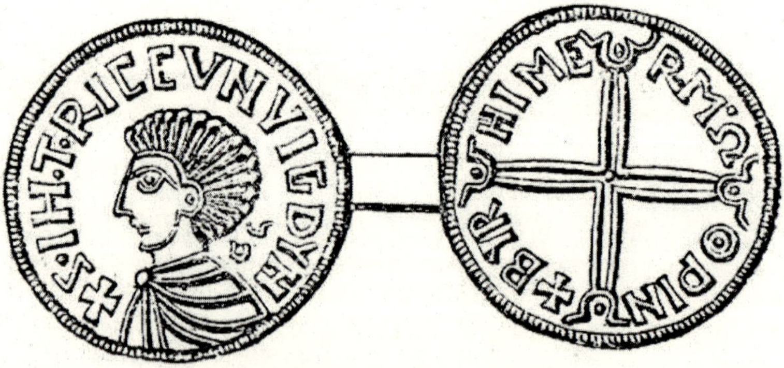

[+ +] Coin: Sigtryg King of Dublin

On the obverse is the legend "Sihtric cunuic dyn," or Sigtryg king of Dublin; and on the reverse, "Byrhtmer mo on Vin;" whence we see that the coiner had an Anglo-Saxon name, and was certainly an Anglo-Saxon, particularly since he is said to have been "on Vin," that is, of Winchester. Among the coiners' names on the Norwegian-Irish coins, we meet, indeed, with several Scandinavian names, such as Stirbirn (Styrbjörn), Azcetel (Asketil), Ivore (Ivar), Colbrand, Tole (Tule), and Oadin (Odin?); whence we may reasonably conclude that the Norwegians in Ireland soon learned to coin, and were not, therefore, always compelled to avail themselves of foreign coiners. But most of Sigtryg's coiners were Anglo-Saxons; and not a few of his coins are, like that above delineated, even struck by coiners in England; as, for instance, in "Efrweec," or "Eofer (wick)" (York), "Veced" (Watchet, in Somersetshire), "Vilt" (Wilton), "Vint" (Winchester), and "Luni" (London). This admits of two explanations; either that these comers at Sigtryg's request minted coins for him, or that Sigtryg, who at one time was driven from his kingdom, resided in some at least of the above-named places, and caused coins to be minted there(?). The origin of several coins minted in Dublin about Sigtryg's time by the Anglo-Saxon king Ethelred the Second—as well as by the Danish-English king Canute the Great, and which for the most part are struck by the same Dublin coiner, Færemin, who minted most of Sigtryg's own coins—is involved in no less obscurity. Although history is silent, we might be almost tempted to believe that Ethelred and Canute were acknowledged by Sigtryg as his liege lords, or that possibly they ruled in Dublin for a short time; but in weighing these probabilities it must be remembered that neither Ethelred nor Canute calls himself on these coins king of Dublin, but simply "Rex Anglorum," or king of the English.

The great number and variety in which Sigtryg's coins appear, and the comparatively good stamp that distinguishes them from the rest of the Norwegian-Irish coins, seem to show that the years of Sigtryg's reign must have been a period very favourable to Scandinavian trade and power in Ireland. In later times the Norwegian-Irish coins became worse, as the coiners did not confine themselves to imitating coins of the older Norwegian-Irish kings, and of the later English kings, Canute the Great, Hardicanute, Edward the Confessor, William the Conqueror, and others, but even copied copies to such a degree that the stamp and inscriptions of the original coins were very frequently not to be recognised. Of the coins current in Ireland in the last half of the eleventh, and in the whole of the twelfth, century, pretty large quantities have been dug up, both in and out of Ireland, and particularly in the neighbouring Isle of Man.

It must, however, be regarded as very doubtful how far this deterioration of the coins affords any reasonable confirmation of the justness of the usual conviction among the Irish, that after Sigtryg's time, or rather after his defeat in the battle of Clontarf, the power of the Norwegians in Ireland was completely broken. For, in that case, we might expect, among other things, that the victorious Irish kings, during the long period of more than a hundred and fifty years, which elapsed from the time of the battle of Clontarf until the English conquest of Ireland, would have minted their own coins. But during the whole of this period there are very few coins that can possibly be regarded as having been minted for native Irish kings. For the rest, the whole of the coins minted in Ireland, from the commencement of minting there (at latest in 950) till the English conquest (1171), seem to owe their existence exclusively to the kings and bishops of the Ostmen, who ruled in the most important trading towns of Ireland[10].

10. See Appendix, No. II.

SECTION VI

The Battle of Clontarf.—Power of the Ostmen after the Battle.—Their

Churches and Bishops.—Their Land and Sea Forces.—The English

Conquest.—Remains of the Ostmen.—Their Importance for Ireland.

The cause of the battle of Clontarf, so celebrated in song and legend, or, as it is called in the Sagas, "Briánsbardagi" (Brian's battle, after King Brian, who fell in it in 1014), is not precisely known. All that we are acquainted with is, that Brian, who was connected by very close ties of relationship with the Norwegian royal family in Dublin, had long availed himself of the assistance of the Norwegians to subdue other Irish princes, until, at length, after gaining victories in that manner, he came to a rupture with King Sigtryg of Dublin. The prospects of Sigtryg, and of the Norwegian power in Ireland, seem really to have been threatening enough; at least it is said that Scandinavian warriors hastened in numbers to Sigtryg's assistance from the Scandinavian kingdoms in England, the Isle of Man, the Syder Isles, and Orkneys. From the last, in particular, came Jarl Sigurd the Stout, with a chosen force, in the midst of which waved a flag with the image of Odin's holy raven. Sigurd's own mother had woven this raven, which, with fluttering wings, had often before led the warriors to victory and glory.

This time, however, the raven was checked in its flight. After many of the standard bearers had been killed, Sigurd Jarl himself took the flag from the staff, and wrapt it about his body. He seemed to foresee, what really happened shortly afterwards, that the raven-flag would be his winding-sheet. The Norwegians were at length forced to give way, even if the battle was not so entirely lost as the exaggerated Irish accounts represent. The Scandinavian auxiliaries withdrew to their ships, and King Sigtryg retired with the remnant of his army to Dublin.

But, as the Irish chronicles contain nothing about Sigtryg and his men having been afterwards expelled from Dublin, or about the Norwegian dominion there having been entirely destroyed, we cannot conclude from them that the power of the Ostmen in the rest of the Irish cities was annihilated in consequence of Sigtryg's defeat in the battle of Clontarf. It would, besides, have been singular enough if the power of the Norwegians in Ireland had been perfectly destroyed so early as the year 1014, since it was just after that time that the Northmen in the neighbouring countries acquired their greatest power by means of their victories. Instead of the Norwegian influence in Ireland having ceased, we not only find, long after this battle, King Sigtryg of Dublin fighting bravely with his Ostmen, though at times with varying fortune, against several Irish kings and chiefs, but we further behold the Ostmen displaying a very remarkable degree of strength and independence in various places in Ireland.

About five-and-twenty years after the battle of Clontarf (say the Irish chroniclers themselves), Sigtryg, king of the Ostmen in Dublin, and Donat (Dunan), their bishop, built, in the middle of that city, the church of the Holy Trinity, also called Christ Church. That the Ostmen should then have founded one of the principal churches of Dublin, which even lay without their own town (Ostmantown), in the very heart of ancient Dublin, is highly significant. After the church was built, Bishop Donat presented several relics to it, amongst which are mentioned "pieces of the clothes of King Olaf the Saint." The great respect in which the name of the Norwegian Saint Olaf was held in Dublin is also manifest from the circumstance that a church consecrated to St. Olave, or, as the Irish common people gradually corrupted the name, to "Tulloch" (compare the name of Tooley Street in London, corrupted from St. Olave Street), was to be found there till at least far into the sixteenth century. This church adjoined the northern end of Fishshamble Street, near Wood-Quay; but originally, perhaps, it was just outside the city.

In the same year (1038) that Christ Church was, partly through the exertions of Bishop Donat, erected in Dublin, he likewise built the chapel of St. Michael. Half a century later (1095) another "Ostman" built Saint Michan's Church in the "Ostmen's" town in Dublin; and about the same time the cathedral in Waterford, dedicated to the Holy Trinity, was founded and erected by the Ostmen there.

The "Ostmen" in Ireland thus possessed not only their own churches, but likewise, as the Irish records also mention, their own bishops, who were consecrated in England by the archbishop of Canterbury; whilst the Irish bishops were consecrated in Ireland itself by the Irish archbishop of Armagh. The Dublin "Ostmen's" first bishop Donat, or Dunan, died in the year 1074, and was buried in Christ Church, to the erection of which he had himself so considerably contributed. After him, by desire of the Dublin king Godred, or Godfred, another "Ostman," Patrick, was chosen bishop of the Ostmen in Dublin, but perished by shipwreck on his voyage home from Canterbury (1084). He was succeeded by the "Ostman" Donat O'Haingly (+1095); whose cousin, Samuel O'Haingly, previously a monk in the convent of St. Alban's in England, afterwards filled the see of the "Ostmen" in Dublin until the year 1121. His successor, Gregorius, was the first of these Ostmen's bishops in Dublin who was made archbishop. This probably arose from the circumstance of the "Ostmen" in the other Irish towns having in the meantime obtained bishops, who were now to have a common superior in the Archbishop of Dublin. In the year 1096 the "Ostmen" in Waterford are said to have obtained a bishop, Malchus, who is stated to have been a native of Ireland. In the year 1136 Waterford had an "Ostman" named Toste (Tuistius, or Tostius) for its bishop. A few years later (1140) Gille, or Gilbert, the "Ostmen's" bishop of Limerick, died; after whom the "Ostmen" chose a certain Patrick. In the year 1151 the "Ostman" Harald, bishop of Limerick, died, and was succeeded by his countryman Thorgils ("Thorgesius"). Twenty years previously (1131) the death of the "Ostman" Everard, or Eberhard, abbot of the convent of St. Mary, near Dublin, is mentioned; which confirms, what is indeed almost a matter of course, that the Ostmen, who had their own churches and bishops, must also have had their own convents partly filled with Scandinavian monks and abbots. At length, in the year 1161, Gregorius, archbishop of Dublin, died; and from his time until the present Dublin has constantly been the seat of one of Ireland's principal archbishops. But precisely because this archbishopric was originally founded by Ostmen, or foreigners, the archbishop of Dublin did not afterwards become the primate of all Ireland, as, from the importance of Dublin, we might otherwise have expected. That dignity, on the contrary, has constantly been reserved for the genuine old Irish archbishopric of Armagh, in the north-east of Ireland. Even Gregorius' successor in the archiepiscopal see is said to have been consecrated in Dublin by the archbishop of Armagh. It has lately been discovered (compare P. Chalmers in the Journal of the Brit. Archæol. Assoc., Oct., 1850, p. 323, &c.) that these archbishops of Dublin not only administered their own diocese, but, at least at times, acted as superintendents of the Norwegian bishoprics in the Isle of Man and the Sudreyjar. There is a letter of Pope Honorius of the beginning of the thirteenth century, from which it appears that the archbishop of Dublin at that time consecrated a bishop of Man and the Sudreyjar, a privilege which in more ancient times belonged to the archbishops of York, and afterwards (from 1181 to 1334) to the archbishops of Trondhjem. It is quite certain that this was a result of the lively intercourse which undoubtedly took place between the successors of the Ostmen in Ireland and their near kinsmen in the Norwegian kingdoms in Man and the Sudreyjar.

It was, above all, a very fortunate circumstance for the independence of the Irish Ostmen that such powerful Norwegian kingdoms continued to exist on the west coast of Scotland. From these they could usually obtain assistance in their battles with the Irish; and by means of them they also kept up a constant connection with their Norwegian fatherland. That they were able to maintain their peculiar independent position in Ireland for more than a century after the Danish dominion in England had ceased to exist, was clearly not so much owing to their military skill and compact force, in comparison with the internal dissensions and perfect want of union among the Irish, as to the considerable wealth and power which they constantly derived from their extensive trade and navigation, and the influence which by such means they must necessarily have exercised in Ireland. The Irish chronicles and pedigrees teach us that friendly connections and reciprocal marriages increased more and more between the Irish and the Ostmen, both in Ireland and Norway, so that the Irish aristocracy became mixed in a considerable degree with Norwegian blood. We also learn from the same documents that the Ostmen and their kings constantly continued to ally themselves with Irish princes, whose power they often essentially contributed to support. The Irish king Konofögr gained a naval battle in Ulfreksfjord against Einar, jarl of Orkney, because, as it is stated, the Norwegian Viking, Eyvind Urarhorn, had joined the former with his ships. When King Magnus Barfod of Norway undertook his expedition to Ireland, he concluded an alliance with Myrjartak, King of Connaught (*O. N.*, "Kunnáktir"), whose daughter, Biadmynja, was married to Magnus Barfod's son Sigurd. But when Magnus fell in Ulster (in 1103), Sigurd abandoned Biadmynja. Yet the connections formed in Ireland by Magnus through this expedition produced important results for Norway. An Irishman named Harald Gille came forward and passed himself off for a son of that monarch by an Irishwoman; and after proving his descent by walking over red-hot iron, actually became king of Norway, and left its throne as an inheritance to his family.

The Ostmen settled in Dublin and other places in Ireland were more and more induced to form connections with native Irish princes, nay, even sometimes to submit to them, as the support which they derived from their own country continually decreased during the eleventh and twelfth centuries. Shortly after the battle of Clontarf, Christianity was introduced into the Scandinavian North, and thus an end was put to the Vikings' expeditions, which had hitherto incessantly brought colonists and auxiliary forces into Ireland. Even the reinforcements which the Ostmen were able to obtain from their countrymen in Man, the Sudreyjar, and the Orkneys, were naturally not so important as before; since on these islands also Christianity gradually annihilated the bold Viking spirit of the people.

Under such circumstances it is surprising that Godfred (or Godred) Merenagh, king of the Ostmen in Dublin, had in the year 1095 a naval force of not fewer than ninety ships in the harbour of Dublin; and that the land forces of the Ostmen in that city were proportionately powerful. The Irish chronicles mention many battles in the eleventh and twelfth centuries in which the Dublin Ostmen brought numerous warriors into the field, and in which they often suffered very considerable loss, without, however, being entirely annihilated or driven out of the town. Even in the year 1167, and consequently a hundred and fifty years after the battle of Clontarf, a great meeting of the Irish people was held by Athboy of Tlactga, at which, the Irish themselves say, *thousands* of the first Ostmen in Dublin were present.

That this account is not exaggerated, and that the number of Ostmen in Dublin, as well as in the other Irish cities, was really very considerable at the close of the twelfth century, is clearly shown by the notorious fact, that when the English, under Earl Strongbow and Miles de Cogan, obtained, in the years 1170 and 1171, a firm footing in Ireland, the Ostmen in Dublin, Limerick, and Cork, were able to offer a very powerful resistance. Respecting the conquest of Dublin by the English we find the following statement in the "Dublin Annals" (by O'Donovan):—

"The year 1170. The Danes of Dublin were treacherously slaughtered in their own garrison by Mac Morough and the English; and they carried away their cattle and their riches. Asgal, the son of Reginald, king of the Danes in Dublin, fled from them.

"1171. A battle was fought at Dublin, between Miles de Cogan and Asgal, son of Reginald, king of the Danes of Dublin. Many fell on both sides, both of the English archers and of the Danes; among whom was Asgal himself, and Hoan, a Dane from the Orkney Isles."

On this occasion Asgal, or "Hasculph," is said to have returned to the city with sixty ships. His warriors, say the chronicles, were accoutred, according to the usual custom of the Danes, in armour and coats of mail, and had red circular shields bound with iron. But though these men were "just as steeled in soul as in arms" (homines tam animis ferrei quam armis), and though, as well as their brethern in Limerick and Cork, they fought the fight of desperation in defence of their property and liberties, yet they were not able to withstand the English. Thus these new conquerors succeeded in annihilating the dominion of the Ostmen in Ireland, or rather in the most important cities of that country, after it had lasted above three hundred years.

Nevertheless we must not believe that the Ostmen were even now wholly expelled from Ireland, or that their influence there was entirely at an end. After the taking of Dublin by the English, so many Ostmen still remained in the city that "the Galls of Dublin" continued to have their own separate army, which even seems to have acted pretty independently of the English conquerors. An Irish chronicle (Annals of the Four Masters) states that Mulrony O'Keary, Lord of Carbury, was treacherously slain by the "Dublin Ostmen" in the year 1174, and consequently some years after the taking of Dublin. In the same year the English themselves were forced to obtain the assistance of the "Dublin Ostmen" against the Irish; and it is expressly stated that in a subsequent attack of the Irish on this united Anglo-Norwegian army not far from Dublin, there fell no fewer than "*four hundred* Ostmen." The contemporary author, Giraldus Cambrensis, to whom we owe this account, also speaks of the Ostmen, after the conquest of Ireland, as a peculiar and decidedly separate people, who carried on trade and navigation ("gens igitur hæc, quæ nunc Ostmannica gens vocatur," &c).

Even more than a century afterwards we can still trace many Ostmen in the chief cities of Ireland, where, it seems, they continued to preserve those Scandinavian characteristics which distinguished them from the Irish and English. In the year 1201 a verdict was pronounced by twelve Irishmen, twelve Englishmen, and twelve Ostmen in Limerick, concerning the lands, churches, and other property belonging to the church of Limerick; which shows that the Ostmen were sufficiently numerous there to be placed on an equal footing with the English and Irish. There is in the Tower of London a document of the year 1283, issued by the English king Edward I., ordering that the Ostmen in Waterford ("Custumanni," Oustumanni, Austumanni?) should, pursuant to King Henry the Second's ordinance, have, and be judged by, the same laws as the English settled in Ireland, which clearly indicates that the Ostmen at that time still formed a distinct and separate people. We might almost believe that the Ostmen in Waterford had even refused to observe the English laws, or that at least there was a doubt how far these laws could be applied to them; since King Edward found it necessary to enforce Henry the Second's ordinance, and to enjoin his chief justice and magistrates in Ireland that the three men named in the document should, "like other Ostmen in Waterford," be judged, and as far as possible ("quantum in vobis est"), punished, according to the laws in force for Englishmen in Ireland. (See the Latin document in the

Appendix.) The striking historical account that in the year 1263 the Irish applied to the Norwegian king Hakon Hakonsön, then lying with his fleet on the south-west coast of Scotland, for assistance against the English, will now no longer be inexplicable or improbable; for it is placed beyond all doubt that amongst the Irish who thus in vain implored King Hakon for help, there must have been a number of the Ostmen still living in Ireland, who naturally continued to maintain a connection with their countrymen in the Norwegian kingdoms on the south-west coast of Scotland, until these kingdoms also were destroyed in the middle ages.

But from this time forward the "Ostmen" do not play any prominent part in the history of Ireland. Their political independence was annihilated; and their national characteristics were not sufficiently supported by fresh arrivals from the mother-country, to enable them in the long run to maintain a distinct position in face of the rapidly advancing English nationality. Their descendants continued, nevertheless, to dwell in Ireland; where they gradually became amalgamated partly with the English conquerors and partly with the native Irish. The Irish chronicles point out various clans in Ireland which were either of Norwegian descent, or at all events had been much mixed with Norwegian blood. In the annals and pedigrees of the middle ages we also meet with both laymen and clergy in Ireland bearing Scandinavian names. For instance, in Christ Church in Dublin, built by the Norwegians, canons and monks are spoken of in the fourteenth and fifteenth centuries called "Harrold," "Olof," "Siwird" (Sivard), "Regenald," (Ragnvald) "Iwyr," &c., names entirely unknown in Ireland previously to the arrival of the Norwegians. The often-mentioned Irish chronicler makes use of a highly remarkable expression. In stating that most of the merchants' families in Dublin in his time (about the year 1650) were descendants of the Norwegian-Irish king Olaf Kvaran, by Brian Boroihma's (Boru's) daughter Save, he adds: "and the descendants of that Amlave Cuaran are still in Dublin opposing the Gadelians of Erinn:" whence we clearly see that national distinctions and national disputes between the descendants of the Irish and of the Norwegians, were still very prominent only two hundred years ago, or full six hundred years after the battle of Clontarf (1014).

Even to the present day we can follow, particularly in Leinster, the last traces of the Ostmen through a similar series of peculiar family names, which are by no means Irish, but clearly original Norwegian names; for instance, Mac Hitteric or Shiteric (son of Sigtryg), O'Bruadair (son of Broder), Mac Ragnall (son of Ragnvald), Roailb (Rolf), Auleev (Olaf), Mánus (Magnus), and others. It is even asserted that among the families of the Dublin merchants are still to be found descendants of the old Norwegian merchants formerly so numerous in that city. The names of families adduced in confirmation of this, as Harrold (Harald), Iver (Ivar), Cotter or Mac Otter (Ottar), and others, which are genuine Norwegian names, corroborate the assertion that Norwegian families appear to have propagated themselves uninterruptedly in Dublin down to our times, as living evidences of the dominion which their forefathers once exercised there.

It is thus satisfactorily proved, by notorious facts of the most various kinds, that for more than three hundred years the Norwegians lived according to their own manners and customs, and under their own bishops and kings, in the most important towns of Ireland, which they in part ruled, down to the time of the English conquest (1170); that they were the first who minted coins, and carried on any considerable trade and navigation in Ireland; and lastly, that great numbers of their descendants continued to reside in that country even after it had long been conquered by the English. No impartial person, therefore, will be able any longer to deny that the settlements of the Ostmen, although commenced by the frequent demolition of churches and convents, were ultimately in the most essential matters particularly fortunate for Ireland; since, by introducing trade and navigation to an extent before unknown, they opened for that sequestered country channels of animated communication and intercourse with the rest of Europe and its continually advancing civilization. The Irish towns occupied by the Ostmen, which have continued to be the principal depots for foreign merchandise, and consequently also the central points of intercourse with foreign countries, may with justice be said to be indebted chiefly to that people for their present greatness, wealth, and power.

Nor, on a larger historical survey, will it appear less evident that, as the Norwegians first opened the way for peaceful connections between Ireland and the rest of Europe, so they also facilitated the English conquest. In consequence both of their frequent wars, and of their frequent alliances with Irish kings, party feeling had rather increased than diminished among the Irish chiefs; whilst numerous Irish families, even the greatest in the land, had by degrees become so much mixed with Norwegian blood, that the strength of the Irish as a nation was not a little weakened and divided. This was particularly the case in those districts of the east coast of Ireland where the English or Norman power afterwards obtained its chief seat. Add to this that the Irish, through the long dominion of the Norwegians in their chief towns, and the advantages which they reaped from it, had become more and more accustomed to behold with indifference the sway of strangers in their country; a circumstance which contributed to the powerful support given to the English on their first invasion of Ireland by several of the native chiefs.

It may possibly be said that the Norwegians in Ireland, by thus preparing the way for the Norman or English conquest, rendered a far greater service to England than to subjugated Ireland. But all the chronicles, it must be recollected, bear witness that the Irish were neither strong enough to govern their own country independently, nor capable of keeping pace with the advance of European civilization by means of an active commerce. We have seen that even in later times the same baleful and sanguinary spirit of dissension which weakened Ireland in ancient days is yet scarcely extinct among the original Irish race. It is manifest, therefore, that Ireland, which would otherwise have been divided from the rest of Europe, and devastated by terrible intestine contentions, has been much benefited by being united to so great and powerful a country as England, which has both the ability and the will to promote the true welfare of the Irish people. England will, by degrees, employ the great advantages afforded by the excellent soil and situation of Ireland, and thus conduct that country, torn as it is by all possible distresses and misfortunes, to a happier existence.

SECTION VII

Conclusion.—Warlike and Peaceful Colonizations.—Resemblances and Differences.—Before and Now.

Denmark and Norway, as is known, are not distinguished by any remarkable extent of fertile and densely-populated country. The whole population in both those kingdoms does not at present amount to three millions; and in ancient times it scarcely seems to have been greater, even when the southern portion of the present kingdom of Sweden still belonged to Denmark.

Nevertheless, Denmark and Norway were able, in ancient times, to send forth great multitudes of people to other countries. Not only were Greenland, Iceland, the Shetland Isles, the Orkneys, and Faroe Isles, colonized from Norway, but also considerable districts in Scotland and Ireland. Many Norwegians, moreover, settled in England and Normandy. At the same time Danes emigrated in great numbers to Normandy, North Holland, and especially England, where they colonized, we may say, the whole of the extensive district to the north of Watlinga-Stræt, or almost half England.

We are not informed that Denmark and Norway were emptied of their population in consequence of these great emigrations, or even that there was any sensible want of inhabitants to supply agricultural labourers and soldiers. In the immediately following centuries Denmark was powerful enough to make the Baltic a Danish lake. We can hardly, therefore, assume, like the monkish chronicles of antiquity, which naturally breathe both fear and hatred of the Scandinavian heathens, that the Norwegians and Danes were merely barbarous Vikings, who procured themselves a footing in the western countries only through brute force. On such grounds we should be perfectly unable to explain satisfactorily how Denmark and Norway, with a proportionately small population, should have been able (without becoming too depopulated) to send out at once such a host of people as were evidently required to take possession, by force of arms, of those rich western lands, and also, it must be observed, to maintain their conquests for centuries. If, instead of blindly following these partial and prejudiced chroniclers, we adhere to what the traces of the nature and importance of the Scandinavian emigrations clearly prove, namely, that from the eighth to the twelfth century, and contemporary with the destructive Viking expeditions, peaceful emigrations from the North constantly took place—which, in reality, were just as effective, perhaps even more so, than the purely warlike expeditions of conquest—this matter will be placed in a far more probable and intelligible point of view. As we have seen, sagacity and the arts of peace, together with navigation and trade, in no slight degree assisted the Danes and Norwegians to procure a footing in the British Islands, and especially in England and Ireland. By perseverance and ability in the occupations of peace as well as war, they were soon enabled to gain the ascendancy in the most important seaport towns; whence, by means of various connections of trade, friendship, and family alliances, they extended their influence and dominion over the adjacent towns and districts. They gradually multiplied themselves, and were joined by fresh immigrants; and thus the foundations were almost imperceptibly laid of Scandinavian colonies, which awaited only the coming of some bold military adventurer to appear as independent, nay, even as dominant states. The great warriors to whom history assigns the honour of the conquests in England and Ireland—and, we may also add, Normandy—would scarcely have been able to obtain them with the generally inferior numbers under their command, had not the Scandinavian merchants, and other peaceful colonists, both opened the way for them, and afterwards supported the conquests they had achieved. It is, on the whole, obvious that the ancient Northmen possessed a very great talent for colonization, which their kinsmen, the English of modern times, seem to have inherited from them.

But as the Scandinavian colonies in the British Islands varied greatly in importance, so also must the effects which they produced have been somewhat different. In Ireland, as well as in Scotland, where the Norwegians met with tribes who, in spite of their apparent Christianity, stood rather below them in civilization, they kept themselves more apart from the natives. In Ireland, especially, they dwelt in their own strongly-fortified towns; where, until late in the middle ages, they maintained their own characteristic language, manners, customs, and laws. But in consequence of this, their Norwegian institutions had no real influence on the development of the national life or institutions of Ireland. At most they merely contributed to facilitate the introduction and establishment of the analogous Anglo-Norman institutions into the Irish cities. In England, on the contrary, where the Scandinavian colonies were far more numerous and powerful than in Scotland and Ireland, the Danish colonists certainly sought, after the Scandinavian fashion, to maintain in the midst of a foreign country their pure Danish laws, manners, and customs. Yet here the Danes, owing to the superior civilization which prevailed among the earlier Anglo-Saxon inhabitants of England, were soon influenced by their language and culture, and became more and more amalgamated with them. Nevertheless the Danes in England were sufficiently numerous and independent to maintain the most important of their free Scandinavian characteristics, which coalesced with, and by degrees visibly impressed themselves upon, the more modern English manners and institutions.

The Danish colonies in England, and the Norwegian colonies in Scotland and Ireland, had so far the same historical importance that they essentially conduced to found a new life, both externally and internally, in the British Islands, partly by extending trade and navigation, partly by subduing, or at least weakening, the power of the Anglo-Saxons, the Scotch, and the Irish, and thus in general preparing for a kindred race (the Normans) the dominion over all these people. It is well known that the Norman sway and the Norman spirit established themselves in Scotland and Ireland far later than in England—a circumstance chiefly owing to the conquests and settlements of the Norwegians in those countries having been far less extensive and important than the Danish conquests in England. Yet that the Danish-Norman spirit predominating in England has been able to maintain to our times its dominion in Scotland and Ireland also, is no slight evidence of the excellent and solid manner in which the Norwegians must originally have prepared the way.

I have shown that the memorials of the great exploits performed by the Danes and Norwegians in the British Islands still appear as fresh and vivid as if they were of modern date. In this respect, the national pride of those nations will find complete gratification. Still, however, it is possible that a general view of the mighty achievements of the ancient Northmen in the western lands may awaken mingled feelings in many a Scandinavian of the present day. The thought may involuntarily arise in him of what the North *was*, when its victorious fleets appeared in the north, south, east, and west, and when Scandinavians exercised dominion far and wide, and what it *is now*—confined within narrow boundaries, menaced from many quarters, and without any preponderating influence on the state of Europe. Beyond the precincts of the North, he will no longer hear his native language, which in former times frequently resounded on foreign shores. The North was forced to shed some of its best and noblest blood; and yet the Northman must now be content, if he can succeed in tracing out, by means of a few words in the popular language, by the names of towns and districts, or by half-erased runic inscriptions on bauta stones, where it was that the "Danish tongue" once prevailed, and where the barrows still rise which cover the race that spoke it.

But such morbid complaints will necessarily vanish when the Scandinavian considers how vividly the ancient power of his race has again displayed itself to the world, and how mighty have been the results of the Norman expeditions; but especially when he ponders on the notorious fact, that the North sent out the flower of its youth and strength, not merely to destroy and plunder, but rather to lay the foundations of a fresher life in the western lands, and thus to impart a new and powerful impulse to human civilization. In our times, besides, it is not chiefly in conquests and the lustre of external greatness that the true happiness and glory of a nation should be sought.

A people, which, like the Scandinavian, have preserved—together with the memorials of former great achievements, and of conquests bringing blessings in their train—enough of the character and courage of their forefathers, not only to maintain the freedom and independence of their country, but also, in comparison with other nations, an honourable place in science and art, cannot justly be said to want either glory or happiness.

APPENDIX I

DOCUMENT OF EDWARD I., OF THE YEAR 1283,

CONCERNING THE OSTMEN IN WATERFORD AND IRELAND.

(*From a Register in the Tower of London; Patent Roll II. Edward I.*

Memb. 9. Communicated by Mr. Duffus Hardy.)

"Pro Custumannis[11] Waterfordi in Hibernia. Rex Justiciario suo Hibernie et omnibus aliis Ballivis et fidelibus suis Hibernie ad quos, &c., salutem. Quia per inspeccionem carte Domini Henrici Regis, filii Imperatricis, quondam Domini Hibernie preavi nostri, nobis constat quod Custumanni nostri Waterford legem Anglicorum in Hibernia habere et secundum ipsam legem judicari et deduci debent. Vobis mandamus quod Gillecrist Makgillemory, William Makgillemory, et Johannem Makgillemory, et alios Custumannos de Civitate et Communitate Waterford, qui de predictis Custumannis predicti domini regis preavi nostri originem duxerunt legem Anglicorum in partibus illis juxta tenorem carte predicte habere et eos secundum ipsam legem quantum in vobis est deduci faciatis, donec aluid de consilio nostro inde duximus ordinandum. In cujus, &c. ... v. die Octobr."

11. This is undoubtedly an old fault in the way of writing or reading for "Oustumannis," "Austumannis." That the word is at all events meant to signify the Ostmen is also assumed in Sir John Davies' "Reports" (fol. 236).

APPENDIX II

COINAGE OF THE NORWEGIANS IN DUBLIN.

(*See page 338.*)

While this work was going through the press, a silver coin, forming an entirely new and highly remarkable contribution to our knowledge of the early Norwegian coinage in the capital of Ireland, was discovered among the collection bequeathed by the late Mr. Devegge to the Royal Cabinet of Coins in Copenhagen. It is represented in the annexed woodcut.

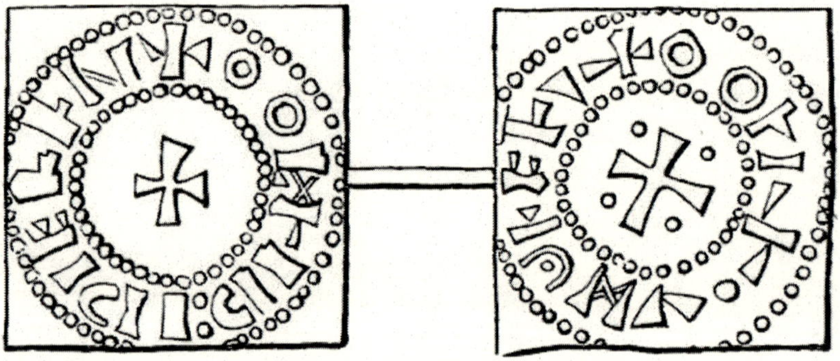

[++] Coin: Olaf in Dublin.

The legend on the obverse is "*Oolaf i divielin*," or "Olaf in Dublin." That on the reverse almost seems to be "*Oolafn me feci(t)*," or "Olaf made me;" in which case the coiner must have had the same Scandinavian name as the king. However this may be, it is clear enough that the coin owes its origin to a Norwegian or Scandinavian king Olaf in Dublin; and, as the stamp shows, it must have been struck in the tenth century. It thus forms a link between the runic coin of Canute in Dublin, and the somewhat later coins of Sigtryg, before described. (See p. 338, *et seq.*)

A great number of coins have been mentioned as minted in Ireland by Scandinavian kings named Olaf; but that above delineated is in reality the first, and, as far as is known, the only one on which we can with certainty read "Olaf in Dublin."

Kings of that name are mentioned in the Irish chronicles in the years 853, 934, 954, 962, &c. (See the list of Norwegian Kings in Ireland, p. 317.)

G. Woodfall & Son, Printers, Angel Court, Skinner Street.

50, ALBEMARLE STREET, LONDON.

April, 1852.

MR. MURRAY'S GENERAL LIST OF WORKS

ABBOTT'S (REV. J.) Philip Musgrave; or Memoirs of a Church of England Missionary in the North American Colonies. Post 8vo. 2*s.* 6*d.*

ABELL'S (MRS.) Recollections of the Emperor Napoleon during the First Three Years of his Captivity on the Island of St. Helena. *Second Edition.* Woodcuts. Post 8vo. 10*s.* 6*d.*

ABERCROMBIE'S (JOHN, M.D.) Enquiries concerning the Intellectual Powers and the Investigation of Truth. *Thirteenth Edition.* Fcap. 8vo. 6*s.* 6*d.*

—— Philosophy of the Moral Feelings. *Eighth Edition.* Fcap. 8vo. 4*s.*

—— Pathological and Practical Researches on the Diseases of the Stomach, the Intestinal Canal, the Liver, and other Viscera of the Abdomen. *Third Edition.* Fcap. 8vo. 6*s.*

ACLAND'S (REV. CHARLES) Popular Account of the Manners and Customs of India, Illustrated with Numerous Anecdotes. Post 8vo. 2*s.* 6*d.*

ÆSCHYLUS. (The Agamemnon and Choephoræ). *A New Edition* of the Text, with Notes, Critical, Explanatory, and Philological, for the Use of Students. By Rev. W. Peile, D.D., Head Master of Repton School. *Second Edition.* 2 Vols. 8vo. 9*s.* each.

ÆSOP'S FABLES, for Old and Young. A New Version. By Rev. THOMAS JAMES, M.A. Illustrated with 100 Woodcuts, by JOHN TENNIEL. Post 8vo. 2*s.* 6*d.*
⁂ *A Few Copies on fine Paper.* 8*vo.* 16*s.*

AGRICULTURAL (THE) JOURNAL. Published (half-yearly) by the Royal Agricultural Society of England. 8vo. 10*s.*

AMBER-WITCH (THE). The most interesting Trial for Witchcraft ever known. Edited by DR. MEINHOLD. Translated from the German by LADY DUFF GORDON. Post 8vo. 2*s.* 6*d.*

ARABIAN NIGHTS (THE). A New Translation arranged for Family Reading. With Explanatory Notes. *New Edition.* With 600 Woodcuts, by WILLIAM HARVEY. One Volume. Medium 8vo.

ARAGO'S (M.) Historical Eloge on James Watt. Translated from the French, with Notes by J. P. MUIRHEAD. Portrait. 8vo. 8*s.* 6*d.*, and 4to. 21*s.*

ARTHUR'S (LITTLE) History of England. By LADY CALLCOTT. *Fifteenth Edition.* Woodcuts. 18mo. 2*s.* 6*d.*

AUNT IDA'S Walks and Talks; a Story Book for Children. By a Lady. Woodcuts. 16mo. 5*s.*

ADMIRALTY PUBLICATIONS (THE); Published by direction of the Lords Commissioners of the Admiralty:—

1. A MANUAL OF SCIENTIFIC ENQUIRY, for the Use of Officers in H.M. Navy and Travellers in General. By Various Hands. Edited by SIR J. F. HERSCHEL, Bart. *Second Edition.* Post 8vo. 10*s.* 6*d.*

2. AIRY'S ASTRONOMICAL OBSERVATIONS MADE AT GREENWICH. 1836 to 1847. Royal 4to. 50*s.* each.

3. —— APPENDIX TO THE ASTRONOMICAL OBSERVATIONS. 1836, 1837, 1842, 8*s.* each; and 1847, 14*s.* Royal 4to.

CONTENTS.

1836.—Bessel's Refraction Tables. Table for converting Errors of R. A. and N.P.D. into Errors of Longitude and Ecliptic P. D.

1837.—Logarithms of Sines and Cosines to every Ten Seconds of Time. Table for converting Sidereal into Mean Solar Time.

1842.—Catalogue of 1439 Stars.

1847.—Twelve Years' Catalogue of Stars.

4. —— MAGNETICAL AND METEOROLOGICAL OBSERVATIONS. 1840 to 1847. Royal 4to. 50*s.* each.

5. —— ASTRONOMICAL, MAGNETICAL AND METEOROLOGICAL OBSERVATIONS, 1848, 49, & 50. Royal 4to. 50*s.* each.

6. —— REDUCTION OF THE OBSERVATIONS OF PLANETS. 1750 to 1830. Royal 4to. 50*s.*

7. —— —— LUNAR OBSERVATIONS. 1750 to 1830. 2 Vols. Royal 4to. 50*s.* each.

8. BERNOULLI'S SEXCENTENARY TABLE. *London*, 1779. 4to. 5*s.*

9. BESSEL'S AUXILIARY TABLES FOR HIS Method OF CLEARING LUNAR DISTANCES. 8vo.

10. —— FUNDAMENTA ASTRONOMIÆ: *Regiomonti*. 1818. Folio. 60*s.*

11. BIRD'S Method OF CONSTRUCTING MURAL QUADRANTS. *London*, 1768. 4to. 2*s.* 6*d.*

12. —— Method OF DIVIDING ASTRONOMICAL INSTRUMENTS. *London*, 1767. 4to. 2*s.* 6*d.*

13. COOK, KING, AND BAYLY'S ASTRONOMICAL OBSERVATIONS. *London*, 1782. 4to. 21*s.*

14. EDWARDS ON THE COMPOSITION OF METALS FOR REFLECTING TELESCOPES. 8vo. 2*s.* 6*d.*

15. EIFFE'S ACCOUNT OF IMPROVEMENTS IN CHRONOMETERS. 4to. 2*s.*

16. ENCKE'S BERLINER JAHRBUCH, FOR 1830. *Berlin*, 1828. 8vo. 9*s.*

17. GROOMBRIDGE'S CATALOGUE OF CIRCUMPOLAR STARS. 4to. 10*s.*

18. HARRISON'S PRINCIPLES OF HIS TIME-KEEPER. PLATES. 1767. 4to. 5*s.*

19. HUTTON'S TABLES OF THE PRODUCTS AND POWERS OF NUMBERS. 1781. Folio. 7*s.* 6*d.*

20. LAX'S TABLES FOR FINDING THE LATITUDE AND LONGITUDE. 1821. 8vo. 10*s.*

21. LUNAR OBSERVATIONS at GREENWICH. 1783 to 1819. Compared with the Tables, 1821. 4to. 7*s.* 6*d.*

22. —— DISTANCES of the MOON'S CENTRE from the PLANETS, 1822, 3*s.*; 1823, 4*s.* 6*d.* 1824 to 1835. 8vo. 4*s. each.*

23. MASKELYNE'S ACCOUNT OF THE GOING OF HARRISON'S WATCH. 1767. 4to. 2*s.* 6*d.*

24. MAYER'S THEORIA LUNÆ JUXTA SYSTEMA NEWTONIANUM. 4to. 2*s.* 6*d.*

25. —— TABULÆ MOTUUM SOLIS ET LUNÆ. 1770. 4to. 5*s.*

26. —— ASTRONOMICAL OBSERVATIONS MADE AT GOTTINGEN, from 1756 to 1761. 1826. Folio. 7*s.* 6*d.*

27. NAUTICAL ALMANACS, from 1767 to 1855. 8vo. 2*s.* 6*d.* each.

28. —— SELECTIONS FROM THE ADDITIONS, up to 1812. 8vo. 5*s.*—1834-54. 8vo. 5*s.*

29. —— SUPPLEMENTS, 1828 to 1833, 1837 and 1838. 8vo. 2*s.* each.

30. —— TABLES requisite to be used with the N. A. 1766. 8vo. 2*s.* 6*d.*

31. —— *Second Edition, enlarged.* 1781. 8vo. 5*s.*

32. —— *Third Edition, corrected.* 1802. 8vo. 5*s.*

33. POND'S ASTRONOMICAL OBSERVATIONS. 1811 to 1835. 4to. 21*s.* each.

34. RAMSDEN'S ENGINE for DIVIDING MATHEMATICAL INSTRUMENTS. 4to. 5*s.*

35. —— ENGINE for DIVIDING STRAIGHT LINES. 4to. 5*s.*

36. SABINE'S PENDULUM EXPERIMENTS to DETERMINE THE FIGURE of the EARTH. 1825. 4to. 40*s.*

37. SHEPHERD'S TABLES for CORRECTING LUNAR DISTANCES. 1772. Royal 4to. 21*s.*

38. —— TABLES, GENERAL, of the MOON'S DISTANCE from the SUN, and 10 STARS. 1787. Folio. 5*s.* 6*d.*

39. TAYLOR'S SEXAGESIMAL TABLE. 1780. 4to. 15*s.*

40. —— TABLES of LOGARITHMS. 4to. 3*l.*

41. TIARK'S ASTRONOMICAL OBSERVATIONS for the LONGITUDE of MADEIRA. 1822. 4to. 5*s.*

42. —— CHRONOMETRICAL OBSERVATIONS for DIFFERENCES of LONGITUDE between DOVER, PORTSMOUTH, and FALMOUTH. 1823. 4to. 5*s.*

43. VENUS and JUPITER: OBSERVATIONS of, compared with the TABLES. *London*, 1822. 4to. 2*s.*

44. WALES and BAYLY'S ASTRONOMICAL OBSERVATIONS. 1777. 4to. 21*s.*

45. WALES' REDUCTION of ASTRONOMICAL OBSERVATIONS MADE IN THE SOUTHERN HEMISPHERE. 1764-1771. 1788. 4to. 10*s.* 6*d.*

AUSTIN'S (MRS.) Fragments from German Prose Writers. Translated, with Biographical Notes. Post 8vo. 10*s.*

—— Translation of Ranke's Political and Ecclesiastical History of the Popes of Rome. *Third Edition.* 2 Vols. 8vo. 24*s.*

BABBAGE'S (CHARLES) Economy of Machinery and Manufactures. *Fourth Edition.* Fcap. 8vo. 6*s.*

—— Table of the Logarithms of the Natural Numbers from 1 to 108000. *Fourth Edition.* Royal 8vo. 6*s.*

—— Ninth Bridgewater Treatise. *Second Edition.* 8vo. 9*s.* 6*d.*

—— Reflections on the Decline of Science in England, and on some of its Causes. 4to. 15*s.*

—— Exposition of 1851; or, Views of the Industry, the Science, and the Government of England. *Second Edition.* 8vo. 7*s.* 6*d.*

BARROW'S (SIR JOHN) Autobiographical Memoir, including Reflections, Observations, and Reminiscences at Home and Abroad. From Early Life to Advanced Age. Portrait. 8vo. 16*s.*

—— Voyages of Discovery and Research within the Arctic Regions, from the earliest periods to the present time. Under the Command of the several Naval Officers, employed by Sea and Land, in search of a North-west Passage from the Atlantic to the Pacific: with Two Attempts to reach the North Pole. Abridged and arranged from the Official Narratives, with Occasional Remarks. Maps. 2 Vols. 8vo. Vol. 1, 12*s.* Vol. 2, 15*s.*

—— (JOHN) Memoirs of the Naval Worthies of Queen Elizabeth's Reign, their Gallant Deeds, Daring Adventures and Services in the infant state of the British Navy, with brief Biographical Notices of the respective Commanders. 8vo. 14*s.*

—— Life and Voyages of Sir Francis Drake. With numerous Original Letters from him and the Lord High Admiral to the Queen, and great Officers of State. Post 8vo. 2*s.* 6*d.*

—— Tour in Austrian Lombardy, the Northern Tyrol, and Bavaria. Woodcuts. Post 8vo. 10*s.* 6*d.*

BASSOMPIERRE'S Memoirs of his Embassy to the Court of England in 1626. Translated with Notes. 8vo. 9*s.* 6*d.*

BEES AND FLOWERS. Two Essays reprinted from the "Quarterly Review." Fcap. 8vo. 1*s. each.*

BELISARIUS (LIFE OF). By LORD MAHON. *Second Edition.* Post 8vo. 10*s.* 6*d.*

BELL (SIR CHARLES) on the Anatomy and Philosophy of Expression as connected with the Fine Arts. *Fourth Edition.* Plates. Impl. 8vo. 21*s.*

—— On The Mechanism and Vital Endowments of the Hand as evincing Design. *Fifth Edition.* Portrait and Woodcuts. Post 8vo. 7*s.* 6*d.*

BENEDICT'S (JULES) Sketch of the Life and Works of Felix Mendelssohn Bartholdy. 8vo. 2*s.* 6*d.*

BERTHA'S Journal during a Visit to her Uncle in England. Containing a Variety of Interesting and Instructive Information, adapted for Young Persons. *Seventh Edition.* Woodcuts. 12mo. 7*s.* 6*d.*

—— Heiress in her Minority; or, the Progress of Character. By Author of "BERTHA'S JOURNAL." 2 Vols. 12mo. 18*s.*

BETHUNE'S (J. E. D.) Specimens of Swedish and German Poetry. Part I. Poems of Esaias Tegner. Part II. Schiller's Maid of Orleans. Crown 8vo. 12*s.*

BIRCH'S (SAMUEL) History of Ancient Pottery: Egyptian, Asiatic, Greek, Roman, Etruscan, and Celtic. With Illustrations. 8vo. (*Nearly Ready.*)

BIRT'S (W. R.) Hurricane Guide. Being an Attempt to connect the Rotatory Gale, or Revolving Storm, with Atmospheric Waves. Including Instructions for observing the Phenomena of the Waves and Storms; with Practical Directions for avoiding the Centres of the latter. With Circles on Cards. Post 8vo. 3*s.*

BIOSCOPE (THE); or, the Dial of Life explained. To which is added a Translation of St. Paulinus' Epistle to Celantia on the Rule of Christian Life and an Elementary View of general Chronology. By GRANVILLE PENN. *Second Edition.* With Plate. 12mo. 12*s.*

BLUNT'S (REV. J. J.) Undesigned Coincidences in the Writings of the Old and New Testament, an Argument of their Veracity: with an Appendix containing Undesigned Coincidences between the Gospels, Acts, and Josephus. *Third Edition.* 8vo. 9*s.*

—— Principles for the proper understanding of the Mosaic Writings, stated and applied, together with an incidental argument for the truth of the Resurrection of our Lord. Being the HULSEAN LECTURES for 1832. Post 8vo. 6*s.* 6*d.*

BOOK OF COMMON PRAYER. Illuminated with 1000 Illustrations of Borders, Initials, and Vignettes. Medium 8vo. 21*s.* cloth, 31*s.* 6*d.* calf, or 42*s.* morocco.

BORROW'S (GEORGE) Lavengro; The Scholar—Gipsy—and Priest. With Portrait. 3 Vols. Post 8vo. 30*s.*

—— Bible in Spain; Or the Journeys, Adventures, and Imprisonments of an Englishman in an Attempt to circulate the Scriptures in the Peninsula. 3 Vols. Post 8vo. 27*s.*, or *Cheap Edition*, 16mo. 5*s.*

—— Zincali, or the Gipsies of Spain; their Manners, Customs, Religion, and Language. 2 Vols. Post 8vo. 18*s.*, or *Cheap Edition*, 16mo. 5*s.*

BOSWELL'S (JAMES) Life of Dr. Samuel Johnson. Including the Tour to the Hebrides, with Notes by Sir W. SCOTT. Edited by the Right Hon. JOHN WILSON CROKER. *A New and Cheaper Edition, thoroughly revised (1848), with much additional Matter.* With Portraits. Complete in one Volume Royal 8vo. 15*s.*
⁂ *Be particular in ordering* CROKER'S BOSWELL, *in* ONE VOLUME.

BRAY'S (MRS.) Life of Thomas Stothard, R.A. With Personal Reminiscences. Illustrated with Portrait and 60 Woodcuts of his chief works. 4to. 21*s.*

BREWSTER'S (SIR DAVID) Martyrs of Science, or the Lives of Galileo, Tycho Brahe, and Kepler. *Second Edition.* Fcap. 8vo. 4*s.* 6*d.*

BRITISH ASSOCIATION REPORTS. Held at York and Oxford 1831-32, 13*s.* 6*d.* Cambridge, 1833, 12*s.* Edinburgh, 1834, 15*s.* Dublin, 1835, 13*s.* 6*d.* Bristol, 1836, 12*s.* Liverpool, 1837, 16*s.* 6*d.* Newcastle, 1838, 15*s.* Birmingham, 1839, 13*s.* 6*d.* Glasgow, 1840, 15*s.* Plymouth, 1841, 13*s.* 6*d.* Manchester, 1842, 10*s.* 6*d.* Cork, 1843, 12*s.* York, 1844, 20*s.* Cambridge, 1845, 12*s.* Southampton, 1846, 15*s.* Oxford, 1847, 18*s.* Swansea, 1848, 9*s.* Birmingham, 1849, 10*s.* Edinburgh, 1850, 15*s.* 8vo.

BROGDEN'S (REV. JAS.) Illustrations of the Liturgy and Ritual of the United Church of England and Ireland. Being Sermons and Discourses Selected from the Works of eminent Divines of the 17th Century. 3 Vols. Post 8vo. 27*s.*

—— Catholic Safeguards against the Errors, Corruptions, and Novelties of the Church of Rome. Being Sermons and Tracts selected from the Works of eminent Divines of the 17th Century. *Second Edition.*

With Preface and Index. 3 Vols. 8vo. 36*s.*
⁂ *The Second and Third Volume may be had separately, 14s. each.*

—— Records of the Supremacy of the Crown, and of the Civic and Religious Liberties of the People of England. 8vo.

BROOKE'S (SIR JAMES) Journals of Events in Borneo, including the Occupation of Labuan, and a Visit to the Celebes. Together with the Expedition of H.M.S. Iris. By CAPT. RODNEY MUNDY, R.N. Plates. 2 Vols. 8vo. 32*s.*

BUBBLES FROM THE BRUNNEN OF NASSAU. By an OLD MAN. *Sixth Edition.* 16mo. 5*s.*

BUNBURY'S (C. J. F.) Journal of a Residence at the Cape of Good Hope; with Excursions into the Interior, and Notes on the Natural History and Native Tribes of the Country. Woodcuts. Post 8vo. 9*s.*

BUNYAN (JOHN) and Oliver Cromwell. Select Biographies. By ROBERT SOUTHEY. Post 8vo. 2*s.* 6*d.*

BURGHERSH'S (LORD) Memoir of the Operations of the Allied Armies under Prince Schwarzenberg and Marshal Blucher during the latter end of 1813-14. 8vo. 21*s.*

—— Early Campaigns of the Duke of Wellington in Portugal and Spain. 8vo. 8*s.* 6*d.*

BURN'S (COL.) Naval and Military Technological Dictionary of the French Language. English and French—French and English. For the Use of Soldiers, Sailors, and Engineers. Crown 8vo.

BURNES' (SIR ALEXANDER) Journey to and Residence in the City of Cabool. *Second Edition.* Plates. 8vo. 18*s.*

BURNS' (ROBERT), Life. By JOHN GIBSON LOCKHART. *Fifth Edition.* Fcap. 8vo. 3*s.*

BURR'S (G. D.) Instructions in Practical Surveying, Topographical Plan-drawing, and on sketching ground without Instruments. *Second Edition.* Woodcuts. Post 8vo. 7*s.* 6*d.*

BUXTON'S (SIR FOWELL) Memoirs. With Selections from his Correspondence. By his Son. *Fourth Edition.* Portrait. 8vo, 16*s.*; or *Cheap Edition*, 16mo. 8*s.* 6*d.*

BYRON'S (LORD) Life and Letters. By THOMAS MOORE. Collected and Arranged, with Notes and Illustrations. *Library Edition.* Plates. 6 Vols. Fcap. 8vo. 18*s.*

—— Complete in One Volume. *New Edition.* Portraits. Royal 8vo. 12*s.*

—— Poetical Works. Collected and Arranged, with Notes and Illustrations. *Library Edition.* Plates. 10 Vols. Fcap. 8vo. 30*s.*

—— Complete in One Volume. *New Edition.* Portrait. Royal 8vo. 12*s.*

—— Poetical Works. *New Edition.* 8 Vols. 16mo. 20*s.*, or separately, at 2*s.* 6*d.* each. (*Nearly Ready.*)

Vol. 1.—Childe Harold.

Vol. 2.—Tales and Poems.

Vol. 3.—Dramas, Vol. 1.

Vol. 4.—Dramas, Vol. 2.

Vol. 5.—Miscellanies, Vol. 1.

Vol. 6.—Miscellanies, Vol. 2.

Vol. 7.—Don Juan, Vol. 1.

Vol. 8.—Don Juan, Vol. 2.

—— Childe Harold. Illustrated with Vignettes. *A New Edition.* Crown 8vo. *In the Press.*

BUTTMAN'S LEXILOGUS; or, a Critical Examination of the Meaning and Etymology of numerous Greek Words and Passages, intended principally for Homer and Hesiod. Translated, and edited with Explanatory Notes, and copious Indexes, by REV. J. R. FISHLAKE. *Third Edition.* 8vo. 14*s.*

—— Catalogue of Irregular Greek Verbs; With all the Tenses extant—their Formation, Meaning, and Usage, accompanied by an

Index. Translated, with Notes, by REV. J. R. FISHLAKE. *Second Edition.* 8vo. 7*s.* 6*d.*

CALVIN'S (JOHN) Life. With Extracts from his Correspondence. By THOMAS H. DYER. Portrait. 8vo. 15*s.*

CALLCOTT'S (LADY) Little Arthur's History of England. *Fifteenth Edition.* Woodcuts. 18mo. 2*s.* 6*d.*

CAMPBELL'S (LORD) Lives of the Lord Chancellors and Keepers of the Great Seal of England. From the Earliest Times to the death of Lord Eldon in 1838. *Third Edition.* 7 Vols. 8vo. 102*s.*

The Work may also be had as follows.

Vols. 1-3. From the Earliest Times to the Revolution of 1688.

Vols. 4-5. From the Revolution of 1688 to the Death of Lord Thurlow in 1806.

Vols. 6-7. From the Birth of Lord Loughborough, in 1733, to the Death of Lord Eldon, in 1838.

—— Lives of the Chief Justices of England. From the Norman Conquest till the Death of Lord Mansfield. 2 Vols. 8vo. 30*s.*

—— (GEORGE.) Modern India. A Sketch of the System of Civil Government. With some Account of the Natives and Native Institutions. 8vo. 16*s.*

—— (THOS.) Specimens of the British Poets. With Biographical and Critical Notices, and an Essay on English Poetry. *Third Edition.* Revised by PETER CUNNINGHAM. Portrait and Vignette. Royal 8vo. 15*s.*

—— Short Lives of the British Poets. With an Essay on English Poetry. Post 8vo. 6*s.*

CARÈME'S FRENCH COOKERY. Translated by W. HALL. *Second Edition.* Plates. 8vo. 15*s.*

CARMICHAEL'S (A. N.) Greek Verbs. Their Formations, Irregularities, and Defects. *Second Edition.* Post 8vo. 8*s.* 6*d.*

CARNARVON'S (LORD) Portugal, Gallicia, and the Basque Provinces. From Notes made during a Journey to those Countries. *Third Edition.* Post 8vo. 6*s.*

CATHCART'S (MAJOR GENERAL) Commentaries on the War in Russia and Germany, 1812-13. With 28 Diagrams and Plans. 8vo. 14*s.*

CHARMED ROE (THE); or, The Story of the Little Brother and Sister. With Illustrations. By OTTO SPECKTER. 16mo. 5*s.*

CLARENDON GALLERY (THE); or, Lives of the Friends and Contemporaries of the Lord Chancellor Clarendon, illustrative of Portraits in his Gallery. With an Introduction, containing a Descriptive Catalogue of the Pictures, and an Account of the Origin of the Collection. By Lady THERESA LEWIS. Portraits. 3 Vols. 8vo. 42*s.*

CLARK (SIR JAMES) On the Sanative Influence of Climate, with an Account of the Best Places for Invalids in the South of Europe, &c. *Fourth Edition.* Post 8vo. 10*s.* 6*d.*

CLAUSEWITZ'S (GENERAL CARL VON) Campaign of 1812, in Russia. Translated from the German. With a Map. 8vo. 10*s.* 6*d.*

CLIVE'S (LORD) Life. By REV. G. R. GLEIG, M.A. Post 8vo. 5*s.*

COLERIDGE'S (SAMUEL TAYLOR) Table-Talk. *Fourth Edition.* Portrait. Fcap. 8vo. 6*s.*

—— (HENRY NELSON) Introductions to the Study of the Greek Classic Poets. *Third Edition.* Fcap. 8vo. 5*s.* 6*d.*

COLONIAL LIBRARY. [See Home and Colonial Library.]

COMBER'S (DEAN) Friendly and Seasonable Advice to the Roman Catholics of England. Edited, with a Preface and Notes, by Rev. Dr. HOOK. Fcap. 8vo. 3*s.*

CONDÉ'S (LOUIS PRINCE) Life, surnamed the Great. By LORD MAHON. Post 8vo. 5*s.*

CRABBE'S (REV. GEORGE) Life and Letters. By his SON. Portrait. Fcap. 8vo. 3*s.*

—— Life and Poetical Works. Collected and Arranged, with Notes and Illustrations. *Library Edition.* Plates. 8 Vols. Fcap. 8vo. 24*s.*

—— Complete in One Volume. *New Edition.* Portrait and Vignette. Royal 8vo. 10*s.* 6*d.*

CROKER'S (RIGHT HON. J. W.) Progressive Geography for Children. *Fourth Edition.* 18mo. 1*s.* 6*d.*

—— Stories for Children Selected from the History of England. *Fourteenth Edition.* Woodcuts. 16mo. 5*s.*

—— Boswell's Life of Johnson. Including the Tour to the Hebrides, with Notes by Sir W. SCOTT. *A New and Cheaper Edition*, Revised (1848), with much additional Matter. Complete in One Volume. Portraits. Royal 8vo. 15*s.*

—— Memoirs of the Reign of George the Second, from his accession to the death of Queen Caroline. By JOHN LORD HERVEY. Portrait. 2 Vols. 8vo. 36*s.*

CROMWELL (OLIVER) and John Bunyan. Select Biographies. By ROBERT SOUTHEY. Post 8vo. 2*s.* 6*d.*

CUMMING'S (R. GORDON) Five Years of a Hunter's Life in the Far Interior of South Africa. With Anecdotes of the Chase in hunting the Wild Beasts of the Deserts and Forests. *Third Edition.* With Woodcuts. 2 Vols. Post 8vo. 24*s.*

COOKERY (MODERN DOMESTIC). Founded on Principles of Economy and Practical Knowledge, and adapted for Private Families. *A New Edition.* BY A LADY. Woodcuts. Fcap. 8vo. 6*s.*

CURZON'S (HON. ROBERT) Visits to the Monasteries of the Levant. *Fourth Edition.* Woodcuts. Post 8vo. 15*s.*

CUNNINGHAM'S (ALLAN) Life of Sir David Wilkie. With his Journals, Tours, and Critical Remarks on Works of Art, and a Selection from his Correspondence. Portrait. 3 Vols. 8vo. 42*s.*

CUNNINGHAM'S (ALLAN) Poems and Songs. Now first collected and arranged, with an Introduction and Biographical Notice of the Author, by PETER CUNNINGHAM. 24mo. 2*s.* 6*d.*

—— (CAPT. J. D.) History of the Sikhs. From the Origin of the Nation to the Battle of the Sutlej. *Second Edition.* With Memoir of the Author. Maps. 8vo. 15*s.*

—— (PETER) London—Past and Present. A Handbook to the Antiquities, Curiosities, Churches, Works of Art, Public Buildings, and Places connected with interesting and historical associations. *Second Edition.* Post 8vo. 16*s.*

—— Modern London. A complete Guide for Visitors to the Metropolis. Map. 16mo. 5*s.*

—— Environs of London. Including a circle of 30 miles round St. Paul's. With Hints for Excursions by Rail,—Road,—and River. Post 8vo. *In the Press.*

—— Westminster Abbey. Its Art, Architecture, and Associations. Woodcuts. Fcap. 8vo. 1*s.*

DARWIN'S (CHARLES) Journal of Researches into the Natural History and Geology of the Countries visited during a Voyage round the World. Post 8vo. 8*s.* 6*d.*

DATES AND DISTANCES; Showing what may be done in a Tour of Sixteen Months upon the Continent of Europe. Post 8vo. 8*s.* 6*d.*

DAVY'S (SIR HUMPHRY) Consolations in Travel; or, Last Days of a Philosopher. *Fifth Edition.* Woodcuts. Fcap. 8vo. 6*s.*

—— Salmonia; or, Days of Fly Fishing. With some Account of the Habits of Fishes belonging to the genus Salmo. *Fourth Edition.* Woodcuts. Fcap. 8vo. 6*s.*

DENNIS' (GEORGE) Cities and Cemeteries of Etruria; or, the extant Local Remains of Etruscan Art. Plates. 2 Vols. 8vo. 42*s.*

DEVEREUX'S (HON. CAPT., R.N.) Lives of the three Devereux, Earls of Essex, from 1540 to 1646. 1. The Earl Marshal of Ireland. 2. The Favourite. 3. The General of the Parliament. Founded upon letters and documents chiefly unpublished. 2 Vols. 8vo. *In the Press.*

DE VERE'S (AUBREY) English Misrule and Irish Misdeeds. Four Letters from Ireland, addressed to an English M.P. *Second Edition.* Post 8vo. 7*s.* 6*d.*

DODGSON'S (REV. C.) Controversy of Faith; or, Advice to Candidates for Holy Orders. Containing an Analysis and Exposition of the Argument by which the Catholic Interpretation of the Baptismal Services is to be vindicated. 12mo. 3*s.*

DOG-BREAKING; the Most Expeditious, Certain, and Easy Method, whether great excellence or only mediocrity be required. By LIEUT.-COL. HUTCHINSON. *Second Edition.* Woodcuts. Fcap. 8vo. 7*s.* 6*d.*

DOMESTIC MODERN COOKERY. Founded on Principles of Economy and Practical Knowledge, and adapted for Private Families. *A New Edition.* BY A LADY. Woodcuts. Fcap. 8vo. 6*s.*

DOUGLAS'S (GENERAL SIR HOWARD) Treatise on Improved Gunnery: with Descriptions of the New Guns introduced since the War. *Third Edition, revised.* Plates. 8vo. 21*s.*

DRAKE'S (SIR FRANCIS) Life, Voyages, and Exploits, by Sea and Land. By JOHN BARROW. *Third Edition.* Post 8vo. 2*s.* 6*d.*

DRINKWATER'S (JOHN) History of the Siege of Gibraltar. 1779-1783. With a Description and Account of that Garrison from the Earliest Periods. Post 8vo. 2*s.* 6*d.*

DRUMMOND'S (HENRY) Abstract Principles of Revealed Religion. Post 8vo. 9*s.* 6*d.*

DUDLEY'S (EARL OF) Letters to the late Bishop of Llandaff. *Second Edition.* Portrait. 8vo. 10*s.* 6*d.*

DURHAM'S (ADMIRAL SIR PHILIP) Naval Life and Services. By his Nephew, CAPT. ALEXANDER MURRAY. 8vo. 5*s.* 6*d.*

DYER'S (THOMAS H.) Life and Letters of John Calvin. Compiled from authentic Sources. Portrait. 8vo. 15*s.*

EASTLAKE (SIR CHARLES, P.R.A.) The Schools of Painting in Italy. From the earliest Times. From the German of KUGLER. Edited, with Notes. *Second Edition.* Illustrated with 100 Engravings from the Old Masters. 2 Vols. Post 8vo. 24*s.*

—— Contributions to the Literature of the Fine Arts. 8vo. 12*s.*

EDWARDS' (W. H.) Voyage up the River Amazon, Including a Visit to Para. Post 8vo. 2*s.* 6*d.*

EGERTON'S (CAPTAIN FRANCIS) Winter's Tour in India; with a Visit to Nepaul. Woodcuts. 2 Vols. Post 8vo.

ELDON'S (LORD CHANCELLOR) Public and Private Life, with Selections from his Correspondence and Diaries. By HORACE TWISS. *Third Edition.* Portrait. 2 Vols. Post 8vo. 21*s.*

ELLESMERE'S (LORD) Two Sieges of Vienna by the Turks. Translated from the German. Post 8vo. 2*s.* 6*d.*

—— Military Events in Italy, 1848-9. Translated from the German. With a Map. Post 8vo. 9*s.*

—— Second Campaign of Radetzky in Piedmont. The Defence of Temeswar and the Camp of the Ban. From the German. Post 8vo. 6*s.* 6*d.*

ELPHINSTONE'S (HON. MOUNTSTUART) History of India—the Hindoo and Mahommedan Periods. *Third Edition.* Map. 8vo. 18*s.*

ENGLAND (HISTORY OF) From the Peace of Utrecht to the First Years of the American War, 1763-80. By LORD MAHON. *Second Edition.* 6 Vols. 8vo. 82*s.*

—— Under the House of Lancaster. With an Introductory View of the Early Reformation. 8vo. 15*s.*

—— From the First Invasion by the Romans, down to the 14th year of Queen Victoria's Reign. By MRS. MARKHAM. *New Edition.* Woodcuts. 12mo. 7*s.* 6*d.*

—— AS IT IS: Social, Political, and Industrial, in the Middle of the 19th Century. By W. JOHNSTON. 2 Vols. Post 8vo. 18*s.*

ESKIMAUX (THE) and English Vocabulary, for the use of Travellers in the Arctic Regions. 16mo. 3*s.* 6*d.*

ESSAYS FROM "THE TIMES." Being a Selection from the LITERARY PAPERS which have appeared in that JOURNAL. Reprinted by permission. *New Edition.* Fcap. 8vo. 4*s.*

ESSEX (THE EARLS OF) Lives of; 1540-1646. 1. The Earl Marshal of Ireland. 2. The Favourite. 3. The General of the Parliament. Founded upon Letters and Documents chiefly unpublished By HON. CAPT. DEVEREUX, R.N. 2 Vols. 8vo. *In the Press.*

EXETER'S (BISHOP OF) Letter to the Archbishop of Canterbury on the Gorham Case. *20th Edition.* 8vo. 3*s.* 6*d.*

—— Pastoral Letter to his Clergy on the Present State of the Church (1851). *Eighth Edition.* 8vo. 4*s.*

—— Acts of the Diocesan Synod, held in the Cathedral Church of Exeter. By the BISHOP OF EXETER, on June 25th, 26th, and 27th, 1851. *Third Edition.* 8vo. 2*s.* 6*d.*

—— Letters to the late Charles Butler, on the Theological parts of his Book of the Roman Catholic Church; with Remarks on certain Works of Dr. Milner and Dr. Lingard, and on some parts of the Evidence of Dr. Doyle. *Second Edition.* 8vo. 16*s.*

—— Sermons. Preached during the Visitation of the Bishop of Exeter in 1845. Published by Request. 12mo. 6*s.*

FAIRY RING (THE), A Collection of TALES and STORIES for Young Persons. From the German of Grimm, now first rendered into English. By J. E. TAYLOR. Illustrated by RICHARD DOYLE. *Second Edition.* Fcap. 8vo. 7*s.* 6*d.*

FALKNER'S (FRED.) Muck Manual for the Use of Farmers. A Treatise on the Nature and Value of Manures. *Second Edition*, with a Glossary of Terms and an Index. Fcap. 8vo. 5*s.*

FAMILY RECEIPT-BOOK. A Collection of a Thousand Valuable and Useful Receipts. Fcap. 8vo. 5*s.* 6*d.*

FARADAY'S (MICHAEL) Chemical Manipulation; Being Instructions to Students in Chemistry, on the Methods of performing Experiments of Demonstration or Research, with Accuracy and Success. *Third Edition.* 8vo. 18*s.*

FARINI'S (LUIGI CARLO) History of the Roman State, 1815-50. Translated from the Italian. By Right Hon. W. E. GLADSTONE. 3 Vols. 8vo. 36*s.*

FEATHERSTONHAUGH'S (G. W.) Tour through the Slave States of North America, from the river Potomac, to Texas and the Frontiers of Mexico. Plates. 2 Vols. 8vo. 26*s.*

FELLOWS' (SIR CHARLES) Excursion in Asia Minor, and Discoveries in Lycia. During two Visits to those Countries. With numerous Plates. 2 vols. Imperial 8vo. Also a *Popular Edition*. Post 8vo.

—— Xanthian Marbles now in the British Museum; their Acquisition and Transmission to England. Plates. 8vo. 5*s.*

—— Account of the Ionic Trophy Monument excavated at Xanthus. Imperial 8vo. 5*s.*

FERGUSON'S (ROBERT, M. D.) Essays on the Diseases of Women. Part I. Puerperal Fever. Post 8vo. 9*s.* 6*d.*

FERGUSSON'S (JAMES) Palaces of Nineveh and Persepolis Restored: an Essay on Ancient Assyrian and Persian Architecture. With 45 Woodcuts. 8vo. 16*s.*

—— Illustrated Handbook of Architecture. Being a Concise and Popular Account of the Different Styles prevailing in all Ages and Countries in the World. With a Description of the most remarkable Buildings. With Illustrations. 8vo. *In the Press.*

FEUERBACH'S Remarkable German Crimes and Trials. Translated from the German by Lady DUFF GORDON. 8vo. 12*s.*

FISHER'S (REV. GEORGE) Elements of Geometry, for the Use of Schools. *Third Edition.* 18mo. 3*s.* (*Published by order of the Admiralty.*)

—— First Principles of Algebra, for the Use of Schools. *Third Edition.* 18mo. 3*s.*

FISHLAKE'S (REV. J. R.) Translation of Buttman's Lexilogus; A Critical Examination of the Meaning and Etymology of numerous Greek Words and Passages, intended principally for Homer and Hesiod. With Explanatory Notes and Copious Indexes. *Third Edition.* 8vo. 14*s.*

—— Translation of Buttman's Catalogue of Irregular Greek Verbs; with all the Tenses extant—their Formation, Meaning, and Usage.

With Explanatory Notes, and accompanied by an Index. *Second Edition.* 8vo. 7*s.* 6*d.*

FLOWER GARDEN (THE). An Essay reprinted from the "Quarterly Review." Fcap. 8vo. 1s.

FORD'S (RICHARD) Handbook for Spain, Andalusia, Ronda, Valencia, Catalonia, Granada, Gallicia, Arragon, Navarre, &c. Post 8vo. 16*s.*

—— Gatherings from Spain. Post 8vo. 6*s.*

FORSYTH'S (WILLIAM) Hortensius, or the Advocate: an Historical Essay on the Office and Duties of an Advocate. Post 8vo. 12*s.*

FORTUNE'S (ROBERT) Three Years' Wanderings in the Northern Provinces of China. With a Visit to the Tea and Cotton Countries. *Second Edition.* Woodcuts. 8vo. 15*s.*

—— Journey to Sung-Lo-Shan and the Bohea Mountains; with some Account of the British Tea Plantations in the North-west Provinces of India. Woodcuts. 8vo.

FRANCE (HISTORY OF). From the Conquest by the Gauls to the Death of Louis Philippe. By MRS. MARKHAM. *New Edition.* Woodcuts. 12mo. 7*s.* 6*d.*

FRENCH (THE) in Algiers; The Soldier of the Foreign Legion—and the Prisoners of Abd-el-Kadir. Translated by Lady DUFF GORDON. Post 8vo. 2*s.* 6*d.*

GEOGRAPHICAL (THE) Journal. Published (occasionally) by the Royal Geographical Society of London. 8vo.

GERMANY (HISTORY OF). From the Invasion by Marius, to the Battle of Leipsic. By MRS. MARKHAM. Woodcuts. 12mo. 7*s.* 6*d.*

GIBBON'S (EDWARD) Decline and Fall of the Roman Empire. Edited with Notes. By the DEAN OF ST. PAUL'S and M. GUIZOT. *Second Edition.* Maps. 6 Vols. 8vo. 3*l.* 3*s.*

—— Life and Correspondence. Edited, with Notes, by the DEAN OF ST. PAUL'S. Portrait. 8vo. 9*s.*

GIFFARD'S (EDWARD) Short Visit to the Ionian Islands, Athens, and the Morea, Plates. Post 8vo. 12*s.*

—— Deeds of Naval Daring; or, Anecdotes of the British Navy. Fcap. 8vo. 2*s.* 6*d.*

GILLY'S (REV. DR.) Romaunt Version of the Gospel of St. John, originally in use among the old Waldenses. Edited from the MSS., with Notes. 8vo. 5*s.* 6*d.*

GLADSTONE'S (RIGHT HON. W. E.) Prayers arranged from the Liturgy for Family Use. *Second Edition.* 12mo. 2*s.* 6*d.*

—— Translation of Farini's History of the Roman State, 1815-1850. 3 Vols. 8vo. 36*s.*

GLEIG'S (REV. G. R.) Campaigns of the British Army at Washington and New Orleans. Post 8vo. 2*s.* 6*d.*

—— Story of the Battle of Waterloo. Compiled from Public and Authentic Sources. Post 8vo. 5*s.*

—— Narrative of Sir Robert Sale's Brigade in Afghanistan, with an Account of the Seizure and Defence of Jellalabad. Post 8vo. 2*s.* 6*d.*

—— Life of Robert Lord Clive. Post 8vo. 5*s.*

—— Life and Letters of General Sir Thomas Munro. Post 8vo. 5*s.*

GODLEY'S (JOHN ROBERT) Letters from America. 2 Vols. Post 8vo. 16*s.*

GOLDSMITH (OLIVER); a Biography. By Washington Irving. Post 8vo. 5*s.*

GOOCH (ROBERT, M.D.), On the most Important Diseases peculiar to Women. *Second Edition.* 8vo. 12*s.*

GORDON'S (SIR ALEX. DUFF) Sketches of German Life, and Scenes from the War of Liberation. Translated from the German of Varnhagen Von Ense. Post 8vo. 5*s.*

—— (LADY DUFF), Amber-Witch: the most interesting Trial for Witchcraft ever known. Translated from the German of MEINHOLD. Post 8vo. 2*s.* 6*d.*

—— French in Algiers. 1. The Soldier of the Foreign Legion. 2. The Prisoners of Abd-el-Kadir. Translated from the French. Post 8vo. 2*s.* 6*d.*

—— Remarkable German Crimes and Trials. Translated from the German of Feuerbach. 8vo. 12*s.*

GOSPEL STORIES FOR CHILDREN. An Attempt to render the Chief Events of the Life of Our Saviour intelligible and profitable. *Second Edition.* 18mo. 3*s.* 6*d.*

GRANT'S (ASAHEL), Nestorians, or the Lost Tribes; containing Evidence of their Identity, their Manners, Customs, and Ceremonies; with Sketches of Travel in Ancient Assyria, Armenia, and Mesopotamia; and Illustrations of Scripture Prophecy. *Third Edition.* Fcap. 8vo. 6*s.*

GRENVILLE PAPERS (THE); being the Private Correspondence of Richard Grenville, Earl Temple, and his Brother, George Grenville,—their Friends and Contemporaries;—Including MR. GRENVILLE'S DIARY OF POLITICAL EVENTS; 1763-65. Edited by WILLIAM JAMES SMITH. Vols. 1 & 2. 8vo. 32*s.* (To be completed in 4 Vols.)

GREEK GRAMMAR FOR SCHOOLS. Abridged from MATTHIÆ. By the BISHOP OF LONDON. *Seventh Edition*, revised by Rev. J. EDWARDS. 12mo. 3*s.*

—— ACCIDENCE FOR SCHOOLS. Abridged from MATTHIÆ. By the BISHOP OF LONDON. *Fourth Edition*, revised by Rev. J. EDWARDS. 12mo. 2*s.*

GROTE'S (GEORGE) History of Greece. From the Earliest Period to the Accession of Philip of Macedon (B.C. 403-359). Maps. Vols. I. to X. 8vo. 16*s.* each. *The Work may be had as follows*:—

VOLS. I.-II.—Legendary Greece. Grecian History to the Reign of Peisistratus at Athens.

VOLS. III.-IV.—History of Early Athens, and the Legislation of Solon. Grecian Colonies. View of the Contemporary Nations surrounding Greece. Grecian History down to the first Persian Invasion, and the Battle of Marathon.

VOLS. V.-VI.—Persian War and Invasion of Greece by Xerxes. Period between the Persian and the Peloponnesian Wars. Peloponnesian War down to the Expedition of the Athenians against Syracuse.

VOLS. VII.-VIII.—The Peace of Nikias down to the Battle of Knidus (B.C. 421 to 394). Socrates and the Sophists.

VOLS. IX.-X.—From the Restoration of the Democracy at Athens down to the Accession of Philip of Macedon (B.C. 403-359).

GUIZOT (M.), On the Causes of the Success of the English Revolution of 1640-1688. 8vo. 6*s.*; or *Cheap Edition*, 12mo, 1*s.*

—— Democracy in France. *Sixth Edition.* 8vo. 3*s.* 6*d.*

GURWOOD'S (COL.) Despatches of the Duke of Wellington during his various Campaigns in India. Denmark. Portugal, Spain, the Low Countries, and France. Compiled from Official and other Authentic Documents. *A New and Enlarged Edition.* 8 Vols. 8vo. 21*s.* each.

—— Selections from the Wellington Despatches and General Orders. *New and Cheaper Edition.* 1 Vol. 8vo. 18*s.*

GUSTAVUS VASA (History of), King of Sweden. With Extracts from his Correspondence. Portrait. 8vo. 10*s.* 6*d.*

HALFORD'S (the late SIR HENRY) Popular Essays and Orations. *Third Edition.* Fcap. 8vo. 6*s.* 6*d.*

—— Nugæ Metricæ. Fcap. 8vo. 3*s.* 6*d.*

HALLAM'S (HENRY) Constitutional History of England, from the Accession of Henry the Seventh to the Death of George the Second. *Sixth Edition.* 2 Vols. 8vo. 24*s.*

—— View of the State of Europe during the Middle Ages. *Tenth Edition.* Including all the Additional Notes. 3 Vols. 8vo.

—— Introduction to the Literary History of Europe, during the 16th, 17th, and 18th Centuries. *Third Edition.* 3 Vols. 8vo. 36*s.*

HAMILTON'S (WALTER) Facts to Assist the Memory in various Sciences. *Second Edition.* Fcap. 8vo. 6*s.* 6*d.*

—— Hindostan, Geographically, Statistically, and Historically. Map. 2 Vols. 4to. 4*l.* 14*s.* 6*d.*

—— (W. J.) Researches in Asia Minor, Pontus, and Armenia; with some Account of the Antiquities and Geology of those Countries. Plates. 2 Vols. 8vo. 38*s.*

HAMPDEN'S (BISHOP OF HEREFORD) Essay on the Philosophical Evidence of Christianity, or the Credibility obtained to a Scripture Revelation, from its Coincidence with the Facts of Nature. 8vo. 9*s.* 6*d.*

HARCOURT'S (EDWARD VERNON) Sketch of Madeira; with Map and Plates. Post 8vo. 8*s.* 6*d.*

HART'S ARMY LIST. (*Published Quarterly and Annually.*) 8vo. 5*s.* and 20*s.* each.

HASE'S ANCIENT GREEKS; their Public and Private Life, Manners, and Customs. Translated from the German. By Mrs. AUSTIN. Fcap. 8vo. 5*s.* 6*d.*

HAND-BOOK OF TRAVEL-TALK; or, Conversations in English, German, French, and Italian. Intended for Englishmen travelling abroad, or Foreigners visiting Great Britain. 18mo. 5*s.*

—— NORTH GERMANY—Holland, Belgium, Prussia, and the Rhine. Map. Post 8vo. 12*s.*

—— SOUTH GERMANY—Bavaria, Austria, Salzburg, the Austrian and Bavarian Alps, the Tyrol, and the Danube, from Ulm to the Black Sea. Map. Post 8vo. 12*s.*

—— PAINTING—the German, Flemish, and Dutch Schools. From the German of KUGLER. Edited, with Notes, by SIR EDMUND HEAD. Post 8vo. 12*s.*

—— SWITZERLAND—the Alps of Savoy, and Piedmont. Map. Post 8vo. 10*s.*

—— FRANCE—Normandy, Brittany, the French Alps, the Rivers Loire, Seine, Rhone, and Garonne, Dauphiné, Provence, and the Pyrenees. Maps. Post 8vo. 12*s.*

—— SPAIN—Andalusia, Ronda, Granada, Valencia, Catalonia, Gallicia, Arragon, and Navarre. Maps. Post 8vo. 16*s.*

—— PAINTING—the Spanish and French Schools. By SIR EDMUND HEAD. Post 8vo. 12*s.*

—— MADEIRA. Maps and Woodcuts. Post 8vo. 8*s.* 6*d.*

—— NORTH ITALY—Florence, Sardinia, Genoa, the Riviera, Venice, Lombardy, and Tuscany. Map. Post 8vo. 12*s.*

—— CENTRAL ITALY—Rome, the Papal States, and Cities of Etruria. Maps. Post 8vo. 16*s.*

—— SOUTHERN ITALY—Sicily, and Naples. Map. Post 8vo. *Nearly ready.*

—— PAINTING—the Italian Schools. From the German of KUGLER. Edited, by Sir CHARLES EASTLAKE. With Woodcuts from the Old Masters. 2 Vols. Post 8vo. 24*s.*

—— MALTA—the Ionian Islands, Turkey, Asia Minor, and Constantinople. Maps. Post 8vo. 15*s.*

—— EGYPT—Thebes, the Nile, Alexandria, Cairo, the Pyramids, Mount Sinai, &c. Map. Post 8vo. 15*s.*

—— SYRIA AND THE HOLY LAND. Maps. Post 8vo. (*Preparing.*)

—— DENMARK, NORWAY, AND SWEDEN. Maps. Post 8vo. 12*s.*

—— RUSSIA AND FINLAND. Maps. Post 8vo. 12*s.*

—— ENGLAND AND WALES. Giving an account of the most remarkable *Places* and most frequented Roads and Railways in England—describing the objects calculated to *interest strangers and passing travellers*, showing how they may be accessible with the least expenditure of time and money. Map. Post 8vo. *Nearly ready.*

—— DEVON AND CORNWALL. Maps. Post 8vo. 6*s.*

HAND-BOOK FOR LONDON, PAST AND PRESENT. Being an Alphabetical Account of all the Antiquities, Curiosities, Churches, Works of Art, Places, and Streets connected with Interesting and Historical Associations. Post 8vo. 16*s.*

—— MODERN LONDON. A Guide to all the Sights and objects of interest in the Metropolis. Map. 16mo. 5*s.*

—— ENVIRONS OF LONDON. Including a Circle of 30 Miles round St. Paul's. Alphabetically arranged. Maps. Post 8vo. *Nearly ready.*

—— ANTIQUITIES AND SCULPTURE IN THE BRITISH MUSEUM. With 300 Woodcuts. Post 8vo. 7*s.* 6*d.*

—— PICTURE GALLERIES IN LONDON AND ITS NEIGHBOURHOOD. With Historical, Biographical, and Critical Notices. Post 8vo. 10*s.*

—— WINDSOR AND ETON. 16mo. 1*s.*

—— WESTMINSTER ABBEY—its Art, Architecture, and Associations. Woodcuts. 16mo. 1*s.*

—— CHRONOLOGY, Alphabetically arranged as a Book of easy reference. 8vo. (*Nearly Ready.*)

—— (OFFICIAL) OF CHURCH AND STATE. Giving a detailed and Historical Account of the Duties attached to the various Departments of the Government and the Church. Post 8vo. 6*s.*

—— FAMILIAR QUOTATIONS. From English Authors. Fcap. 8vo. *In the Press.*

—— ARCHITECTURE. Illustrated. Being a Concise and Popular Account of the Different Styles prevailing in all Ages and Countries in the World. With a Description of the most remarkable Buildings. By JAMES FERGUSSON. Illustrations. 8vo. *In the Press.*

—— CATHEDRALS OF ENGLAND. Containing a short Description of each. By Rev. G. A. POOLE. With Plates. Post 8vo. *In the Press.*

—— MEDIÆVAL ART. Translated from the French of M. Jules Labarthe, and Edited, with Notes and Illustrations, by MRS. PALLISER. With Illustrations. 8vo. *In the Press.*

HAWKSTONE; a Tale of England in the Year 184—. *Third Edition.* 2 Vols. Fcap. 8vo. 12*s.*

HAY'S (J. H. DRUMMOND) Western Barbary, its wild Tribes and savage Animals. Post 8vo. 2*s.* 6*d.*

HEBER'S (BISHOP) Parish Sermons; on the Lessons, the Gospel, or the Epistle, for every Sunday in the Year, and for Week-day Festivals. *Sixth Edition.* 2 Vols. Post 8vo. 16*s.*

—— Sermons Preached in England. *Second Edition.* 8vo. 9*s.* 6*d.*

—— Hymns written and adapted for the weekly Church Service of the Year. *Twelfth Edition.* 16mo. 2*s.*

—— Poetical Works. *Fifth Edition.* Portrait. Fcap. 8vo. 7*s.* 6*d.*

—— Journey through the Upper Provinces of India. From Calcutta to Bombay, with a Journey to Madras and the Southern Provinces. 2 Vols. Post 8vo. 12*s.*

HEAD'S (SIR FRANCIS) Rough Notes of some Rapid Journeys across the Pampas and over the Andes. Post 8vo. 2*s.* 6*d.*

—— Bubbles from the Brunnen of Nassau. By an OLD MAN. *Sixth Edition.* 16mo. 5*s.*

—— Emigrant. *Fifth Edition.* Post 8vo. 12*s.*

—— Stokers and Pokers, or the London and North-Western Railway. Post 8vo. 2*s.* 6*d.*

—— Defenceless State of Great Britain. A series of Descriptive Sketches, containing—1. Military Warfare. 2. Naval Warfare. 3. The Invasion of England. 4. The Capture of London by a French Army. 5. The Treatment of Women in War. 6. How to Defend Great Britain. Post 8vo. 12*s.*

—— Paris in 1851—a Faggot of French Sticks. *Second Edition.* 2 Vols. Post 8vo. 24*s.*

—— (SIR GEORGE) Forest Scenes and Incidents in Canada. *Second Edition.* Post 8vo. 10*s.*

—— Home Tour through the Manufacturing Districts of England, Scotland, and Ireland, including the Channel Islands, and the Isle of Man. *Third Edition.* 2 Vols. Post 8vo. 12*s.*

—— (SIR EDMUND) Handbook of Painting—the Spanish and French Schools. Post 8vo. 12*s.*

—— Handbook of Painting—the German, Flemish, and Dutch Schools. From the German of KUGLER. Edited, with Notes. Post 8vo. 12*s.*

HEIRESS (THE) in Her Minority; or, The Progress of Character. By the Author of "BERTHA'S JOURNAL." 2 Vols. 12mo. 18*s.*

HERODOTUS. A New English Version. Translated from the Text of GAISFORD, and Edited with Notes, illustrating the History and Geography of Herodotus, from the most recent sources of information, embodying the chief Results, Historical and Ethnographical, which have been arrived at in the progress of Cuneiform and Hieroglyphical Discovery. By Rev. GEORGE RAWLINSON, M.A. Assisted by COLONEL RAWLINSON and SIR J. G. WILKINSON. 4 Vols. 8vo. *In Preparation.*

HERSCHEL'S (SIR J. W. F.) Manual of Scientific Enquiry, for the Use of Officers and Travellers in general. By various Writers. *Second Edition.* Post 8vo. 10*s.* 6*d.* (*Published by Authority of the Lords of the Admiralty.*)

HERVEY'S (LORD) Memoirs of the Reign of George the Second, from his Accession to the Death of Queen Caroline. Edited, with Notes, by Right Hon. J. W. CROKER. Portrait. 2 Vols. 8vo. 36*s.*

HICKMAN'S (WM.) Treatise on the Law and Practice of Naval Courts Martial. 8vo. 10*s.* 6*d.*

HILL'S (GENERAL LORD) Life. By REV. EDWIN SIDNEY. *Second Edition.* Portrait. 8vo. 12*s.*

HISTORY OF ENGLAND UNDER THE HOUSE OF LANCASTER. With an Introductory View of the Early Reformation. 8vo. 15*s.*

—— THE LATE WAR: With Sketches of Nelson, Wellington, and Napoleon. By J. G. LOCKHART. 18mo. 2*s.* 6*d.*

HOLLAND'S (REV. W. B.) Psalms and Hymns, selected and adapted to the various Solemnities of the Church. 24mo. 1*s.* 6*d.*

HOLMES' (MRS. DALKEITH) Ride on Horseback through France and Switzerland to Florence. 2 Vols. Post 8vo. 18*s.*

HOME AND COLONIAL LIBRARY. Complete in 76 Parts. Post 8vo. 2*s.* 6*d.* each, sewed, or bound in 37 Volumes, 6*s.* each, cloth. (*Vols. 12 and 36 are 8s. 6d. each.*) The Series contains the following works.

THE BIBLE IN SPAIN. By GEORGE BORROW. (*Two Parts.*)

JOURNALS IN INDIA. By BISHOP HEBER. (*Four Parts.*)

TRAVELS IN THE HOLY LAND. By CAPTAINS IRBY and MANGLES.

THE SIEGE OF GIBRALTAR. By JOHN DRINKWATER.

MOROCCO AND THE MOORS. By J. DRUMMOND HAY.

LETTERS FROM THE BALTIC. By a LADY.

THE AMBER WITCH. By LADY DUFF GORDON.

OLIVER CROMWELL & JOHN BUNYAN. By ROBERT SOUTHEY.

NEW SOUTH WALES. By MRS. MEREDITH.

LIFE OF SIR FRANCIS DRAKE. By JOHN BARROW.

FATHER RIPA'S MEMOIRS OF THE COURT OF CHINA.

A RESIDENCE IN THE WEST INDIES. By M. G. LEWIS.

SKETCHES OF PERSIA. By SIR JOHN MALCOLM. (*Two Parts.*)

THE FRENCH IN ALGIERS. By LADY DUFF GORDON.

BRACEBRIDGE HALL. By WASHINGTON IRVING. (*Two Parts.*)

VOYAGE OF A NATURALIST. By CHARLES DARWIN. (*Three Parts.*)

HISTORY OF THE FALL OF THE JESUITS.

LIFE OF LOUIS PRINCE CONDE. By LORD MAHON. (*Two Parts.*)

GIPSIES OF SPAIN. By GEORGE BORROW. (*Two Parts.*)

TYPEE; OR THE MARQUESAS. By HERMANN MELVILLE. (*Two Parts.*)

LIVONIAN TALES. By a LADY.

MISSIONARY LIFE IN CANADA. By REV. J. ABBOTT.

SALE'S BRIGADE IN AFFGHANISTAN. By REV. G. R. GLEIG.

LETTERS FROM MADRAS. By a LADY.

HIGHLAND SPORTS. By CHARLES ST. JOHN. (*Two Parts.*)

JOURNEYS ACROSS THE PAMPAS. By SIR F. B. HEAD.

GATHERINGS FROM SPAIN. By RICHARD FORD. (*Two Parts.*)

SIEGES OF VIENNA BY THE TURKS. By LORD ELLESMERE.

SKETCHES OF GERMAN LIFE. By SIR A. GORDON. (*Two Parts.*)

OMOO; OR THE SOUTH SEAS. By HERMANN MELVILLE. (*Two Parts.*)

STORY OF BATTLE OF WATERLOO. By REV. G. R. GLEIG. (*Two Parts.*)

A VOYAGE UP THE RIVER AMAZON. By W. H. EDWARDS.

THE WAYSIDE CROSS. By CAPT. MILMAN.

MANNERS & CUSTOMS OF INDIA. By REV. C. ACLAND.

CAMPAIGNS AT WASHINGTON. By REV. G. R. GLEIG.

ADVENTURES IN MEXICO. By G. F. RUXTON. (*Two Parts.*)

PORTUGAL AND GALLICIA. By LORD CARNARVON. (*Two Parts.*)

LIFE OF LORD CLIVE. By REV. G. R. GLEIG. (*Two Parts.*)

TALES OF A TRAVELLER. By WASHINGTON IRVING. (*Two Parts.*)

SHORT LIVES OF THE POETS. By THOMAS CAMPBELL. (*Two Parts.*)

HISTORICAL ESSAYS. By LORD MAHON. (*Two Parts.*)

LONDON & NORTH-WESTERN RAILWAY. By SIR F. B. HEAD.

ADVENTURES IN THE LYBIAN DESERT. By BAYLE ST. JOHN.

A RESIDENCE AT SIERRA LEONE. By a LADY. (*Two Parts.*)

LIFE OF GENERAL MUNRO. By REV. G. R. GLEIG. (*Two Parts.*)

MEMOIRS OF SIR FOWELL BUXTON. (*Three Parts.*)

OLIVER GOLDSMITH. By WASHINGTON IRVING. (*Two Parts.*)

HONEY BEE (THE). An Essay Reprinted from the "Quarterly Review." Fcap. 8vo. 1*s.*

HOOK'S (REV. DR.) Means of rendering more effectual the Education of the People. *Tenth Edition.* 8vo. 2*s.* 6*d.*

—— Church Dictionary. *Sixth Edition.* 8vo. *In the Press.*

—— Friendly and Seasonable Advice to the Roman Catholics of England. By DEAN COMBER. *A New Edition.* Edited with Notes and a Preface. Fcap. 8vo. 3*s.*

—— Three Reformations; Lutheran—Roman—and Anglican. *Third Edition.* 8vo. 3*s.*

—— Nonentity of Romish Saints, and Inanity of Roman Ordinances. *Third Edition.* 8vo. 2*s.* 6*d.*

—— Sermons, chiefly historical. 8vo. *In the Press.*

—— (THEODORE) Life. A Sketch. Reprinted from the "Quarterly Review." Fcap. 8vo. 1*s.*

HORACE (LIFE AND WORKS OF). Edited with Notes by the Rev. H. H. MILMAN, Dean of St. Paul's. Illustrated with Views, Vignettes, &c., taken chiefly from the Antique. 300 Woodcuts. Crown 8vo. 42*s.*

HOSKING'S (WM.) Guide to the Regulation of Buildings in Towns as a Means of Securing the Health of the Inhabitants. *Second Edition.* Post 8vo. 3*s.* 6*d.*

HOUSTOUN'S (MRS.) Yacht Voyage to Texas and the Gulf of Mexico. Plates. 2 Vols. Post 8vo. 21*s.*

HUMBOLDT'S (ALEX.) Cosmos; or, a Physical Description of the World. Translated by COL. and MRS. SABINE. *Seventh Edition.* 3 Vols. Post 8vo. 10*s.* 6*d.*

—— Aspects of Nature in different Lands and in different Climates. Translated by COL. and MRS. SABINE. 2 Vols. Post 8vo. 5*s.*

HUTCHINSON (COLONEL) on Dog-Breaking; the most expeditious, certain, and easy Method, whether great Excellence or only Mediocrity be required. *Second Edition.* Woodcuts. Fcap. 8vo. 7*s.* 6*d.*

INKERSLEY'S (THOS.) Gothic Architecture in France; Being an Inquiry into the Chronological Succession of the Romanesque and Pointed Styles; with Notices of some of the principal Buildings, and an Index. 8vo. 12*s.*

IRBY AND MANGLES' (CAPTAINS) Travels in Egypt, Nubia, Syria, and the Holy Land, including a Journey round the Dead Sea, and through the Country east of the Jordan. Post 8vo. 2*s.* 6*d.*

IRVING'S (WASHINGTON) Oliver Goldsmith. Post 8vo. 5*s.*

—— Lives of Mahomet and his Successors. 2 Vols. 8vo. 21*s.*

JAMES' (REV. THOMAS) Fables of Æsop. A New Version, for Old and Young. With 100 Original Designs, by JOHN TENNIEL. Post 8vo, 2*s.* 6*d.* *Also a few Copies on Fine Paper*, 8vo. 16*s.*

JAMESON'S (MRS.) Handbook to the Picture Galleries in and near London. With Historical, Biographical, and Critical Notices. Post 8vo. *Second Edition.* 10*s.*

JAPAN AND THE JAPANESE. Described from the Accounts of Recent Dutch Travellers. Post 8vo. 9*s.* 6*d.*

JERVIS'S (LIEUT.) Manual of Field Operations, adapted for the Use of Officers of the Army. Post 8vo. 9*s.* 6*d.*

JESSE'S (EDWARD) Visits to Spots of Interest in the Vicinity of Windsor and Eton. Woodcuts. Post 8vo. 12*s.*

—— Gleanings in Natural History, adapted for Schools. With Anecdotes of the Sagacity and Instinct of Animals. *Sixth Edition.* Fcap. 8vo. 6*s.* 6*d.*

JOCELYN'S (LORD) Six Months with the Chinese Expedition; or, Leaves from a Soldier's Note-Book. *Seventh Edition.* Fcap. 8vo. 5*s.* 6*d.*

JOHNSON'S (DR. SAMUEL) Life: By James Boswell. Including the Tour to the Hebrides, with Notes by SIR W. SCOTT. Edited by the Right Hon. JOHN WILSON CROKER. *A New and Cheaper Edition*, revised (1848), with much additional matter. Complete in One Volume. Portraits. Royal 8vo. 15*s.*

JOHNSTON'S (WM.) England as it is: Social, Political, and Industrial, in the Middle of the 19th Century. 2 Vols. Post 8vo. 18*s.*

JONES'S (REV. RICHARD) Essay on the Distribution of Wealth, and on the Sources of Taxation. Part I.—RENT. *Second Edition.* Post 8vo. 7*s.* 6*d.*

JOURNAL OF A NATURALIST. *Fourth Edition.* Woodcuts. Post 8vo. 9*s.* 6*d.*

JUKES' (J. B.) Excursions in and about Newfoundland during the Years 1839-40. Map. 2 Vols. Post 8vo. 21*s.*

KING EDWARD VITH'S Latin Grammar; or an Introduction to the Latin Tongue, for the Use of Schools. *Seventh Edition.* 12mo. 3*s.* 6*d.*

—— Latin Accidence; or Elements of the Latin Tongue, for the Use of Junior Classes. 12mo. 2*s.*

KINNEAR'S (JOHN G.) Cairo, Petra, and Damascus, described from Notes made during a Tour in those Countries: with Remarks on the Government of Mehemet Ali, and on the present prospects of Syria. Post 8vo. 9*s.* 6*d.*

KUGLER'S (DR. FRANZ) Handbook to The Italian Schools of Painting. Translated from the German. By a LADY. Edited, with Notes, by SIR CHARLES EASTLAKE. *Second Edition.* Illustrated with 100 Woodcuts from the Old Masters. 2 Vols. Post 8vo. 24*s.*

—— Handbook to the German, Flemish, and Dutch Schools of Painting. Translated from the German. By a LADY. Edited, with Notes, by SIR EDMUND HEAD, Bart. Post 8vo. 12*s.*

—— Handbook to the Spanish and French Schools of Painting. On the Plan of Kugler. By SIR EDMUND HEAD, Bart. Post 8vo. 12*s.*

LABORDE'S (LEON DE) Journey through Arabia Petræa, to Mount Sinai, and the Excavated City of Petræa,—the Edom of the Prophecies. *Second Edition.* With Plates. 8vo. 18*s.*

LAMBERT'S (Miss) Church Needlework. With Practical Remarks on its Preparation and Arrangement. Plates. Post 8vo. 9*s.* 6*d.*

—— My Knitting Book. Woodcuts. *Two Parts.* 16mo. 3*s.*

—— My Crochet Sampler. Woodcuts. *Two Parts.* 16mo. 4*s.*

—— Hints on Decorative Needlework. 16mo. 1*s.* 6*d.*

LATIN GRAMMAR (KING EDWARD THE VITH'S.) For the Use of Schools. *Seventh Edition.* 12mo. 3*s.* 6*d.*

LATIN ACCIDENCE; or, Elements of the Latin Tongue, for the Use of Junior Classes. 12mo. 2*s.*

LAYARD'S (AUSTEN H.) Nineveh and its Remains. Being a Narrative of Researches and Discoveries amidst the Ruins of Assyria. With an Account of the Chaldean Christians of Kurdistan; the Yezedis, or Devil-worshippers; and an Enquiry into the Manners and Arts of the Ancient Assyrians. *Fifth Edition.* Plates and Woodcuts. 2 Vols. 8vo. 36*s.*

—— Popular Account of his Researches at Nineveh, arranged by himself. From the above Work. Woodcuts. Post 8vo. 5*s.*

—— Monuments of Nineveh. Illustrated by One Hundred Engravings. Imperial Folio, 10*l.* 10*s.*; or Colombier folio, 14*l.* 14*s.*

—— Fresh Discoveries at Nineveh, and Researches at Babylon; being the Results of the Second Expedition to Assyria; also a Journey to the Khabour, the Desert, Lake Van, ancient Armenia, Kurdistan, and the Borders of the Euphrates. With many Woodcuts. 2 Vols. 8vo. *In the Press.*

—— Illustrations of Vases, Sculptures, and Bronzes recently Discovered at Nineveh and Babylon, principally Bas-Reliefs of the Wars and Exploits of Sennacherib from his Palace at Kouyunjik. Imperial Colombier. Folio. *Nearly Ready.*

LETTERS FROM THE SHORES OF THE BALTIC. By a LADY. Post 8vo. 2*s.* 6*d.*

—— MADRAS; or, First Impressions of Life and Manners in India. By a LADY. Post 8vo. 2*s.* 6*d.*

—— SIERRA LEONE, written to Friends at Home. By a LADY. Edited by Mrs. NORTON. Post 8vo. 5*s.*

LEWIS' (G. CORNEWALL) Essay on the Government of Dependencies. 8vo. 12*s.*

—— History and Antiquities of the Doric Race. Translated from MULLER by HENRY TUFNELL and GEORGE CORNEWALL LEWIS. *Second Edition.* Maps. 2 Vols. 8vo. 26*s.*

—— Glossary of Provincial Words used in Herefordshire and some of the adjoining Counties. 12mo. 4*s.* 6*d.*

—— Essay on the Origin and Formation of the Romance Languages: containing an Examination of M. Raynouard's Theory on the Relation of the Italian, Spanish, Provençal, and French to the Latin. *Second Edition.* 8vo. 12*s.*

—— (LADY THERESA) Lives of the Friends and Contemporaries of the Lord Chancellor Clarendon, illustrative of Portraits in his Gallery. With an Introduction, containing a Descriptive Catalogue of the Pictures, and an Account of the Origin of the Collection. Portraits. 3 Vols. 8vo. 42*s.*

—— (M. G.) Journal of a Residence among the Negroes in the West Indies. Post 8vo. 2*s.* 6*d.*

LEXINGTON (THE) PAPERS; or, Some Account of the Courts of London and Vienna at the end of the 17th Century. Extracted from the Official and Private Correspondence of ROBERT SUTTON (LORD LEXINGTON) while Minister at Vienna, 1694-1698. Edited by HON. H. MANNERS SUTTON. 8vo. 14*s.*

LINDSAY'S (LORD) Sketches of the History of Christian Art. 3 Vols. 8vo. 31*s.* 6*d.*

—— Lives of the Lindsays; or, a Memoir of the Houses of Crawford and Balcarres. To which are added, Extracts from the Official

Correspondence of Alexander, sixth Earl of Balcarres, during the Maroon War; together with Personal Narratives, by his Brothers, the Hon. Robert, Colin, James, John, and Hugh Lindsay; and by his Sister, Lady Anne Barnard. 3 Vols. 8vo. 42*s.*

—— Progression by Antagonism. A Theory, involving Considerations touching the Present Position, Duties, and Destiny of Great Britain. 8vo. 6*s.*

—— (Rev. HENRY) Practical Lectures on the Historical Books of the Old Testament. 2 Vols. 16mo. 10*s.*

LITTLE ARTHUR'S HISTORY OF ENGLAND. By LADY CALLCOTT. *Fifteenth Edition.* 18mo. 2*s.* 6*d.*

LIVONIAN TALES.—The Disponent.—The Wolves.—The Jewess. By the Author of "Letters from the Baltic." Post 8vo. 2*s.* 6*d.*

LOCH'S (CAPT. G. C.) Events of the Closing Campaign in China. Map. Post 8vo. 8*s.* 6*d.*

LOCKHART'S (J. G.) Ancient Spanish Ballads, Historical and Romantic, Translated, with Notes. *New Edition*, with Illuminated Titles, Borders, &c. 4to.

—— Life of Robert Burns. *Fifth Edition.* Fcap. 8vo. 3*s.*

—— History of the Late War: with Sketches of Nelson, Wellington, and Napoleon. 18mo. 2*s.* 6*d.*

LONG'S (GEORGE) Essays on the Conduct of Life, and Moral Nature of Man. 2 Vols. Post 8vo. 6*s.* each.

LOUDON'S (MRS.) Instructions in Gardening for Ladies. With Directions for Every Month in the Year, and a Calendar of Operations. *Eighth Edition.* Woodcuts. Fcap. 8vo. 5*s.*

—— Modern Botany for Ladies; or, a Popular Introduction to the Natural System of Plants. *Second Edition.* Woodcuts. Fcap. 8vo. 6*s.*

LOWE'S (GENERAL SIR HUDSON) Letters and Journals, revealing the true History of Napoleon during his Captivity at St. Helena. Portrait. 3 Vols. 8vo. *In Preparation.*

LUSHINGTON'S (*Mrs.*) Narrative of a Journey from Calcutta to Europe, by way of Egypt. *Second Edition.* Post 8vo. 8*s.* 6*d.*

LYELL'S (SIR CHARLES) Principles of Geology; or, the Modern Changes of the Earth and its Inhabitants considered as illustrative of Geology. *Eighth Edition.* Woodcuts. 8vo. 18*s.*

—— Manual of Elementary Geology; or, the Ancient Changes of the Earth and its Inhabitants illustrated by its Geological Monuments. *Fourth Edition.* Woodcuts. 8vo. 12*s.*

—— Travels in North America, 1841-2; with Observations on the United States, Canada, and Nova Scotia. Plates. 2 Vols. Post 8vo. 21*s.*

—— Second Visit to the United States of North America, 1845-6. *Second Edition.* 2 Vols. Post 8vo. 18*s.*

MACCULLOCH'S (J. R.) Edition of RICARDO'S Political Works. With a Notice of his Life and Writings. 8vo. 16*s.*

MACFARLANE'S (CHARLES) Travels in Turkey during the Years 1847-8, made on purpose to examine into the True State of that Country. 2 Vols. 8vo. 28*s.*

MAHON'S (LORD) History of England, from the Peace of Utrecht to the First Years of the American War, 1763-80. *Second Edition.* 6 Vols. 8vo. 82*s.*

—— "Forty-Five;" or, a Narrative of the Rebellion in Scotland. Post 8vo. 3*s.*

—— History of the War of the Succession in Spain. *Second Edition.* Map. 8vo. 15*s.*

—— Spain under Charles the Second; or, Extracts from the Correspondence of the Hon. ALEXANDER STANHOPE, British Minister at Madrid from 1690 to 1700. *Second Edition.* Post 8vo. 6*s.* 6*d.*

—— Life of Louis Prince of Condé, surnamed the Great. Post 8vo. 5*s.*

—— Life of Belisarius. *Second Edition.* Post 8vo. 10*s.* 6*d.*

MALCOLM'S (SIR JOHN) Sketches of Persia. *Third Edition.* Post 8vo. 5*s.*

MANTELL'S (GIDEON A.) Thoughts on Animalcules; or, the Invisible World, as revealed by the Microscope. *Second Edition.* Plates. 16mo. 6*s.*

MANUAL OF SCIENTIFIC ENQUIRY, Prepared for the Use of Officers and Travellers in general. By various Writers. Edited by SIR J. HERSCHEL, Bart. *Second Edition.* Maps. Post 8vo. 10*s.* 6*d.* (*Published by order of the Lords of the Admiralty.*)

MARKHAM'S (MRS.) History of England. From the First Invasion by the Romans, down to the fourteenth year of Queen Victoria's Reign. *New Edition.* Woodcuts. 12mo. 7*s.* 6*d.*

—— History of France. From the Conquest by the Gauls, to the Death of Louis Philippe. *New Edition.* Woodcuts. 12mo. 7*s.* 6*d.*

—— History of Germany. From the Invasion by Marius, to the Battle of Leipsic. Woodcuts. 12mo. 7*s.* 6*d.*

—— History of Rome and Greece. 12mo. *In Preparation.*

—— Sermons for Children. *Second Edition.* Fcap. 8vo. 3*s.*

MARKLAND'S (J. H.) Remarks on English Churches, and on rendering Sepulchral Memorials subservient to pious and Christian Uses. *Fourth Edition.* Woodcuts. Fcap. 8vo. 6*s.* 6*d.*

—— Reverence due to Holy Places. *Third Edition.* Woodcuts. Fcap. 8vo. 2*s.*

MARRYAT'S (JOSEPH) Collections towards a History of Pottery and Porcelain, in the Fifteenth, Sixteenth, Seventeenth, and Eighteenth Centuries. With a Description of the Manufacture, a Glossary, and a List of Monograms. With Coloured Plates and Woodcuts. 8vo. 31*s.* 6*d.*

⁂ *A few copies on India Proofs, mounted on Large Paper.* 4to. 5*l.* 5*s.*

MARRYAT'S ANCIENT POTTERY; Egyptian, Asiatic, Greek, Roman, Etruscan, and Celtic. By SAMUEL BIRCH. Woodcuts. 8vo. *Nearly Ready.*

MATTHIÆ'S (AUGUSTUS) Greek Grammar for Schools. Abridged from the Larger Grammar. By the BISHOP OF LONDON. *Seventh Edition* revised by Rev. J. EDWARDS. 12mo. 3*s.*

MATTHIÆ'S Greek Accidence for Schools. Abridged by the BISHOP OF LONDON. *Fourth Edition*, revised by Rev. J. EDWARDS. 12mo. 2*s.*

—— Index of Quotations from Greek Authors contained in Matthiæ's Greek Grammar. *Second Edition.* 8vo. 7*s.* 6*d.*

MAWE'S (H. L.) Journal of a Passage from the Pacific to the Atlantic, crossing the Andes in the Northern Provinces of Peru, and descending the great River Maranon. 8vo. 12*s.*

MAXIMS AND HINTS for an Angler, and the Miseries of Fishing. To which is added, Maxims and Hints for a Chess-Player. By RICHARD PENN. *Second Edition.* Woodcuts. 12mo. 5*s.*

MAYO'S (DR.) Elements of the Pathology of the Human Mind. Fcap. 8vo. 5*s.* 6*d.*

MELVILLE'S (HERMANN) Typee and Omoo; or, Adventures amongst the Marquesas and South Seas. 2 Vols. Post 8vo. 10*s.*

MENDELSSOHN'S (FELIX BARTHOLDY) Life. By JULES BENEDICT. 8vo. 2*s.* 6*d.*

MERRIFIELD (MRS.) Original Treatises on the Arts of Painting in Oil, Miniature, Mosaic, and on Glass; of Gilding, Dyeing, and the Preparation of Colours and Artificial Gems, described in several unpublished Manuscripts, dating from the Twelfth to the Eighteenth Centuries. Preceded by a General Introduction, with Translations, Preface, and Notes. 2 Vols. 8vo. 30*s.* (*Published by Authority.*)

MEREDITH'S (MRS. CHARLES) Notes and Sketches of New South Wales, during a Residence from 1839 to 1844. Post 8vo. 2*s.* 6*d.*

MILFORD'S (JOHN) Norway and her Laplanders in 1841; with a Few Hints to the Salmon Fisher. 8vo. 10*s.* 6*d.*

MITCHELL'S (THOMAS) Plays of Aristophanes. With English Notes. 8vo.—1. CLOUDS, 10*s.*—2. WASPS, 10*s.*—3. FROGS, 15*s.*

MODERN DOMESTIC COOKERY. Founded on Principles of Economy and Practical Knowledge, and adapted for Private Families. Woodcuts. Fcap. 8vo. 6*s.*

MILMAN'S (DEAN OF ST. PAUL'S) History of Christianity, from the Birth of Christ to the Extinction of Paganism in the Roman Empire. 3 Vols. 8vo. 36*s.*

—— Edition of Gibbon's Decline and Fall of the Roman Empire. Edited with Notes, by M. GUIZOT. *Second Edition.* Maps. 6 Vols. 8vo. 3*l.* 3*s.*

—— Life and Correspondence of Edward Gibbon. Portrait. 8vo. 9*s.*

—— Life and Works of Horace. Illustrated with Statues, Coins, Views, &c., chiefly from the Antique. With 300 Woodcuts. Crown 8vo. 42*s.*

—— Poetical Works. *Second Edition.* Plates. 3 Vols. Fcap. 8vo. 18*s.*

—— Character and Conduct of the Apostles Considered as an Evidence of Christianity. (Being the Bampton Lecture, 1827.) 8vo. 10*s.* 6*d.*

MILMAN'S (CAPT. E. A.) Wayside Cross; or, the Raid of Gomez. A Tale of the Carlist War. Post 8vo. 2*s.* 6*d.*

MOORE'S (THOMAS) Life and Letters of Lord Byron. With Notes and Illustrations. *Library Edition.* Plates. 6 Vols. Fcap. 8vo. 18*s.*

—— Complete in One Volume. Portrait and Vignette. Royal 8vo. 12*s.*

MUCK MANUAL (The) for the Use of Farmers. A Practical Treatise on the Chemical Properties, Management, and Application of Manures. By FREDERICK FALKNER. *Second Edition.* Fcap. 8vo. 5*s.*

MUIRHEAD (J. P.). James Watt, an Historical Eloge. By M. ARAGO. Translated, with Notes. 8vo, 8*s.* 6*d.*; or 4to. 21*s.*

—— Correspondence of James Watt on his Discovery of the Theory of the Composition of Water, with a Letter from his Son. Portrait. 8vo, 10*s.* 6*d.*; or 4to. 24*s.*

MULLER'S DORIANS; The History and Antiquities of the Doric Race. Translated by HENRY TUFNELL and GEORGE CORNEWALL LEWIS. *Second Edition.* Maps. 2 Vols. 8vo. 26*s.*

MUNDY'S (CAPT. RODNEY) Events in Borneo, including the Occupation of Labuan and Visit to the Celebes. Plates. 2 Vols. 8vo. 32*s.*

MUNRO'S (GENERAL SIR THOMAS) Life and Letters. By the REV. G. R. GLEIG. Post 8vo. 5*s.*

MURCHISON'S (SIR RODERICK) Russia in Europe and the Ural Mountains; Geologically Illustrated. With Coloured Maps, Plates, Sections, &c. 2 Vols. Royal 4to. 8*l.* 8*s.*

—— Earlier Forms of Life, as disclosed in the Older Rocks. With Plates and Woodcuts. 8vo. *In the Press.*

MURRAY'S (CAPT. A.) Naval Life and Services of Admiral Sir Philip Durham. 8vo. 5*s.* 6*d.*

MURRAY'S RAILWAY READING. Or CHEAP BOOKS in large readable type. To be published occasionally, varying in size and price.

Already published:

NIMROD ON THE CHACE. 1*s.*

LITERARY ESSAYS FROM "THE TIMES." 4*s.*

MUSIC AND DRESS. 1*s.*

LAYARD'S POPULAR ACCOUNT OF NINEVEH. 5*s.*

NIMROD ON THE ROAD. 1*s.*

MAHON'S HISTORY OF THE "FORTY-FIVE." 3*s.*

LIFE OF THEODORE HOOK. 1*s.*

DEEDS OF NAVAL DARING. 2*s.* 6*d.*

THE FLOWER GARDEN. 1*s.*

JAMES' FABLES OF ÆSOP. 100 Woodcuts. 2*s.* 6*d.*

THE HONEY BEE. 1*s.*

NIMROD ON THE TURF. 1*s.* 6*d.*

MUSIC AND THE ART OF DRESS. Two Essays reprinted from the "Quarterly Review." Fcap. 8vo. 1*s.*

NAUTICAL ALMANACK (The). (*Published by Order of the Lords Commissioners of the Admiralty.*) Royal 8vo. 2*s.* 6*d.*

NAVY LIST (The Royal). (*Published Quarterly, by Authority.*) 12mo. 2*s.* 6*d.*

NEALE'S (E. V.) Feasts and Fasts: an Essay on the Rise, Progress, and Present State of the Laws relating to Sundays and other Holidays, &c. Fcap. 8vo. 9*s.* 6*d.*

NEVILLE'S (HON. RICHARD CORNWALLIS) Anglo-Saxon Remains, discovered 1851, at Little Wilbraham, Cambridgeshire. With short Descriptions. Illustrated by 40 Plates. 4to. *Nearly Ready.*

NEWBOLD'S (LIEUT.) Straits of Malacca Penang, Malacca, and Singapore. 2 Vols. 8vo. 26*s.*

NIMROD On the Chase—The Turf—and The Road. Reprinted from the "Quarterly Review." Woodcuts. Fcap. 8vo. 1*s.* each; or bound in 1 Vol., 3*s.* 6*d.*

NORTON (THE HON. MRS.) Letters from Sierra Leone, written to Friends at Home. By a LADY. Edited by Mrs. NORTON. Post 8vo. 5*s.*

O'BYRNE'S (W. R.) Naval Biographical Dictionary, comprising the Life and Services of every Living Officer in H. M. Navy, from the Rank of Admiral of the Fleet to that of Lieutenant. Compiled from Authentic and Family Documents. Royal 8vo. 42*s.*

O'CONNOR'S (R.) Field Sports of France; or, Hunting, Shooting, and Fishing on the Continent. Woodcuts. 12mo. 7*s.* 6*d.*

OLIPHANT'S (LAURENCE) Journey to Katmandu, (Capital of Nepaul,) with the Camp of JUNG BAHADER; including a Sketch of the Nepaulese Ambassador at Home. Fcap. 8vo.

OXENHAM'S (REV. W.) English Notes for Latin Elegiacs; designed for early Proficients in the Art of Latin Versification, with Prefatory Rules of Composition in Elegiac Metre. *Second Edition.* 12mo. 4*s.*

OXFORD'S (BISHOP OF) Charge to his Clergy at his Second Visitation in Nov., 1851. *Second Edition.* 8vo. 3*s.* 6*d.*

PAGET'S (JOHN) Hungary and Transylvania. With Remarks on their Condition, Social, Political, and Economical. *Second Edition.* Woodcuts. 2 Vols. 8vo. 24*s.*

PARISH'S (SIR WOODBINE) Buenos Ayres and the Provinces of the Rio de la Plata. Their First Discovery and Conquest, Present State, Trade, Debt, &c. Map and Woodcuts. 8vo.

PARIS'S (T. C.) Letters from the Pyrenees during Three Months' Pedestrian Wanderings amidst the Wildest Scenes of the French and Spanish Pyrenees. Woodcuts. Post 8vo. 10*s.* 6*d.*

PARKYNS' (MANSFIELD) Personal Narrative of a Residence in Abyssinia. With Maps and Woodcuts. 2 Vols. 8vo. *In the Press.*

PEILE'S (REV. DR.) Agamemnon of Æschylus. A New Edition of the Text, with Notes, Critical, Explanatory, and Philological, for the Use of Students. *Second Edition.* 8vo. 9*s.*

—— Choephoræ of Æschylus. A New Edition of the Text, with Notes, Critical, Explanatory, and Philological, for the Use of Students. *Second Edition.* 8vo. 9*s.*

PELLEW'S (DEAN OF NORWICH) Life of Lord Sidmouth, with his Correspondence. Portraits. 3 Vols. 8vo. 42*s.*

PENN'S (RICHARD) Maxims and Hints for an Angler, and the Miseries of Fishing. To which is added, Maxims and Hints for a Chess-player. *Second Edition.* Woodcuts. Fcap. 8vo. 5*s.*

—— (GRANVILLE) Bioscope; or, Dial of Life Explained. To which is added, a Translation of St. Paulinus' Epistle to Celantia, on the Rule of Christian Life; and an Elementary View of General Chronology. *Second Edition.* With Dial Plate. 12mo. 12*s.*

PENROSE'S (REV. JOHN) Lives of Vice-Admiral Sir C. V. Penrose, and Captain James Trevenen. Portraits. 8vo. 10*s.* 6*d.*

—— Sermons for Households, or Fifty-four Sermons Written for Sunday Reading in Families. 8vo, 10*s.* 6*d.*

—— (F. C.) Principles of Athenian Architecture, and the Optical Refinements exhibited in the Construction of the Ancient Buildings at Athens, from a Survey. With 40 Plates. Folio. 5*l.* 5*s.* (*Published under the direction of the Dilettanti Society*).

PENNINGTON (G. J.) On the Pronunciation of the Greek Language. 8vo. 8*s.* 6*d.*

PHILLIPS' (JOHN) Memoirs of William Smith, LL.D., (the Geologist). Portrait. 8vo. 7*s.* 6*d.*

—— Geology of Yorkshire. The Yorkshire Coast, and the Mountain-Limestone District. Part I., 31*s.* 6*d.* Part II., 52*s.* 6*d.* Plates, 4to.

PHILOSOPHY IN SPORT MADE SCIENCE IN EARNEST; or, the First Principles of Natural Philosophy inculcated by aid of the Toys and Sports of Youth. *Sixth Edition.* Woodcuts. Fcap. 8vo. 8*s.*

PHILPOTT'S (BISHOP OF EXETER) Letters to the late Charles Butler, on the Theological parts of his Book of the Roman Catholic Church; with Remarks on certain Works of Dr. Milner and Dr. Lingard and on some parts of the Evidence of Dr. Doyle. *Second Edition.* 8vo. 16*s.*

—— Letter to the Archbishop of Canterbury. *Twenty-First Edition.* 8vo. 3*s.* 6*d.*

—— Pastoral Letter, addressed to his Clergy, on the Present State of the Church. *Eighth Edition.* 8vo. 4*s.*

—— Acts of the Diocesan Synod, held in the Cathedral Church of Exeter. On June 25th, 26th, and 27th, 1851. *Third Edition.* 8vo. 2*s.* 6*d.*

PHIPPS' (HON. EDMUND) Memoir, Correspondence, Literary and Unpublished Diaries of Robert Plumer Ward. Portrait. 2 Vols. 8vo. 28*s.*

POOLE'S (R. S.) Horæ Egyptiacæ: or the Chronology of Ancient Egypt, discovered from Astronomical and Hieroglyphic Records upon its Monuments. Plates. 8vo. 10*s.* 6*d.*

—— (REV. G. A.) Handbook for the Cathedrals of England. Containing Descriptions of each. Woodcuts. Post 8vo. *In the Press.*

POPE'S (ALEXANDER) Works. A New Edition, containing numerous unpublished Letters, &c. Edited with Notes and a Life. By the Right Hon. John Wilson Croker. Portraits. 8vo. *In the Press.*

PORTER'S (G. R.) Progress of the Nation, in its various Social and Economical Relations, from the beginning of the Nineteenth Century. *Third Edition.* (1851.) 8vo. 24*s.*

POWELL'S (REV. W. P.) Latin Grammar simplified. 12mo. 3*s.* 6*d.*

PRAYER-BOOK (THE), Illuminated with 1000 Illustrations of Borders, Initials, Vignettes, &c. Medium 8vo. Cloth, 21*s.*; Calf, 31*s.* 6*d.* Morocco, 42*s.*

PUSEY (PHILIP) On Improvement in Farming; or What ought Landlords and Farmers to do? 8vo. 1*s.*

PUSS IN BOOTS. Suited to the Tastes of Little and Grown Children. By OTTO SPECKTER. *Second Edition.* Plates. 16mo. 5*s.*

QUARTERLY REVIEW (THE). 8vo. 6*s.*

RANKE'S (LEOPOLD) Political and Ecclesiastical History of the Popes of Rome, during the Sixteenth and Seventeenth Centuries. Translated from the German by MRS. AUSTIN. *Third Edition.* 2 Vols. 8vo. 24*s.*

—— History of Prussia; or, Memoirs of the House of Brandenburgh. Translated from the German by SIR ALEXANDER DUFF GORDON, BART. 3 Vols. 8vo. 36*s.*

—— History of Servia, and the Servian Revolution. Translated from the German by MRS. KERR. *Second Edition.* Map. 8vo. 15*s.*

RAWLINSON'S (REV. GEORGE) Herodotus. A New English Version. Translated from the Text of GAISFORD, and Edited with Notes, illustrating the History and Geography of Herodotus, from the most recent sources of information, embodying the chief Results, Historical and Ethnographical, which have been arrived at in the progress of Cuneiform and Hieroglyphical Discovery. Assisted by COLONEL RAWLINSON, and SIR J. G. WILKINSON. 4 Vols. 8vo. *In the Press.*

REJECTED ADDRESSES (THE). By JAMES AND HORACE SMITH. With Biographies of the Authors, and additional Notes. *Twenty-second Edition.* Portraits. Fcap. 8vo. 5*s.*

RICARDO'S (DAVID) Political Works. With a Notice of his Life and Writings. By J. R. M'CULLOCH. 8vo. 16*s.*

RIDE on Horseback to Florence through France and Switzerland. Described in a Series of Letters. By a LADY. 2 Vols. Post 8vo. 18*s.*

RIPA'S (FATHER) Memoirs during Thirteen Years' Residence at the Court of Peking, in the Service of the Emperor of China. Translated from the Italian. By FORTUNATO PRANDI. Post 8vo. 2*s.* 6*d.*

ROBERTSON'S (LORD) Leaves from a Journal, and other Fragments in Verse. Crown 8vo. 7*s.* 6*d.*

—— (REV. J. C.) History of the Christian Church, for the Use of Students in Theology, and General Readers. Part I.—TO THE REFORMATION. 2 Vols. 8vo. *In the Press.*

ROMILLY'S (SIR SAMUEL) Memoirs and Political Diary. By his SONS. *Third Edition.* Portrait. 2 Vols. Fcap. 8vo. 12*s.*

ROSS'S (SIR JAMES) Voyage of Discovery and Research in the Southern and Antarctic Regions during the years 1839-43. Plates. 2 Vols. 8vo. 36*s.*

ROYAL SOCIETY OF LITERATURE (THE). TRANSACTIONS. Plates. Vols. I. to III. 8vo. 12*s.* each.

RUNDELL'S (MRS.) Modern Domestic Cookery, founded on Principles of Economy and Practice, and adapted for Private Families. Entirely Revised and corrected to the Present Time, by A LADY. Woodcuts. Fcap. 8vo. 6*s.*

RUXTON'S (GEORGE F.) Travels in Mexico; with Adventures among the Wild Tribes and Animals of the Prairies and Rocky Mountains. Post 8vo. 5*s.*

SALE'S (LADY) Journal of the Disasters in Affghanistan. *Eighth Edition.* Post 8vo. 12*s.*

—— (SIR ROBERT) Brigade in Affghanistan. With an Account of the Seizure and Defence of Jellalabad. By REV. G. R. GLEIG. Post 8vo. 2*s.* 6*d.*

SAXON (THE) in Ireland. Being Notes of the Rambles of an Englishman in the West of Ireland, in search of a Settlement. *Second Edition.* Fcap. 8vo.

SENTENCES FROM THE PROVERBS. In English, French, Italian, and German. For the Daily Use of Young Persons. By A LADY. 16mo. 3*s.* 6*d.*

SCROPE'S (WILLIAM) Days of Deer-Stalking in the Forest of Atholl; with some Account of the Nature and Habits of the Red Deer. *Third Edition.* Woodcuts. Crown 8vo. 20*s.*

—— Days and Nights of Salmon Fishing in the Tweed; with a short Account of the Natural History and Habits of the Salmon, and Instructions to Sportsmen, &c. Plates. Royal 8vo. 42*s.*

SCROPE'S (G. P.) Memoir of Lord Sydenham, and his Administration in Canada. *Second Edition.* Portrait. 8vo. 9*s.* 6*d.*

SERMONS. Preached during the Visitation of the Bishop of Exeter in 1845. Published by Request. 12mo. 6*s.*

SEWELL'S (REV. W.) Evidences of Christianity; or, Dialogues between a Brahmin and a Christian. Fcap. 8vo. 7*s.* 6*d.*

SHAW'S (THOS. B.) Outlines of English Literature, for the Use of Young Students. Post 8vo. 12*s.*

SIDMOUTH'S (LORD) Life and Correspondence. By the HON. and REV. GEORGE PELLEW, DEAN OF NORWICH. Portraits. 3 Vols. 8vo. 42*s.*

SIDNEY'S (REV. EDWIN) Life of Lord Hill. *Second Edition.* Portrait. 8vo. 12*s.*

SIERRA LEONE; Described in a Series of Letters to Friends at Home. By A LADY. Edited by MRS. NORTON. Post 8vo. 6*s.*

SMITH'S (WM., LL.D.) Dictionary of Greek and Roman Antiquities. *Second Edition.* With 500 Woodcuts. 8vo. 42*s.*

—— Dictionary of Greek and Roman Biography and Mythology. With Woodcuts. 3 Vols. 8vo. 5*l.* 15*s.* 6*d.*

—— Dictionary of Greek and Roman Geography. Woodcuts. Parts 1 and 2. 8vo. 4*s.* each.

—— New Latin-English Dictionary, founded on the best and most recent authorities. 8vo. *In Preparation.*

—— Classical Dictionary of Biography, Mythology, and Geography, for Schools. 8vo. 21*s.*

—— Smaller Classical Dictionary for Young Persons, abridged from the larger Work. With 200 Woodcuts. Post 8vo. 10*s.* 6*d.*

—— Smaller Dictionary of Antiquities, for Young Persons, abridged from the larger Work. With 200 Woodcuts. Crown 8vo. 10*s.* 6*d.*

—— (WM. JAS.) Grenville Papers, formerly preserved at Stowe—now for the first time made public. Including also MR. GRENVILLE'S DIARY OF POLITICAL EVENTS. 2 Vols. 8vo. 32*s.*

—— (WM., the Geologist) Memoirs. By JOHN PHILLIPS. Portrait. 8vo. 7*s.* 6*d.*

—— (JAMES & HORACE) Rejected Addresses. *Twenty-second Edition.* Portrait. Fcap. 8vo. 5*s.*

SOMERVILLE'S (MARY) Physical Geography. *Third Edition.* Portrait. 2 Vols. Fcap. 8vo. 12*s.*

—— Connexion of the Physical Sciences. *Eighth Edition.* Plates. Fcap. 8vo. 10*s.* 6*d.*

SOUTHEY'S (ROBERT) Book of the Church; with Notes containing References to the Authorities, and an Index. *Sixth Edition.* 8vo. 12*s.*

—— Lives of John Bunyan & Oliver Cromwell. Post 8vo. 2*s.* 6*d.*

SPECKTER'S (OTTO) Puss in Boots; suited to the Tastes of Little and Grown Children. *Second Edition.* Plates. 16mo. 5*s.*

—— Charmed Roe; the Story of the Little Brother and Sister. Plates. 16mo. 5*s.*

SPENCER'S (REV. J. A.) Sketches of Travels in Egypt and the Holy Land. Plates. 8vo. 21*s.*

STANLEY'S (EDWARD, D.D., late Bishop of Norwich) ADDRESSES AND CHARGES. With a Memoir of his Life. By Rev. A. P. STANLEY. *Second Edition.* 8vo. 10*s.* 6*d.*

ST. JOHN'S (BAYLE) Adventures in the Libyan Desert and the Oasis of Jupiter Ammon. Woodcuts. Post 8vo. 2*s.* 6*d.*

—— (CHARLES) Field Notes of a Sportsman and Naturalist in Sutherland. Woodcuts. 2 Vols. Post 8vo. 18*s.*

—— Wild Sports and Natural History of the Highlands. Post 8vo. 5*s.*

STATE PAPERS during the Reign of Henry the Eighth. Vols. I. to V. 4to. 20*s.* each. (*Published by Authority.*)

STAUNTON'S (SIR GEORGE) Miscellaneous Notices relating to China, and our Commercial Intercourse with that Country. Including a few Translations from the Chinese Language, and some Observations on our Intercourse with that Country. *Third Edition.* 8vo. 10*s.* 6*d.*

STEPHENS' (J. L.) Incidents of a First and Second Visit to Central America and Yucatan. *New Edition.* Post 8vo. *In Preparation.*

STEVENS' (WM., M.D.) Observations on the Healthy and Diseased Properties of the Blood. 8vo. 15*s.*

STISTED'S (MRS. HENRY) Letters from the Bye-Ways of Italy. Plates. 8vo. 18*s.*

STOTHARD'S (THOS. R. A.) Life. With Personal Reminiscences. By Mrs. BRAY. With Portrait, and 60 Woodcuts. 4to. 21*s.*

STRIFE FOR THE MASTERY. Two Allegories. With Illustrations. Crown 8vo.

STRONG'S (REV. CHARLES) Specimens of Italian Sonnets. From the most celebrated Poets, with Translations. 8vo. 6*s.*

SUTTON (HON. H. MANNERS). Some Account of the Courts of London and Vienna, at the end of the Seventeenth Century, extracted from the Official and Private Correspondence of Robert Sutton (late Lord Lexington) while British Minister at Vienna, 1694-98. 8vo. 14*s.*

SUVERN'S ARISTOPHANES. The Birds and the Clouds. Translated by W. R. HAMILTON, F.R.S. 2 Vols. Post 8vo. 9*s.*

SYDENHAM'S (LORD) Memoirs. With his Administration in Canada. By G. POULET SCROPE, M.P. *Second Edition.* Portrait. 8vo. 9*s.* 6*d.*

TAIT'S (DEAN OF CARLISLE) Suggestions offered to the Theological Student under present Difficulties. Five Discourses preached before the University of Oxford. Post 8vo. 6*s.* 6*d.*

TALBOT'S (H. FOX) English Etymologies. 8vo. 12*s.*

TAYLOR'S (HENRY) Notes from Life and Books. *Third Edition.* 2 Vols. Post 8vo. 15*s.*

—— (J. E.) Fairy Ring. A Collection of Stories for Young Persons. From the German. With Illustrations by RICHARD DOYLE. *Second Edition.* Woodcuts. Fcap. 8vo. 7*s.* 6*d.*

—— Michael Angelo considered as a Philosophic Poet. *Second Edition.* Post 8vo. 5*s.*

TENNENT'S (SIR JAMES EMERSON) Christianity in Ceylon. Its Introduction and Progress under the Portuguese, Dutch, British, and American Missions. With an Historical Sketch of the Brahmanical and Buddhist Superstitions. Woodcuts. 8vo. 14*s.*

THEOPHILUS' (THE MONK) Essay upon Various Arts; forming an Encyclopædia of Christian Art of the 11th Century. Translated, with Notes, by ROBERT HENDRIE. 8vo. 21*s.*

THORNTON'S (WM. T.) Plea for Peasant Proprietors; with the Outlines of a Plan for their Establishment in Ireland. Post 8vo. 7*s.* 6*d.*

THREE-LEAVED MANUAL OF FAMILY PRAYER; arranged so as to save the trouble of turning the Pages backwards and forwards. Royal 8vo. 2*s.*

TICKNOR'S (GEORGE) History of Spanish. Literature. With Criticisms on particular Works, and Biographical Notices of Prominent Writers. 3 Vols. 8vo. 42*s.*

TREMENHEERE'S (SEYMOUR) Political Experience of the Ancients, in its bearing on Modern Times. Fcap. 8vo. 2*s.* 6*d.*

—— Notes on Public Subjects, made during a Tour in the United States and Canada. Post 8vo.

TURNBULL'S (P. E.) Narrative of Travels in Austria, with Remarks on its Social and Political Condition. 2 Vols. 8vo. 24*s.*

TWISS' (HORACE) Public and Private Life of Lord Chancellor Eldon, with Selections from his Correspondence. Portrait. *Third Edition.* 2 Vols. Post 8vo. 21*s.*

VAUGHAN'S (REV. DR.) Sermons on Various Occasions. 8vo. 12*s.* 6*d.*

—— Preached in Harrow School. 8vo. 10*s.* 6*d.*

—— Nine Sermons. 12mo. 5*s.*

VAUX'S (W. S. W.) Handbook to the Antiquities in the British Museum; being a Description of the Remains of Greek, Assyrian, Egyptian, and Etruscan Art preserved there. With 300 Woodcuts. Post 8vo. 7*s.* 6*d.*

VENABLES' (REV. R. L.) Domestic Manners of the Russians. Described from a Year's Residence in that Country. Post 8vo. 9*s.* 6*d.*

VOYAGE to the Mauritius and back, touching at the Cape of Good Hope, and St. Helena. By Author of "PADDIANA." Post 8vo. 9*s.* 6*d.*

WAAGEN'S (DR.) Treasures of Art in Great Britain. Being an Account of the Chief Collections of Paintings, Sculpture, Manuscripts, Miniatures, &c., &c., in this Country. Obtained from Personal Inspection during Visits to England. 2 Vols. 8vo. *In Preparation.*

WAKEFIELD'S (E. J.) Adventures in New Zealand, 1839-1844. With some Account of the Beginning of the British Colonisation of the Island. Map. 2 Vols. 8vo. 28*s.*

WALKS AND TALKS. A Story-book for Young Children. By AUNT IDA. With Woodcuts. 16mo. 5*s.*

WARD'S (ROBERT PLUMER) Memoir, Correspondence, Literary and Unpublished Diaries and Remains. By the HON. EDMUND PHIPPS. Portrait. 2 Vols. 8vo. 28*s.*

WASHINGTON'S (GENERAL) Life. By WASHINGTON IRVING. Post 8vo. *Nearly Ready.*

WATT (JAMES); an Historical Eloge. By M. ARAGO. Translated, with Notes. By J. P. MUIRHEAD. 8vo, 8*s.* 6*d.*; or 4to. 21*s.*

—— Correspondence on his Discovery of the Theory of the Composition of Water. Edited, with Notes, by J. P. MUIRHEAD. Portrait. 8vo. 10*s.* 6*d.*; or 4to. 24*s.*

WELLESLEY'S (REV. DR.) Anthologia Polyglotta; a Selection of Versions in various Languages, chiefly from the Greek Anthology. 8vo,. 15*s.*; or 4to. 42*s.*

WELLINGTON'S (THE DUKE OF) Despatches during his various Campaigns in India, Denmark, Portugal, Spain, the Low Countries, and France. Compiled from Official and other Authentic Documents. By COL. GURWOOD, C.B. *A New and Enlarged Edition.* 8 Vols. 8vo. 21*s.* each.

—— Selections from the above Work; arranged as a Convenient Manual for Reference. By COL. GURWOOD. *A New and Cheaper Edition.* 1 Vol. 8vo. 18*s.*

WILBERFORCE'S (BISHOP OF OXFORD) Charge to his Clergy in Nov. 1851. *Second Edition.* 8vo. 3*s.* 6*d.*

—— (ARCHDEACON) on Church Courts and Church Discipline. 8vo. 7*s.*

—— Incarnation of our Lord Jesus Christ in its relation to Mankind. *Third Edition.* 8vo. 12*s.*

—— Doctrine of Holy Baptism, with Remarks upon the REV. W. GOODE'S Effects of Infant Baptism. *Third Edition.* 8vo. 7*s.* 6*d.*

—— Sermons on the New Birth of Man's Nature. 8vo. 8*s.*

—— History of Erastianism. *Second Edition.* Post 8vo. 3*s.*

WILKIE'S (SIR DAVID) Life, Journals, Tours, and Critical Remarks on Works of Art, with a Selection from his Correspondence. By ALLAN CUNNINGHAM. Portrait. 3 Vols. 8vo. 42*s.*

WILKINSON'S (SIR J. G.) Private Life, Manners, and Customs of the Ancient Egyptians. With 600 Illustrations. *Third Edition.* 5 Vols. 8vo. 4*l.* 4*s.*

—— Dalmatia and Montenegro; with a Journey to Mostar in Hertzegovina, and Remarks on the Slavonic Nations. Plates and Woodcuts. 2 Vols. 8vo. 42*s.*

—— Handbook for Egypt.—Thebes, the Nile, Alexandria, Cairo, the Pyramids, Mount Sinai, &c. Map. Post 8vo. 15*s.*

—— (MATTHEW, D.D.) School Sermons, preached in the Chapel of Marlborough College. 8vo. 9*s.*

—— (G. B.) Working Man's Handbook to South Australia; with Advice to the Farmer, and Detailed Information for the several Classes of Labourers and Artisans. Map. 8mo. 1*s.* 6*d.*

WOOD'S (LIEUT.) Voyage up the Indus to the Source of the River Oxus, by Kabul and Badakhshan. Map. 8vo. 14*s.*

WORDSWORTH'S (REV. DR.) Athens and Attica. *New Edition.* Plates. Post 8vo. *In the Press.*

—— Fac-Similes of Ancient Writings on the Walls of Pompeii. *Second Edition.* 8vo. 2*s.* 6*d.*

—— King Edward VIth's Latin Grammar, for the Use of Schools. *Seventh Edition*, revised. 12mo. 3*s.* 6*d.*

—— Accidence, *printed separately* for the Use of Junior Classes. 12mo. 2*s.*

WORSAAE'S (J. J. A.) Account of the Danes and Northmen in England, Scotland, and Ireland. Woodcuts. 8vo. 10*s.* 6*d.*

YOUNG'S (DR. THOS.) Miscellaneous Works, now first collected and edited, with a Memoir of his Life. By GEORGE PEACOCK, D.D., Dean of Ely. 4 Vols. 8vo. *In the Press.*

BRADBURY AND EVANS, PRINTERS, WHITEFRIARS.

Made in United States
North Haven, CT
04 November 2026

21774644R00238